T0339411

MICROGRID

MICROGRID
Advanced Control Methods and Renewable Energy System Integration

Edited by

MAGDI S. MAHMOUD
King Fahd University of Petroleum and Minerals, Systems Engineering Department, Dhahran, Saudi Arabia

AMSTERDAM • BOSTON • HEIDELBERG • LONDON
NEW YORK • OXFORD • PARIS • SAN DIEGO
SAN FRANCISCO • SINGAPORE • SYDNEY • TOKYO
Butterworth-Heinemann is an imprint of Elsevier

Butterworth-Heinemann is an imprint of Elsevier
The Boulevard, Langford Lane, Kidlington, Oxford OX5 1GB, United Kingdom
50 Hampshire Street, 5th Floor, Cambridge, MA 02139, United States

Library of Congress Cataloging-in-Publication Data
A catalog record for this book is available from the Library of Congress

British Library Cataloguing-in-Publication Data
A catalogue record for this book is available from the British Library

ISBN: 978-0-08-101753-1

For information on all Butterworth Heinemann publications
visit our website at https://www.elsevier.com/

Working together
to grow libraries in
developing countries

www.elsevier.com • www.bookaid.org

Publisher: Joe Hayton
Acquisition Editor: Sonnini R. Yura
Editorial Project Manager: Ana Claudia Abad Garcia
Production Project Manager: Kiruthika Govindaraju
Cover Designer: Greg Harris

Typeset by SPi Global, India

CONTENTS

CONTRIBUTORS

M.A. Abido
King Fahd University of Petroleum and Minerals, Dhahran, Saudi Arabia

M.I. Abouheaf
Aswan University, Aswan, Egypt

N.M. Alyazidi
King Fahd University of Petroleum and Minerals (KFUPM), Dhahran, Saudi Arabia

M. Angjelichinoski
Aalborg University, Aalborg, Denmark

F. Blaabjerg
Aalborg University, Aalborg, Denmark

T. Caldognetto
University of Padova, Padova, Italy

W.-Y. Chiu
Yuan Ze University, Taoyuan, Taiwan

H. Dagdougui
ETS, Montreal, Canada

T. Dragičević
Aalborg University, Aalborg, Denmark

A. Khorsandi
Science and Research Branch, Islamic Azad University, Tehran, Iran

D. Li
Durham University, Durham, United Kingdom

M.S. Mahmoud
King Fahd University of Petroleum and Minerals (KFUPM), Dhahran, Saudi Arabia

A. Mehrizi-Sani
Washington State University, Pullman, WA, United States

H. Mokhtari
Sharif University of Technology, Tehran, Iran

A. Ouammi
ETS, Montreal, Canada; The National Center for Scientific and Technical Research, Rabat, Morocco

S. Peyghami
Sharif University of Technology, Tehran, Iran

P. Popovski
Aalborg University, Aalborg, Denmark

R. Sacile
University of Genova, Genova, Italy

M. Shahbazi
School of Engineering and Computing Sciences, Durham University, Durham, United Kingdom

C. Stefanovic
Aalborg University, Aalborg, Denmark

H. Sun
Durham University, Durham, United Kingdom

A.H. Syed
King Fahd University of Petroleum and Minerals, Dhahran, Saudi Arabia

P. Tenti
University of Padova, Padova, Italy

ABOUT THE EDITOR

Magdi S. Mahmoud is Distinguished Professor at King Fahd University of Petroleum and Minerals, Dhahran, Saudi Arabia. He obtained a BSc (honors) degree in communication engineering, an MSc degree in electronic engineering, and a PhD degree in systems engineering from Cairo University in 1968, 1972, and 1974 respectively. He has been a professor of engineering since 1984. He has been on the faculty of different universities worldwide, including Cairo University and the American University in Cairo (Egypt), Kuwait University (Kuwait), United Arab Emirates University (United Arab Emirates), the University of Manchester Institute of Science and Technology (United Kingdom), the University of Pittsburg and Case Western Reserve University (United States), Nanyang Technological University (Singapore), and the University of Adelaide (Australia). He has lectured in Venezuela (Central University of Venezuela), Germany (University of Hanover), the United Kingdom (University of Kent), the United States (University of San Antonio), Canada (University of Montreal), and China (Beijing Institute of Technology, Yanshan University). He is the principal author of 37 books and book chapters and the author/coauthor of more than 525 peer-reviewed articles. He is the recipient of two national, one regional, and several university prizes for outstanding research in engineering and applied mathematics. He is a fellow of the IEE, a senior member of the IEEE, the CEI (United Kingdom), and a registered consultant engineer of information engineering and systems (Egypt). He is currently actively engaged in teaching and research in the development of modern methods for distributed control and filtering, networked-control systems, triggering mechanisms in dynamical systems, fault-tolerant systems, and information technology.

PREFACE

Microgrids (MGs) are collections of electrical loads and microsources functioning as a single system that can operate either connected to a larger power grid or completely autonomously. Microgrids should be able to solve energy problems locally and hence increase flexibility. Power electronics plays an important role in achieving this revolutionary technology. The future grid is envisioned as a number of interconnected microgrids in which every user is responsible for the generation and storage of the energy that is consumed and for sharing the energy with neighbors. Looked at in this light, microgrids are key elements to integrate renewable and distributed energy resources as well as distributed energy storage systems (ESSs). Microgrids are essentially the building blocks of a futuristic smarter super grid, which in decades to come will revolutionize the power sector, ushering in an era of energy revolution just like the Internet of energy.

Microgrids have been attracting much research interest because of their potential to increase energy security and reliability, as well as to foster the penetration of distributed renewable resources (e.g., wind, solar) and distributed storage (e.g., plug-in vehicles, community energy storage). The development of microgrids could facilitate the smart grid, which is conceived to improve instantaneous grid power balancing as well as demand response. It requires microgrid control functions such as power balancing, optimization, prediction, and smart grid and end-user interaction.

From a contemporary perspective, microgrids are local grids comprising different technologies, such as power electronic converters, distributed generation (DG) units, ESSs, and telecommunication systems that can not only operate connected to the traditional centralized grid (macrogrid) but can also operate autonomously in islanded mode. Control structures are essential to the proper control of MGs, providing stability and efficient operation. The important roles that can be achieved with these control structures are frequency and voltage regulation, active and reactive power control between DG units and with the main grid, synchronization of the MG with the main grid, energy management, and economic optimization.

This book is about the state of the art of methods and applications of microgrid control. Several pertinent results on the design of control laws as well as stability analysis are presented.

- **Chapter 1:** "Microgrid Control Problems and Related Issues" by Magdi S. Mahmoud.

  This chapter provides an overview of the control methods of microgrid systems.
- **Chapter 2:** "Distributed Control Techniques in Microgrids" by Ali Mehrizi-Sani.

  This chapter provides a review of distributed control and management strategies for the next generation of microgrids and identifies future research directions.
- **Chapter 3:** "Hierarchical Power Sharing Control in DC Microgrids" by Saeed Peyghami, Hossein Mokhtari, and Frede Blaabjerg.

  This chapter provides a comprehensive treatment of power sharing in dc microgrids.
- **Chapter 4:** "Master/Slave Power-Based Control of Low-Voltage Microgrids" by Paolo Tetni and Tommaso Caldognetto.

  This chapter deals with control of low-voltage microgrids with master-slave architecture, where distributed energy resources interface with the grid by means of conventional current-driven inverters, and a voltage-driven grid-interactive inverter governs the interaction between the utility and the microgrid at their point of common coupling.
- **Chapter 5:** "Load-Frequency Controllers for Distributed Power System Generation Units" by Mohamed I. Abouheaf and Magdi S. Mahmoud.

  In this chapter adaptive learning techniques are proposed to control autonomous microgrids in real time. Online value iteration and policy iteration learning techniques are developed for single-agent and multiagent systems respectively. First, value iteration is used to control an autonomous microgrid. The proposed value iteration technique is based on heuristic dynamic programming. The control policy is selected by optimization of the respective Bellman equation. The proposed control strategy is implemented with use of actor-critic networks. The control strategy is designed with use of a dynamic model of islanded microgrids and makes use of an internal oscillator for frequency control. The proposed technique is implemented online with only partial knowledge of the microgrid dynamics. The simulation results show that the proposed control technique stabilizes the system and is robust to the load disturbances. Second, multiagent reinforcement learning techniques are developed to solve dynamic graphical games online in real time. For the multiplayer graphical game, it is desired to determine the optimal noncooperative solutions.

- **Chapter 6:** "An Optimization Approach to Design Robust Controller for Voltage Source Inverters" by Syed H. Asim and Mohamed A. Abido.

  This chapter presents a new method for intelligent robust control design that achieves the best possible convergence rate for the system, making use of the knowledge of the range of uncertain parameters, resulting in enhanced stability and performance. The proposed method is applied to the grid-connected voltage source inverter (VSI) system with uncertainties in grid impedance. Simulation results illustrate the efficacy of the proposed scheme, where excellent transient and steady-state performance of the controller is observed.

- **Chapter 7:** "Demand Side Management in Microgrid Control Systems" by Dan Li, Wei-Yu Chiu, and Hongjian Sun.

  In this chapter demand-side management in microgrid control systems is investigated from various perspectives. First, the history of demand-side management is briefly presented and basic concepts are introduced. Then, associated problems and challenges are discussed. Next, state-of-the-art technologies for demand-side management are presented.

- **Chapter 8:** "Towards a Concept of Cooperating Power Network for Energy Management and Control of Microgrids" by Hanane Dagdougui, Ahmed Ouammi, and Roberto Sacile.

  This chapter provides an outline of the operation and control of smart microgrids as well demand-side management and demand response programs, in addition to energy market trading. Furthermore, it underlines the concept of coordination and interconnection of a set of smart microgrids in a network, showing the benefits, advantages, and outcomes.

- **Chapter 9:** "Power Electronics for Microgrids: Concepts and Future Trends" by Tomislav Dragicevic and Frede Blaabjerg.

  The rapid increase in electrical energy consumption and environmental concerns are putting increasing pressure on the realization of more efficient power processing systems. Both stationary and vehicular microgrids experience similar trends in this regard. However, different types of microgrids generally have diverse mission objectives and performance requirements, and comprise different types of generators, storage devices, and consumers. Power electronics plays a key role here as enabling technology that allows flexible arrangement and control of these components within a generic microgrid. This chapter provides a brief review of currently used concepts and future power electronics trends in various microgrid areas. For each area the functions of application specific to power electronics subsystems and their performance requirements are also described.

- **Chapter 10:** "Power Electronic Converters in Microgrid Applications" by Mahmoud Shahbazi and Amir Khorsandi.

  In a microgrid, DG units, including renewable energy sources and energy storage units, are connected to the system with use of power electronic converters. These converters are key components of the microgrid and provide high controllability of the system. This chapter investigates the use of power electronic converters for microgrid applications and reviews some of the most commonly used structures, including two-level and three-level neutral point-clamped (NPC) and cascaded H-bridge (CHB) converters. Furthermore, the modeling and control of these converters are presented.

- **Chapter 11:** "Power Talk: Communication in a DC Microgrid Through Modulation of the Power Electronics Components" by Marko Angjelichinoski, Cedomir Stefanovic, Petar Popovski, and Frede Blaabjerg.

  The focus of the chapter is on the design of power talk solutions for DC MGs with droop control. Specifically, the chapter presents the power talk implementation through modification of the droop parameters of the primary control loop, and investigates the design of modulation schemes and detection strategies in different MG setups. In comparison with standard communication solutions, power talk has several important advantages. It alleviates the need for an external communication network, as its implementation requires only software modifications in the power electronic interfaces, while its reliability and availability draw on the reliability and availability of the MG power transmission system.

- **Chapter 12:** "Pilot-Scale Implementation of Coordinated Control for Autonomous Microgrids" by Magdi S. Mahmoud and Nezar M. Alyazidi.

  This chapter describes a pilot-scale implementation of a two-level coordinating control approach for the islanded operation of a microgrid system. In the islanded operation of the microgrid, the master DG unit handles the voltage and frequency control of the load, and the slave DG units regulate their power components by use of the conventional dq-current control strategy. On the basis of the simulation environment within MATLAB/SimPowerSystems, three typical experimental studies including laboratory-scale devices and equipment were conducted. The results illustrate effective performance under the control strategy developed.

Magdi S. Mahmoud
King Fahd University of Petroleum and Minerals, Saudi Arabia

ACKNOWLEDGMENTS

Particular thanks are due to the Elsevier team: Sonnini R. Yura for initiating the writing of the book, and Mariana Kuhl and Ana Claudia Garcia for guidance and dedication throughout the publishing process. I greatly appreciate the excellent work of Kiruthika Govindaraju in the production phase and the expert help of Suresh Kumar with the Elsevier LaTeX template. I am also grateful to all the anonymous referees for carefully reviewing and selecting the final papers. This work is supported by the Deanship of Scientific Research (DSR) at King Fahd University of Petroleum and Minerals (KFUPM) through book writing research project no. BW 151004.

Magdi S. Mahmoud

CHAPTER 1

Microgrid Control Problems and Related Issues

M.S. Mahmoud
King Fahd University of Petroleum and Minerals (KFUPM), Dhahran, Saudi Arabia

1. INTRODUCTION

Economic challenges, technological advancements, and environmental impacts are now demanding distributed generation in place of conventional centralized generation [1]. Power supply companies are now confronted with unprecedented difficulties in terms of meeting the load requirements, consumer satisfaction, and environmental considerations. Thus distributed generation has received much attention because of its potential to alleviate pressure from the main transmission system by supplying a few local loads [2]. The waste heat generated from the fuel-to-electricity conversion is used by the distributed generation system with the help of microturbines, reciprocating engines, and fuel cells to provide heat and power to customers. Adding to the system distributed energy sources such as photovoltaic panels, wind turbines, energy storage devices such as batteries and capacitors, generators extracting energy from other renewables, and controllable loads can provide momentous contributions to future energy generation and distribution. Another noteworthy feature is that the carbon dioxide emission is reduced to a large extent, satisfying the commitment of many nations concerning a decrease of carbon footprints [3]. However, distributed generation faces technical issues regarding its connection to intermittent renewable generation and feeble areas of the distribution network. Further, owing to the distinct behavior of distributed generation unlike the conventional load, alteration in power flow results in problems. To counter the irregular behavior and increasing penetration of distributed generation, the microgrid was introduced.

Microgrid
http://dx.doi.org/10.1016/B978-0-08-101753-1.00001-2

The *microgrid* has entered into distributed generation and looks promising for future aspects. It has the ability to respond to changes in the load, while decreasing feeder losses and improving local reliability. Basically designed to cater for the heat and power requirements of local customers, it can serve as an in-interruptible power supply for critical loads.

The concept of a microgrid has received considerable attention owing to its potential to serve as an alternative power source, utilizing unconventional sources and supplying the most critical loads of the main grid in the case of a network failure. Microgrids are low-voltage networks or distributed energy systems that provide heat and power to a particular area by employing generators and loads. They have the ability to operate independently and isolate themselves from the main grid in the case of a fault [4–16].

If proper control techniques are implemented, they may improve the reliability of electrical energy supply. A microgrid can have microturbines, wind turbines, fuel cells, photovoltaic cells, etc., as sources of energy that are interfaced with the help of power electronic converters. All these units are connected to the main grid through a point of common coupling and look like a solitary unit to the distribution network. No additional inertia is added to the system from the distributed generation units. However, because of this, the power balance among generation and load and the network frequency becomes complicated to maintain, especially when the microgrid is in islanded mode [17]. The microgrid operates in two modes: namely, grid-connected mode and islanded mode.

A comparison between a conventional power grid and a microgrid is presented in Table 1.1 to appreciate the role of renewable energy resources.

Table 1.1 Comparison between the conventional grid and a microgrid [18]

Conventional grid	Smart grid
Electromechanical	Digital
One-way communication	Two-way communication
Centralized generation	Distributed generation
Few sensors	Senors throughout
Manual monitoring	Self-monitoring
Manual restoration	Self-healing
Failures and blackouts	Adaptive and islanding
Limited control	Pervasive control
Few customer choices	Many customer choices

2. MICROGRID REVIEW

Following the standards of the Consortium for Electric Reliability Technology Solutions (CERTS) architecture [19, 20], a basic microgrid architecture is shown in Fig. 1.1.

A microgrid is an interconnection of [3]:

- Distributed energy sources, such as microturbines, wind turbines, fuel cells, and photovoltaics.
- Storage devices for energy integration, such as batteries, flywheels, and power capacitors on low-voltage distribution systems.
- A group of radial feeders, which could be part of a distribution system. There are three sensitive-load feeders (feeders A–C) and one nonsensitive-load feeder (feeder D):
  1. The sensitive-load feeders contain sensitive loads that must always be supplied; thus each feeder must have at least a microsource rated to satisfy the load at that feeder.
  2. The nonsensitive-load feeder is the feeder that may be shut down if there is a disturbance or there are power quality problems on the

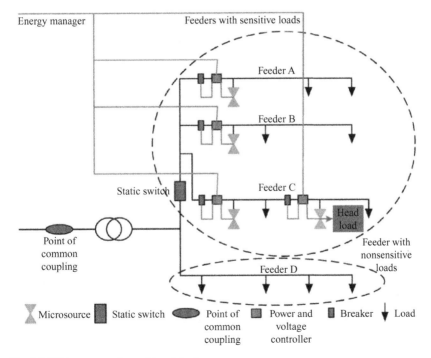

Fig. 1.1 Wide-sense microgrid architecture.

utility; the nonsensitive-load feeder will be left to ride through the disturbance or power quality problems.

3. When there is a problem with the utility supply, feeders A–C can be islanded from the grid with use of the static switch that can separate them in less than a cycle to isolate the sensitive loads from the power grid to minimize disturbance to the sensitive loads.

- The energy manager, which is responsible for managing system operation through power dispatching and voltage setting to each microsource controller. Some possible criteria for the microgrid to fulfill this responsibility are as follows [19]:

1. ensure that the necessary electrical loads and heat are fulfilled by the microsources;
2. ensure that the microgrid satisfies operational contracts with the utility;
3. minimize emissions and/or system losses; and
4. maximize the operational efficiency of the microsources.

Remark 1.1. *In islanded operation, a microgrid will work autonomously; therefore, it must have enough local generation to meet the demands of the sensitive loads [5, 19]. Furthermore, a disturbance requiring a feeder to operate individually may also occur. Each sensitive-load feeder in the microgrid design must have enough local generation to supply its own loads, while the nonsensitive-load feeder will rely on the utility supply.*

Remark 1.2. *After a disturbance the microgrid will reconnect to the utility and work normally as a grid-connected system. In this grid-connected system, excess local power generation, if any, will supply the nonsensitive loads or charge the energy storage devices for later use. The excess power generated by the microgrid may also be sold to the utility; in this case, the microgrid will participate in the market operation or provide ancillary services.*

Remark 1.3. *The disconnection or reconnection processes must be specified by the point of common coupling, a single point of connection to the utility located on the primary side of the transformer. At this point the microgrid must meet the established interface requirements, such as defined in the IEEE Standard 1547 series. Furthermore, the successful disconnection or reconnection processes depend on microgrid controls. The controllers must ensure that the processes occur seamlessly and the operating points after the processes are satisfied.*

Remark 1.4. *In grid-connected mode, the microgrid is supposed to follow the rules of the distribution network without being involved in the operation of the main power system. The microgrid operation based on this approach is significant for the stable operation of the power system. In this mode the microgrid can draw power from* *the main grid or can supply its power to the main grid, thus functioning similarly to a controllable load or source. By supplying or drawing power, the microgrid should be able to control the active and reactive power flows and keep an eye on the energy storage [21, 22]. However, in this mode, owing to the small size of the distribution units, the system dynamics have to be fixed to a wide extent. Another issue is the slow response of the control signals whenever there is a change in output power. Furthermore, because of the lack of synchronous machines connected to the low-power grid, virtual inertia has to be incorporated in the control loops of the power electronic interfaces [23].*

Remark 1.5. *The islanded mode is an operating condition in which the microgrid isolates itself from the main grid in the case of a fault. However, the transition from the grid-connected mode to the islanded mode must be stable [24]. If the microgrid is consuming or supplying power to the main grid before disconnection, a power imbalance occurs. This is compensated by the energy storage units because the microsources have low inertia and slow dynamic response [25–27]. Further research can be found in [18, 28–38].*

3. MICROGRID COMPONENTS

In a basic microgrid architecture (see Fig. 1.2), the electrical system is assumed to be radial with several feeders and a collection of loads. The radial system is connected to the distribution system through a separation device, usually a static switch, called *point of common coupling*. Each feeder has a circuit breaker and a power flow controller. Developed within the EU R&D microgrid project, The Consortium for Electric Reliability Technology Solutions (CERTS), the microgrid concept adopted in this research involves an operational architecture (see Fig. 1.3). It comprises a low-voltage network, loads (some of them interruptible), both controllable and noncontrollable microsources, storage devices, and a hierarchical-type management and control scheme supported by a communication infrastructure used to monitor and control microsources and loads.

The head of the multilevel control system is the microgrid central controller (MGCC). At a second control level, load controllers and microsource controllers exchange information with the MGCC, which manages microgrid operation by providing set points to both load controllers and microsource controllers.

The amount of data to be exchanged between network controllers is small, since it includes mainly messages containing set points to load controllers and microsource controllers, information requests sent by the MGCC to load controllers and microsource controllers about active and

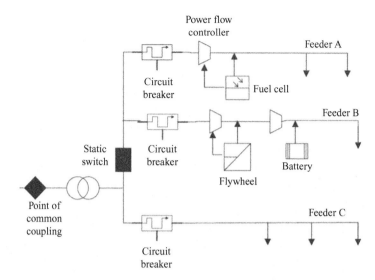

Fig. 1.2 Basic microgrid architecture.

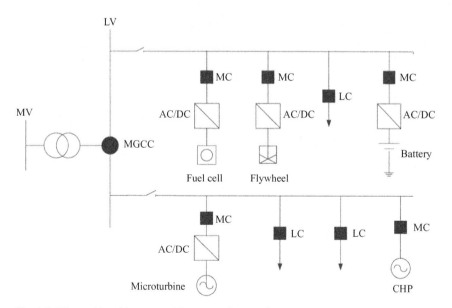

Fig. 1.3 Microgrid architecture with a central controller.

reactive powers, and voltage levels and messages to control microgrid switches.

3.1. Microsources

The microsources of special interest for microgrids are small (around 100-kW) units with power electronic interfaces [39–45]. These sources, including microturbines, wind generators, photovoltaic arrays, photovoltaic panels, and fuel cells, are placed at customer sites. They are low cost, have low voltage, and are highly reliable, with little emission.

Despite this impressive list of benefits, it has been pointed out [1] that distributed energy resource (DER) penetration has not met expectations. Major drawbacks to increased DER utilization are high cost, the need for custom engineering, lack of plug-and-play integration methods, and few successful business models. Many private and public organizations are aggressively addressing these drawbacks, including the Department of Energy (DoE) and other state organizations in US.

Some emerging generation technologies are introduced in detail in [33]. Several microsource models, able to describe their dynamic behavior, have been developed [10]. A simple combination control method for different kinds of DERs to compensate load demand fluctuation in a microgrid is proposed in [11, 12].

The primary functions of the microsource controller are regulation of power flow on a feeder where the operating points of the loads are varying, regulation of voltage at each microsource to accommodate the changing loads on the system, and most importantly to see that every microsource takes its load during islanded operation [19].

3.2. Microturbines

Microturbines are single-shaft, simple mechanical devices consisting of a generator which is a permanent magnet machine functioning at variable speed typically in the range of 50,000–100,000 rpm. The variable-speed generation system is interfaced with the electrical system through power electronics. The microturbines are flexible to operate on different fuels such as natural gas and gasoline. Possessed with good reliability, they are also commercially affordable at economic costs [34].

3.3. Fuel Cells

The microsources include renewable sources of energy such solar, wind, and hydro power. Mini hydro generators, wind turbines, and photovoltaic panels are located geographically in a microgrid. The energy from these renewable sources is harnessed and converted to electricity. The only shortcoming of these sources is their intermittent nature.

Fuel cells are unconventional and produce electricity from hydrogen and oxygen. They release water vapor and have low harmful emissions. Also, they offer higher efficiencies when compared with the microturbines just discussed. However, fuel cells are currently uneconomical and quite expensive relative to other renewable energy sources. From the environmental perspective, renewable sources and fuel cells are well suited for distributed generation in place of their conventional counterparts, including combustion engines (see Fig. 1.2).

3.4. Storage Devices

Energy storage is a vital factor to legitimize renewable energy resources as a reliable contributor to the main sources of energy and to provide successful operation of a microgrid. The energy storage process plays an important role in the balance between the generation of power and the energy demanded [20]. Lasseter [34, 35] points out that a system with clusters of microgrids designed to operate in an islanded mode must provide some form of storage to ensure there is an initial energy balance. Because of the large time constants (from 10 to 200 s) of the responses of some microsources, such as fuel cells and microturbines, storage devices must be able to provide the amount of power required to balance the system following disturbances and/or significant load changes.

These devices act as controllable AC voltage sources to face sudden system changes such as in load-following situations. In spite of acting as voltage sources, these devices have physical limitations and thus a finite capacity for storing energy.

The necessary microgrid storage can come in several forms; batteries or supercapacitors on the DC bus of each microsource; direct connection of AC storage devices (batteries, flywheels, etc.); or use of traditional generation with inertia with the microsource.

The requirements of energy storage in a microgrid are as follows:

1. Balancing power demand between the generation side and the load side is the first priority for energy storage devices (since the sources are intermittent and transient disturbances lack inertia).

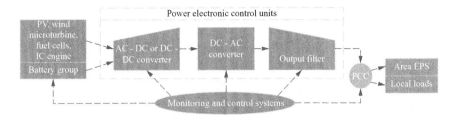

Fig. 1.4 Power electronics of a typical distributed energy resource system [22].

2. Storage of maximum energy demands during off-peak hours and able to supply all loads when required.
3. Elimination of loaded parts from the microgrid that helps to meet unpredicted and sudden demands.
4. Provision of smooth transient conditions from grid-connected to is-landed operation or vice versa.
5. Accommodation of the minute-hour peaks in the daily demand curve.

3.5. DER Interfaces

Power converters allow connection of independent equipment and components on a common system. Distributed generation technologies require specific converters and power electronic interfaces that are used to convert the generated energy to suitable power types directly supplied to a grid or to consumers. The development of an advanced power electronic interface helps meet various power demands with lower cost compared with DER systems since power converters provide similar functions. Thus the stability of the microgrid is maintained, while source variety is also accommodated [22]. A block diagram of a typical DER power electronic system and the power electronic interface in a microgrid are shown in Figs. 1.4 and 1.5, respectively.

4. MICROGRID CONTROLS

In microgrid systems, there are two modes of operation:
• grid-connected operation
• grid-islanded operation

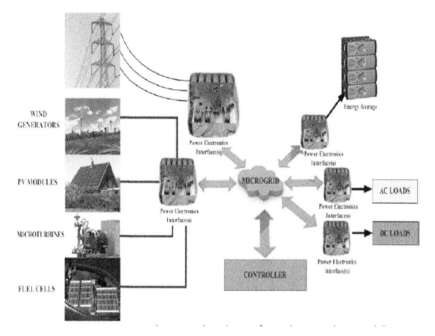

Fig. 1.5 Power electronic interface in a microgrid [18].

4.1. Grid-Connected Operation

For grid-connected operation, the microgrid is required to follow the distribution rules of the network without participating in the operation of the main power system. The microgrid operating on the basis of this approach is important for stable operation of the power system. In the case of grid-connected operation, the microgrid can draw its power from the main grid or it can supply power to the main grid, and it works like a controllable load or source. Fig. 1.6 presents an overview of primary, secondary, and tertiary control of the DER units in the grid-connected operation [11].

The *primary controller* is used to provide reliable operation and to respond to communication failures. It is associated by frequency f and voltage U as inputs and it gives control actions Q_{prim} and P_{prim} to control the deviation of the voltage and frequency.

The *secondary controller* is used to regulate the voltage and frequency on the basis of its response to the load and supply changes. Its control actions P_{sec} and Q_{sec} are given by calculation of the average of primary control actions.

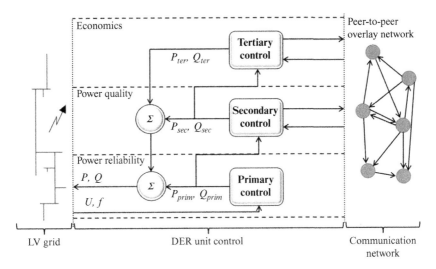

Fig. 1.6 Overview of the primary, secondary, and tertiary control building blocks.

The tertiary controller is used to improve the operation of DER units so they have the optimal cost-efficiency (i.e., as far as the power limits allow the DER units to exchange their power at equal marginal cost).

4.2. Islanded Operation

The microgrid units can be controlled on the basis of a decentralized approach to balance the microgrid components' energy and the demand. When the grid is affected by any abnormal operation or conditions, the microgrid should be disconnected and changed to grid–islanded operation mode. If the microgrid is in grid–islanded mode, it should handle the following considerations [12]:

1. balancing of supply and demand;
2. acceptable power quality;
3. voltage and frequency balance; and
4. communication among the microgrid components.

The control of an isolated microgrid means balancing the generation and the demand power to keep high performance with an acceptable range of frequency and voltage amplitude. In the literature two main control strategies are used for this purpose: a PQ inverter controller to keep the active power constant at a desirable power factor, and a voltage source inverter (VSI) controller to regulate the frequency and voltage amplitude.

These control strategies for an isolated microgrid are discussed in [13] for single-master operation and multimaster operation. The experimental setup results with a VSI controller for the output voltages of the microgrid sources during various loads are given in [14]. The islanded microgrid frequency and the generated power control with big load variations are given in [15]. The controller in [46] automatically controls the frequency drop and restores the frequency to its desired reference.

A proportional-plus-integral (PI) controller is proposed to control the isolated microgrid frequency in [13, 47–52]. The fuzzy controller [53] is used to select optimal PI gains by considering changes in the microsource parameters and particle swarm optimization is used for the tuning PI controller [54]. Also, the power source's reactive current can be used to control the frequency [55]. The case when the frequency depends on the reactive power is presented in [56]. Instead of frequency control, voltage balance control is presented in [57], where a controller is applied to a VSI to improve voltage transient response. Voltage balance controllers are also proposed in [55, 56, 58, 59]. An optimal controller is proposed in [60] to control the microgrid islanded mode frequency and voltage variations, and the laboratory setup for microgrid frequency and voltage control is implemented in [61]. H_∞ robust control is proposed in [62, 63] to control the operation of the connection of two (distributed generation) microgrid units.

The microgrid islanded mode has to continue providing energy to the microgrid loads, so the focus here will be on the islanded operation mode.

4.3. Inverter Controller

Most microsource technologies that can be installed in a microgrid are not suitable for direct connection to the electrical network because of the characteristics of the energy produced. Therefore power electronic interfaces (DC–AC or AC–DC–AC) are required (see Fig. 1.7). The general model for a microsource contains three basic elements:
• prime mover;
• DC interface; and
• VSI.

The microsource couples to the microgrid by means of an inductor. The VSI controls both the magnitude and the phase of its output voltage, $\bar{V}$, so it can control real and reactive powers. The voltage regulation is crucial for a microgrid with integration of large number of microsources so as to overcome oscillation caused by high penetration of microsources. The

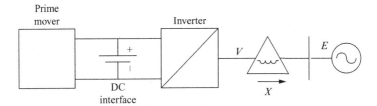

Fig. 1.7 Interface inverter system.

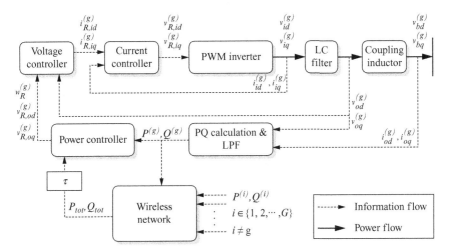

Fig. 1.8 Inverter control.

voltage regulation is also used to ensure that there is no large circulating reactive current between sources [6].

A typical inverter control structure is depicted in Fig. 1.8 identifying all relevant elements.

4.4. Microgrid Classification

Microgrids can be classified in many ways as summarized in Table 1.2.

In addition, microgrids may also be single- or three-phase systems or connected to low- or medium-voltage distribution networks.

4.5. Control Functions

Microgrid controllers have responsibilities to ensure that [19]:

1. microsources work properly at a predefined operating point or at point slightly different from the predefined operating point but that still satisfy the operating limits;

Table 1.2 Classes of microgrids [18]

Power type	Microgrid classes
Application type	DC microgrid
	Hybrid DC-AC microgrid
	High-frequency AC microgrid
	Line-frequency AC microgrid
	Utility microgrid
	Commercial and industrial microgrid
	Isolated microgrid
System structure	Single-stage power-conversion systems
	Two-stage power-conversion systems
Supervisory control	Centralized
	Decentralized
	Distributed
Connection of DER	Electronically coupled
	Conventionally coupled

2. active power and reactive power are transferred according to the needs of the microgrid and/or the distribution system;
3. disconnection and reconnection processes are conducted seamlessly;
4. market participation is optimized by optimizing production of local microsources and power exchanges with the utility;
5. heat utilization for the local installation is optimized;
6. sensitive loads, such as medical equipment and computer servers, are supplied uninterruptedly;
7. in the case of general failure, the microgrid is able to operate through black start; and
8. energy storage systems can support the microgrid and increase the system's reliability and efficiency.

The microgrid must address the following issues when operating in islanded mode:

1. balancing of supply and demand;
2. acceptable power quality;
3. voltage and frequency balance; and
4. communication among the microgrid components.

Controlling the islanded microgrid means balancing the generation and demand power to deliver high performance while maintaining acceptable ranges of frequency and voltage amplitude. The islanded operation of the microgrid will be the focus of this chapter.

Several control strategies for the microgrid have been proposed in the literature, including PI controllers in [64–71]. Robust H_∞ control is presented in [72, 73] for the control of two distributed generation units. An optimal controller is presented for controlling the frequency and voltage fluctuations during islanded mode in [74].

4.6. Control Operations of Microgrids

The major control architectures considered for the operation of microgrids are classified into four categories:

1. Autonomous (islanded) control architecture, in which the microgrid is designed to run autonomously ensuring stable, sustainable, and reliable operation.

   The merits and demerits of droop control are outlined in Table 1.3.

2. Multilevel control architecture, in which microsource controllers and the load controllers are used. The MGCC provides the microsource controllers with the demand requirements among other control functions.

3. Agent-based control architecture, in which the several control functions are represented in terms of agents that can be software or hardware components.

4. Neural network-based energy management system, in which multilayer perceptron neural networks have been used to perform control functions within the microgrid.

4.7. Electronically Coupled Microgrids

Electronically coupled microgrids consist of a number of elements that can operate in parallel either in islanded mode or connected to the main grid. The general structure of an electronically coupled microgrid is depicted in

Table 1.3 Features of droop control

Advantages	Disadvantages
Prevention of communications	Selecting just the voltage regulation or load sharing
Higher flexibility	Hybrid DC–AC microgrid
Increased reliability	Inductance couplings
Free laying	Impact of overall impedance
Various power ratings	Lower response rate

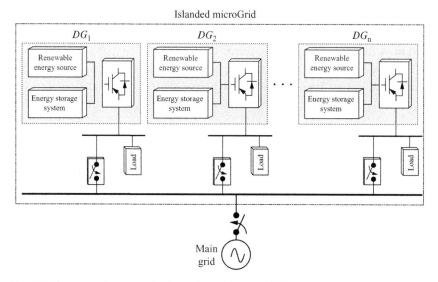

Fig. 1.9 Electronically coupled microgrid architecture [38].

Fig. 1.9, and has *n* distributed generation units. The microgrid is connected to the utility system through a static transfer switch at the point of common coupling. Each distributed generation system comprises a renewable energy source, an energy storage system, and a power electronic interface, which normally consists of a DC-AC inverter. Each distributed generation unit can be connected to a predefined load or to the AC common bus directly to supply power. The DC-AC inverters are classified as VSIs and current source inverters, where the former are commonly used to inject current in grid–connected mode and the latter are used to keep the frequency and voltage stable in autonomous operation. Both can operate in parallel in a microgrid [74–87].

4.8. Basic Control Techniques

There are several control techniques that help to manage the component level of a distribution system:

- *Master and slave control*: The master fixes the voltage and frequency values, while the slaves control the current sources.
- *Current and power flow control*: This method controls the current and power distribution by using control signals.

- *Droop control*: This method is an improved method to be combined with the previous methods since the converters behave as nonideal voltage sources.

4.9. Centralized Control

A centralized control system achieves intelligence from a particular central location, which depends on the network type, and could be a switch, a server, or a controller. It is easy to operate a centrally controlled network as it presents increased control to the operator who maintains the entire system. This feature allows the manager to define broad control strategies so as to meet power requirements (see Fig. 1.10).

Centralized control is best used for microgrids with the following characteristics:

1. The owners of microsources and loads have common goals and seek cooperation to meet their goals.
2. Small-scale microgrids may be feasible to control with the presence of an operator.

Since the primary control is local and does not have intercommunication with other distributed generation units, to achieve global controllability of the microgrid, secondary control is often used. A conventional centralized secondary control loop is implemented in an MGCC [38]. Fig. 1.11 shows the microgrid secondary control architecture consisting of a number

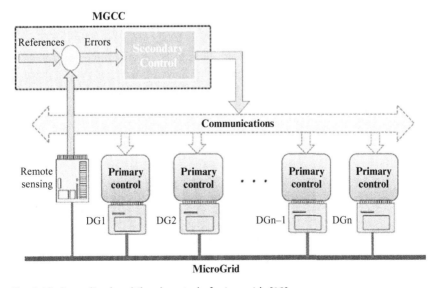

Fig. 1.10 Centralized multilevel control of microgrids [18].

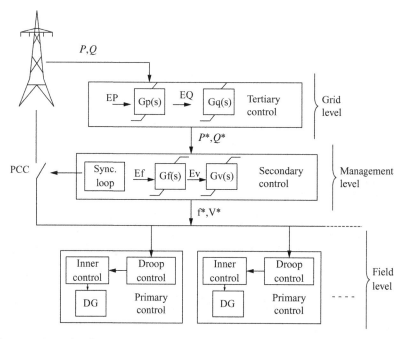

Fig. 1.11 Centralized secondary controller [18].

of distributed generation units locally controlled by a primary controller and a secondary controller, which measures from a remote sensing block a number of parameters to be sent back to the controller by means of a low-bandwidth communication system. Hence those variables are compared with the references so they can be compensated by the secondary controller, which will send the output signal through the communications channel to the primary controller of each distributed generation unit. The advantage of this architecture is that the communication system is not too busy, since only unidirectional messages are sent in only one direction (from the remote sensing platform to the MGCC and from the MGCC to each distributed generation unit). The drawback is that the MGCC is not highly reliable since a failure of this controller is enough to stop the secondary control action.

5. A MICROGRID AS A SYSTEM OF SYSTEMS

It is worth noting, from a control system design viewpoint, that a microgrid is a complex system comprising a variety of systems that are nonlinear in nature and possess strong cross-coupling between them. Hence viewing the microgrid from an intelligent system of systems (SoS) perspective is the need of the hour. Moreover, an efficient control method based on an SoS has to be established *i* to overcome the challenges posed by the microgrid. The concept of an SoS is now widespread and has entered several domains, including defense, IT, health care, manufacturing, energy, space stations, and exploration [88–91].

A control system consisting of a real-time network in its feedback can be termed a *networked control system* [92]. The same concept can be applied to the microgrid operating in islanded mode, where three distributed generation units supplying a load are considered as three subsystems. At the primary level, the islanded system is assumed to be equipped with PI controllers. At the secondary level, the networked control system is designed to control the interconnected distributed generation units, forming a networked microgrid SoS.

As mentioned in [90–93] that the need to design an SoS control system that can tolerate packet loss and delays is one of the prime challenges in SoS networked control, where a network is considered subjected to bounded random packet losses and that the controller stabilizes the system in the presence of packet losses.

There are certain characteristics or features that are unique to an SoS as given in [94]. The SoS is expected to exhibit the following characteristics:

1. *Operational independence*: All the constituent systems within the SoS architecture operate independently and do no interfere with neighboring systems in their functionality.

2. *Managerial independence*: The constituent systems continue to operate on their own unperturbed by the SoS. In other words, they are responsible for their autonomous operation.

3. *Evolutionary development*: The SoS is not designed as a single unit, and is rather flexible, and can accommodate numerous new systems or do away with systems that are no longer necessary.

4. *Emergent behavior*: All the constituent systems function as a collective unit to accomplish a common objective, which cannot be achieved by a single-component system.

5. *Geographic distribution*: The distribution of the subsystems is sequential to facilitate flow of information among them.

5.1. SoS Structure

The microgrid as described so far is a complex system comprising microsources, loads, and energy storage devices. Most of the elements are nonlinear systems and strong cross-coupling exists between them. However, complexity is not the only thing that would be sufficient for the microgrid to qualify as an SoS; it must comply with other features of the SoS presented in [94] mentioned in the previous section.

The microgrid architecture as an SoS is depicted in Fig. 1.12. As can be seen, the subsystems of the microgrid SoS are a photovoltaic system, a wind turbine and microturbines. There could be other distributed generation units such as fuel cells and unconventional sources of generation among the subsystems. The typical characteristics of the microgrid SoS are as follows:

- The subsystems are independently operated and managed.
- The microgrid SoS is evolutionary. It can accommodate new subsystems when required and discard any of them from the structure.
- All subsystems are emergent. Collectively, they form the SoS which accomplishes the overall objective of supplying local loads and providing ancillary services to the main grid.
- The subsystems are geographically distributed.

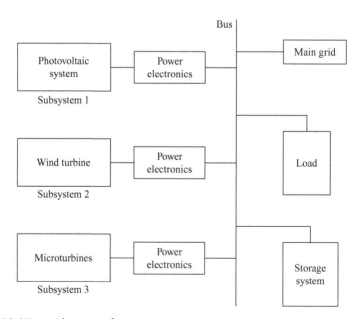

Fig. 1.12 Microgrid system of systems structure.

Owing to the above characteristics that are similar to those of an SoS, the microgrid complies perfectly with the features of an SoS.

5.2. SoS Framework

A framework for a microgrid SoS is presented in Fig. 1.13. The objective of the microgrid is best served when its constituent subsystems are well coordinated, organized, and able to communicate effectively among each other. This calls for an integrated framework that allows each subsystem to operate independently and establishes good communication among the subsystems to accomplish the desired goal. The emergent nature and geographic distribution of the microgrid makes it convenient to

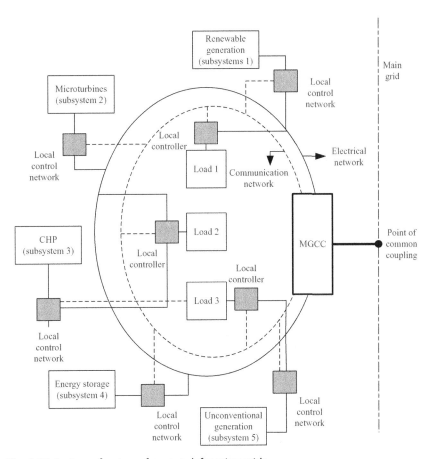

Fig. 1.13 System of systems framework for microgrids.

characterize it in an SoS framework. An SoS framework addresses all the management, communication, and control needs of a microgrid in a systematic manner.

The MGCC is connected to the main grid via a point of common coupling. All the constituent subsystems are integrated with the MGCC through a local control network. An electrical network and a communication network are used for the transfer of control signals and data collection, respectively. Each subsystem has a dedicated local controller. These local controllers form the primary level of the microgrid SoS control hierarchy. Together, all the subsystems constitute the SoS framework.

5.3. Control Hierarchy

The control design for microgrids is a major issue that needs attention. On the basis of the microgrid SoS structure and framework mentioned previously, a control method based on an SoS is proposed for microgrids. A hierarchical control structure for the microgrid SoS is illustrated in Fig. 1.14. It can be seen that subsystems (distributed generation units) of the microgrid are integrated with the primary and secondary level of control. The tertiary level, however, is connected to the SoS through a

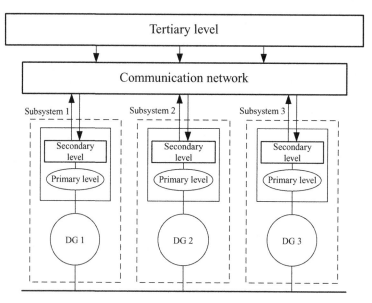

Fig. 1.14 Control structure for a microgrid system of systems.

Table 1.4 Functional levels of the microgrid system of systems

Level	Function
Primary	Voltage and frequency stabilization after islanding
Secondary	Check for any deviations caused by primary control
Tertiary	Govern the power flow between the microgrid and the main grid

Table 1.5 Variables of a microgrid system of systems

Level	Variable
Primary	Voltage (E)
	Frequency (ω)
	Power (P)
	Reactive power (Q)
Secondary	Reference voltage (E^{ref})
	Reference frequency (ω^{ref})
Tertiary	Reference power (P^{ref})
	Reference reactive Power (Q^{ref})

communication network. Table 1.4 summarizes the functions of each level. The variables for each level are described in Table 1.5.

1. *Primary level*: The primary level is responsible for stabilizing the voltage and frequency after islanding. It consists of elementary control hardware. The current and voltage loops of the DERs are included in the control hardware. It adjusts the amplitude and frequency of the reference voltage for instance. Further, it also alleviates circulating currents that are a threat to the power electronic devices of DERs. It also distributes active and reactive power among DERs. Following an islanding event, a microgrid suffers from voltage and frequency instabilities because of power mismatch. Thus voltages and frequencies need to be stabilized. This is accomplished by the primary level [28–32].

2. *Secondary level*: The secondary level is responsible for compensating for any voltage or frequency deviation that is caused by the primary level. It makes sure that the deviations are regulated toward zero if any load or generation changes in the microgrid [95–104].

3. *Tertiary level*: The tertiary control level controls the flow of power from the microgrid to the main grid. This is accomplished by adjustment of the frequencies and amplitudes of the DER voltages. It also ensures economically optimal operation of the microgrid [28–32].

Remark 1.6. *The foregoing guidelines describe the microgrid system as an SoS. The characteristics of the microgrid system are presented that bear remarkable resemblance to an SoS. The structure of the SoS is presented and a framework is proposed for the microgrid. Further, a hierarchical control structure for the microgrid SoS is also presented.*

5.4. Grid Synchronization in Distributed Generation Units

Grid synchronization of all constituent distributed generation units is very critical in a microgrid system [105]. The distributed generation units like microturbines, photovoltaic systems, and wind turbines are connected to the grid through a power conversion unit as shown in Fig. 1.15. This unit can incorporate inverters, DC-DC converters, power conditioning units, and filters. Many synchronization techniques can be found in

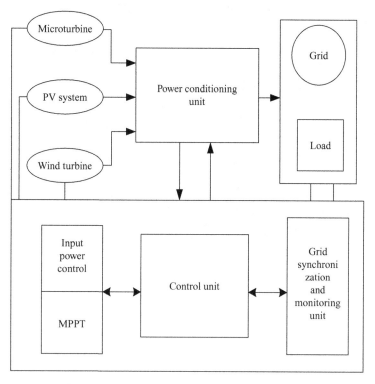

Fig. 1.15 Microgrid structure with synchronization and control units.

the literature [106, 107]. A comparison of various phase-locked-loop techniques to accomplish grid synchronization is presented in [108].

Another important aspect is the control of distributed generation units. When connected to the grid, the control segregates into two divisions: namely, input-side control and grid-side control [109]. While input-side control is primarily concerned with extracting maximum power from the source, grid-side control must achieve the following:

1. synchronization with the grid;
2. active and reactive power control between the distributed generation units and the grid;
3. ensure the quality of power injected is high; and
4. protection from an islanding event and maintaining high efficiency.

5.5. Overall Microgrid Modeling for Control Implementation

A schematic of different distributed generation units in a microgrid system serving as subsystems is presented in Fig. 1.16. These distributed generation units can be integrated with the main grid through a power conditioning unit. Individual modeling of each subsystem was presented earlier. One can obtain a comprehensive model of the entire microgrid system by combining all distributed generation units. As can be observed, all the distributed generation units have a power conditioning unit in common. Hence the dynamic equations of each distributed generation unit coupled with the

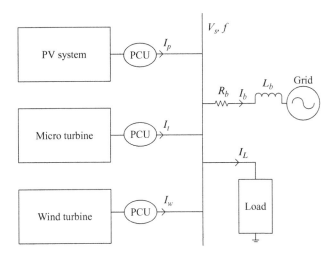

Fig. 1.16 Overall microgrid model.

equations for the currents and voltages of the power conditioning unit will render the complete mathematical model of the distributed generation system.

Remark 1.7. *One of the major challenges in microgrids is the control design. After the microgrid has been drafted in the SoS framework and a survey of modeling methods has been presented, the next task is to explain control paradigms based on an SoS in terms of microgrid control. The next section deals with various control methods for an SoS elucidated for a microgrid.*

6. CONTROL METHODS FOR A MICROGRID SYSTEM OF SYSTEMS

Although extensive research has been conducted in designing control strategies for microgrids, there is still ambiguity regarding the best microgrid control strategies. A survey of the various control techniques developed for microgrids is presented in [17]. Several control strategies were proposed for microgrids in [68–71]. Robust H_∞ control is presented in [72, 73] for the control of two distributed generation units. An optimal controller is presented for control of the frequency and voltage fluctuations during islanded mode in [74].

However, since we are looking at microgrids from an SoS perspective, we need to use the control strategies devised for an SoS. The first hurdle lies in finding a mathematical model for the framework. The control paradigms given in [88] need a mathematical model for their implementation. Nevertheless, a few mathematical models of microgrids are available in the literature that are complex and of high order [95, 96]. Additionally, state space modeling has also been done for microgrids [97, 98].

Once we have obtained the mathematical model, we look at the control strategies that might be applied to microgrids. One interesting control strategy is hierarchical control [23, 100, 110–112].

6.1. Hierarchical Control

Functionally, a microgrid can operate by using the following three main hierarchical control levels:
- Primary control is the droop control used to share load between converters.
- Secondary control is responsible for removing any steady-state error introduced by the droop control.

- Tertiary control concerning more global responsibilities decides on the import or export of energy for the microgrid.

Systems of this kind operate over large synchronous machines with high inertias and inductive networks. However, in power electronics-based microgrids, there are no inertias, and the nature of the networks is mainly resistive. There are major differences between both systems that we have to take into account when designing their control schemes. A three-level hierarchical control can be constructed following the method in [99]. The primary control deals with the inner control of the distributed generation units by adding virtual inertias and controlling their output impedance. The secondary control is conceived to restore the frequency and amplitude deviations produced by the virtual inertias and output virtual impedance. The tertiary control regulates the power flows between the grid and the microgrid at the point of common coupling.

In what follows we look at the basic elements of consensus-based control that can be possibly applied to a microgrid. This is an important class type of the cooperative control paradigm [113]. We focus on the mutual agreement among the different units of a microgrid as subsystems in an SoS.

6.2. Consensus Control

The consensus control problem has been discussed extensively in the literature owing to its applicability in the formation of robots, cooperative control of unmanned aerial vehicles, and communication between sensor networks [114]. Consensus among a group of agents or subsystems means to arrive at a certain agreement concerning a particular value or quantity that is dependent on the states of every agent. A consensus algorithm or protocol is a rule that explains the interaction between an agent and its surrounding subsystems enclosed in the network. However, management of the shared information between the agents is a very critical issue that has to be resolved in order to carry out coordination among the subsystems. To address the consensus problems for dynamic networked agents, a theoretical framework was developed by Olfati-Saber et al. [115, 116]. A consensus protocol is introduced with x_i representing the information state of the ith agent, where the information state is responsible for the information that has to be coordinated among the agents, and the consensus protocol for continuous systems can be formulated as in [117–119].

Consensus control of microgrids is proposed in [120]. Coordinated performance is achieved at the secondary level despite the units being out of synchronization at the beginning. The proposed scheme is tested for an islanded microgrid in various scenarios. Another distributed cooperative control method based on consensus is proposed in [121]. Power sharing is precisely realized among distributed generation units in addition to voltage regulation at the critical bus by the virtue of this method.

6.3. Decentralized Control

With reference to Fig. 1.17, the control of the inverter plus filter interfaces is crucial to the operation of the microgrid. Because of the distributed nature of the system, these interfaces need to be controlled on the basis of local measurements only. The decentralized control of the individual interfaces should address the following basic issues:

1. Interfaces should share the total load (linear or nonlinear) in a desired way.
2. Decentralized control based on local measurement should guarantee stability on a global scale.

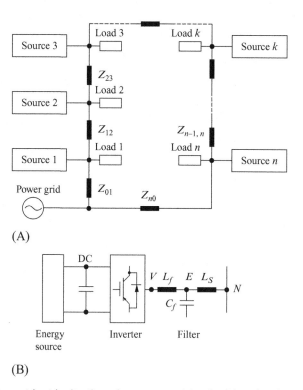

Fig. 1.17 Microgrid with distributed sources and loads. (A) a distributed microgrid structure (B) an inverter interface.

3. Inverter control should prevent any DC voltage offsets on the microgrid.

4. Inverter control should actively damp oscillations between the output filters.

From the viewpoint of decentralized control, it is convenient to classify distributed generation architectures into three classes:

- In *highly dispersed networks* the interconnecting impedances are predominantly inductive, and the voltage magnitude and phase angle at different source interconnects can be very different.

- In *networks spread over a smaller area* the impedances are still inductive, with a significant resistive component. The voltage magnitude does not differ much, but the phase angles can be different for different sources.

- In very small networks the impedances are small and predominantly resistive. Neither magnitude nor phase angle differences are significant at any point.

The main common quantity in all cases is the steady-state frequency, which must be the same for all sources. In grid-connected mode the microgrid frequency is decided by the grid. In islanded mode the frequency is decided by the microgrid control.

In each of these classes, if every source is connected to at most two other sources, as shown in Fig. 1.17A, then the microgrid is radial; otherwise it is meshed. If there is a line connecting source 1 with source k in Fig. 1.17A, then it is a meshed microgrid. It is worth noting that most of the work done on microgrid decentralized control has been for radial–microgrid topologies. The decentralized control of interfaces in meshed topologies is an area that needs further research.

6.4. Networked Control

Another control paradigm based on an SoS that can be extended to a microgrid is networked control. Networked control of an SoS was introduced in [88]. A control system consisting of a real-time network in its feedback can be termed a *networked control system* [92]. As mentioned earlier, the microgrid can operate at multiple levels, forming a control hierarchy. At the primary level there is no need for a communication network, since the control is based on local measurements only. However, at the secondary level, a communication network is required to accomplish global controllability of microgrids. The set points for the voltage and frequency to be maintained are generated by a higher control level and need to be communicated to the primary level for the local controllers.

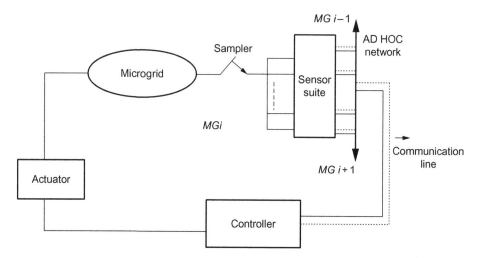

Fig. 1.18 Networked control of a microgrid based on a system of systems.

A networked control structure of a microgrid based on an SoS is shown in Fig. 1.18. The stability of the communication network connecting the distributed generation units is a major concern. Ensuring network stability in this scenario is of paramount importance. Any packet lost or delays in the communication can cause severe power mismatches among the distributed units. On the other hand, the inverters connected to these systems operate under imbalance conditions because of sensitive loads. This leads to switching harmonics and voltage and frequency variations in the microgrid system and disturbs the stability of the system.

The primary challenge in SoS networked control design for a microgrid system is to build a distributed control system that can endure packet losses, delays, and partially decoded packets that affect system stability [88]. In other words, it is expected to add robustness to the system. Alternatively, all characteristics of the ad hoc network must be considered while one is designing a networked control system to check on communication infractions and ensure a robust and stable operation.

7. FUTURE DIRECTIONS

Research interest in microgrids has increased significantly, triggered by the increasing demand for reliable, secure, and sustainable electrical energy. More research into and implementation of microgrids will be conducted to improve the maturity of microgrid technology. Microgrids are rich dynamical systems for modeling, control, optimization, and simulation. Several research problems need to be solved to keep up with planned renewable energy integration in the electrical grid:

- Techniques based on concepts of distributed control and adaptive control strategies are promising, particularly for droop-controlled microgrids including frequency, voltage, and reactive power sharing controllers. The purpose is to generate additional proper signals for the primary control level by use of the measurements of other distributed generation units in each sample time.
- Designing a multiagent-based intelligent microgrid controller, designing a self-reconfigurable microgrid, and handling inherent uncertainties of renewable sources are key issues to be carefully investigated (see Fig. 1.20) [122–139].
- DC distribution-based microgrids and a hybrid distribution network for microgrids are areas of research that call for collaborative efforts.
- The whole idea of using the concept of an SoS is to understand the complex microgrid system, by considering not only the components but also the whole system. It is of interest to start from Fig. 1.19 and build on dynamic modeling of the subsystems.
- Renewable resources are widely distributed and because of the intermittent nature of power, such a new distributed system can be provided by various generation approaches to obtain the maximum potential energy from the sources. The communication system, stability, and control issues of microgrids are worth investigating to obtain concrete results (see Fig. 1.20).

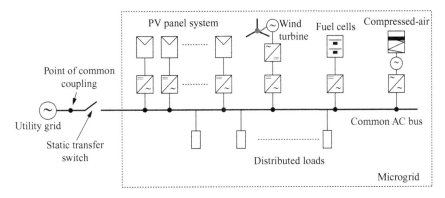

Fig. 1.19 Possible microgrid.

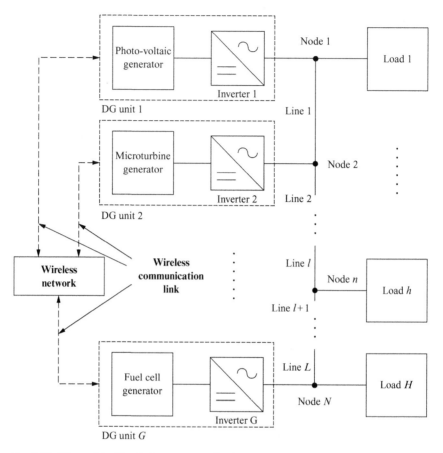

Fig. 1.20 Microgrid with wireless communication.

- Investigation of stability issues for both grid-connected mode and islanded mode for various types of microgrids in term of voltage and frequency has not been fully completed, despite the availability of hardware technology advancement. Investigation of the full-scale development, and experimental evaluation of V/f control methods according to several operation modes is among the significant issues. Determining the transition dynamics between grid-connected and islanded modes on the basis of interactions between the distribution generation and high penetration of distributed generation is another significant issue.
- Definition and formal design of intelligent and robust energy delivery systems in the future by providing significant reliability and security benefits, by considering the microgrid structure in Fig. 1.21.

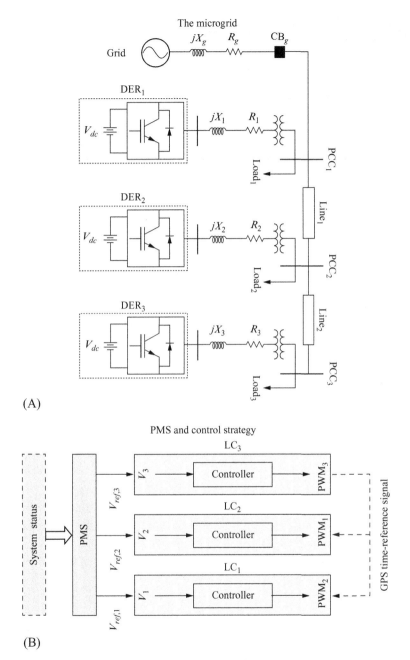

Fig. 1.21 A detailed microgrid architecture. (A) The microgrid. (B) PMS and control strategy.

- Careful examination of the dynamic behavior of microgrids during transition from islanding operation to grid connection or vice versa. This is crucial for the energy sustainability and load balancing of the system. Energy management and load scheduling are issues of prime concern in this regard.
- It is known that a robustly controlled microgrid ensures seamless import/export of active power and reactive power by the main grid and continuous supply of the critical load during islanded operation. These lead to a flexible and smart power system. What type of control strategies guarantee this is a topic for further research; see [140–142].

8. CONCLUSIONS

Microgrid systems facilitate remote applications and allow access to pollution-free energy. They give impetus to the use of renewable sources of energy. Moreover, in the event of a power grid failure, a microgrid is one of the best alternatives. This chapter has provided an overview of microgrid systems and elaborated on several aspects of control, mode of operation, and distributed energy storage applications within microgrids and desired targets. Microgrid research fits very well with ongoing smart grid activities throughout the world, and several challenges need to be overcome before it becomes a reality. In particular, control with limited communication and computing facilities is a challenging problem favoring the adoption of decentralized techniques.

Considering the wide variety of applications of SoS technology, the concept of an SoS was extended to a microgrid. A review of numerous microgrid architectures, models, layouts, and control methods was presented. A unique SoS perspective on microgrids was provided and further elucidated by the proposal of a framework for microgrids. The resemblance of microgrid features to those of an SoS was highlighted, leading to a generalized structure of a microgrid SoS, where the DERs of the microgrid are represented as subsystems.

ACKNOWLEDGMENTS

This work is supported by the Deanship of Scientific Research (DSR) at King Fahd University of Petroleum and Minerals (KFUPM) through book writing research project no. BW 151004.

REFERENCES

[1] R.H. Lasseter, Microgrids and distributed generation, J. Energy Eng. 1 (2007) 1–7.
[2] A. Arulampalam, M. Barnes, A. Engler, A. Goodwin, N. Jenkins, Control of power electronic interfaces in distributed generation microgrids, Int. J. Electron. 91 (2004) 503–523.
[3] N. Hatziargyriou, H. Asano, R. Iravani, C. Marnay, Microgrids: an overview of ongoing research and development and demonstration projects, in: Proc. IEEE, 2007.
[4] European Research Project Microgrids [Online], Available at: http://Microgrids. power.ece.ntua.gr.
[5] N. Hatziargyriou, A. Dimeas, A. Tsikalakis, Centralized and decentralized control of microgrids, Int. J. Dist. Energy Resour. 1 (3) (2005) 97–212.
[6] M. Barnes, A. Engler, C. Fitzer, N. Hatziargyriou, C. Jones, S. Papathanassiou, et al., MicroGrid laboratory facility, in: Proceedings of IEEE International Conference on Future Power Systems, 2005.
[7] G.A. Pagani, M. Aiello, Towards decentralization: a trpological investigation of the medium and low voltage grids, IEEE Trans. Smart Grid 2 (2011) 538–547.
[8] T.L. Vandoorn, B. Renders, L. Degroote, B. Meersman, L. Vandevelde, Active load control in islanded microgrids based on the grid voltage, IEEE Trans. Smart Grid 2 (2011) 139–151.
[9] Q. Yang, J.A. Barria, T.C. Green, Advanced power electronic conversion and control system for universal and flexible power management, IEEE Trans. Smart Grid 2 (2011) 231–243.
[10] G. Pepermans, J. Driesen, D. Haeseldonckx, R. Belmans, W. Dhaeseleer, Distributed generation: definition, benefits and issues, Energy Policy 33 (6) (2005) 787–798.
[11] K. De Brabandere, K. Vanthournout, J. Driesen, G. Deconinck, R. Belmans, Control of microgrids, in: IEEE Power Engineering Society General Meeting, 2007, pp. 1–7.
[12] M.A. Pedrasa, T. Spooner, A survey of techniques used to control microgrid generation and storage during island operation, in: Proceedings of the Australian Universities Power Engineering Conference, 2006.
[13] J.A. Peças Lopes, C.L. Moreira, A.G. Madureira, Defining control strategies for analyzing microgrids islanded operation, in: IEEE Russia Power Tech, 2005, pp. 1–7.
[14] M.H. Ashourian, A.A.M. Zin, A.S. Mokhtar, S.J. Mirazimi, Z. Muda, Controlling and modeling power-electronic interface DERs in islanding mode operation micro grid, in: IEEE Symposium on Industrial Electronics and Applications (ISIEA), 2011, pp. 161–166.
[15] E. Barklund, N. Pogaku, M. Prodanovic, C. Hernandez-Aramburo, T.C. Green, Energy management in autonomous microgrid using stability-constrained droop control of inverters, IEEE Trans. Power Electron. 23 (5) (2008) 2346–2352.
[16] European Research Project More Microgrids [Online], Available at: http:// Microgrids.power.ece.ntua.gr.
[17] Z. Xiao, J. Wu, N. Jenkins, An overview of microgrid control, Intell. Autom. Soft Comput. 16 (2) (2010) 199–212.
[18] M.S. Mahmoud, F.M. Al-Sunni, Control and Optimization of Distributed Generation Systems, Springer-Verlag, UK, 2015.
[19] R. Lasseter, A. Akhil, C. Marnay, J. Stephens, J. Dagle, R. Guttromson, et al., The CERTS microgrid concept, in: White Paper on Integration of Distributed Energy Resources, Prepared for Transmission Reliability Program, Office of Power Technologies, US Department of Energy, 2002, http://certs.lbl.gov/pdf/50829-app.pdf.

[20] J. Eto, R. Lasseter, B. Schenkman, J. Stevens, D. Klapp, H. Volkommer, E. Linton, H. Hurtado, J. Roy, Overview of the CERTS microgrid laboratory test bed, in: PES Joint Symposium Integration of Wide-Scale Renewable Resources into the Power Delivery System, July 29–31, vol. 1, 2009 .

[21] P. Piagi, R.H. Lasseter, Autonomous control of microgrids, in: Proceedings of IEEE Power Engineering Society General Meeting (PES), 2006.

[22] E. Hossain, E. Kabalci, R. Bayindir, R. Perez, A comprehensive study on microgrid technology, Int. J. Renew. Energy Res. 4 (4) (2014) 1094–1107.

[23] J.C. Vasquez, J.M. Guerrero, J. Miret, M. Castilla, L.G.D. Vicuna, Hierarchical control of intelligent microgrids, IEEE Ind. Electron. Mag. 4 (2010) 23–29.

[24] M. Prodanovic, T.C. Green, High-quality power generation through distributed control of a power park microgrid, IEEE Trans. Ind. Electron. 53 (5) (2006) 1471–1482.

[25] E. Jones, C. Fitzer, M. Barnes, Investigation of microgrids, in: Proceedings of the 3rd IET International Conference on Power Electronics, Machines and Drives, 2006, pp. 510–514.

[26] J.A.P. Lopes, C.L. Moreira, A.G. Madureira, Defining control strategies for analyzing microgrids islanded operation, in: Proceedings of IEEE Russia Power Tech, 2007, pp. 1–7.

[27] D. Georgakis, S. Papathanassiou, Operation of a prototype microgrid system based on micro-sources equipped with fast acting power electronics interfaces, in: IEEE 35th Annual Power Electronics Specialists Conference, Germany, vol. 4, 2004, pp. 2521–2526.

[28] A. Bidram, A. Davoudi, Hierarchical structure of microgrids control system, IEEE Trans. Smart Grid 3 (2012) 1963–1976.

[29] S. Baudoin, I. Vechiu, H. Camblong, A review of voltage and frequency control strategies for islanded microgrid, in: Proceedings of IEEE 16th International Conference on System Theory, Control and Computing (ICSTCC), 2012, pp. 1–5.

[30] J.B. Almada, R.P.S. Leao, F.F.D. Montenegro, S.S.V. Miranda, R.F. Sampaio, Modeling and simulation of a microgrid with multiple energy resources, in: Proceedings of IEEE Eurocon, 2013, pp. 1150–1157.

[31] A.S. Dobakhshari, S. Azizi, A.M. Ranjbar, Control of microgrids: aspects and prospects, in: Proceedings of the IEEE International Conference on Networking, Sensing and Control, 2011, pp. 38–43.

[32] S.A. Kale, P.P. Jagtap, J.B. Helonde, Role of micro sources within microgrid, in: Proceedings of the IEEE Fourth International Conference on Emerging Trends in Engineering & Technology, 2011, pp. 174–179.

[33] M.E. Elkhatib, R. El-Shatshat, M.M.A. Salama, Novel coordinated voltage control for smart distributed networks with DG, IEEE Trans. Smart Grid 2 (2011) 598–605.

[34] R.H. Lasseter, P. Paigi, MicroGrid: a conceptual solution, in: IEEE Annual Power Electron Specialists Conference, vol. 1, 2004, pp. 4285–4290.

[35] R.H. Lasseter, MicroGrids, in: IEEE Power Engineering Society Transmission & Distribution Conference, vol. 1, 2002, pp. 305–308.

[36] M. Chris, V. Giri, Microgrids in the evolving electricity generation and delivery infrastructure, in: IEEE Power Engineering Society General Meeting, vol. 18–22, 2006.

[37] J.A.P. Lopes, C.L. Moreira, A.G. Madureira, Defining control strategies for microgrids islanded operation, IEEE Trans Power Syst. 21 (2) (2006) 916–924.

[38] Q. Shafiee, J.M. Guerrero, J.C. Vasquez, Distributed secondary control for islanded microgrids—a novel approach, IEEE Trans. Power Electron. 29 (2) (2014) 1018–1031.

[39] D. Yubing, G. Yulei, L. Qingmin, W. Hui, Modelling and simulation of the microsources within a microgrid, in: IEEE International Conference on Electrical Machines and Systems ICEMS, 2008.

[40] Y. Zhu, K. Tomsovic, Development of models for analyzing load-following performance of microturbines and fuel cells, Electr. Power Syst. Res. 62 (1) (2002) 1–11.

[41] G. Stavrakakis, G. Kariniotakis, A general simulation algorithm for the accurate assessment of isolated diesel-wind turbines systems—part I: a general multi-machine power system model, IEEE Trans. Energy Convers. 10 (3) (1995) 577–583.

[42] S. Roy, O. Malik, G. Hope, An adaptive control scheme for speed control of diesel driven power-plants, IEEE Trans. Energy Convers. 6 (4) (1991) 605–611.

[43] N. Hatziargyriou, G. Kariniotakis, Modelling of microsources for security studies, in: Proceedings of the 2004 CIGRE SESSION, Paris, France, 2004.

[44] G. Kariniotakis, et al., DA1-digital models for microsources, in: Microgrids Project Deliverable DA1, 2003.

[45] F. Mohamed, Microgrid modelling and simulation, Helsinki University of Technology, Finland, 2006.

[46] X. Zhang, J. Liu, T. Liu, L. Zhou, A novel power distribution strategy for parallel inverters in islanded mode microgrid, in: The Twenty-Fifth Annual IEEE Applied Power Electronics Conference and Exposition (APEC), 2010, pp. 2116–2120.

[47] P.K. Ray, S.R. Mohanty, N. Kishor, Dynamic modeling and control of renewable energy based hybrid system for large band wind speed variation, in: Proceedings of IEEE PES Innovative Smart Grid Technologies Conference Europe (ISGT Europe), 2010, pp. 1–10.

[48] N.S. Jayalakshmi, D.N. Gaonkar, Performance study of isolated hybrid power system with multiple generation and energy storage units, in: International Conference on Power and Energy Systems (ICPS), 2011, pp. 1–5.

[49] F. Katiraei, M.R. Iravani, Power management strategies for a microgrid with multiple distributed generation units, IEEE Trans. Power Syst. 21 (4) (2006) 1821–1831.

[50] D.J. Lee, L. Wang, Small-signal stability analysis of an autonomous hybrid renewable energy power generation/energy storage system part I: time-domain simulations, IEEE Trans. Energy Convers. 23 (1) (2008) 311–320.

[51] A.M.O. Haruni, A. Gargoom, M.E. Haque, M. Negnevitsky, Dynamic operation and control of a hybrid wind-diesel stand alone power systems, in: The Twenty-Fifth Annual IEEE Applied Power Electronics Conference and Exposition (APEC), 2010, pp. 162–169.

[52] G.K. Kasal, B. Singh, Voltage and frequency controllers for an asynchronous generator-based isolated wind energy conversion system, IEEE Trans. Energy Convers. 26 (2) (2011) 402–416.

[53] I. Salhi, S. Doubabi, N. Essounbouli, Fuzzy control of micro hydro power plants, in: 5th IET International Conference on Power Electronics, Machines and Drives (PEMD 2010), 2010.

[54] D.C.H. Das, A.K. Roy, N. Sinha, PSO based frequency controller for wind-solar-diesel hybrid energy generation/energy storage system, in: International Conference on Energy, Automation, and Signal (ICEAS), 2011, pp. 1–6.

[55] D. Soto, C. Edrington, S. Balathandayuthapani, S. Ryster, Voltage balancing of islanded microgrids using a time-domain technique, Electr. Power Syst. Res. 84 (1) (2012) 214–223.

[56] C.K. Sao, P.W. Lehn, Intentional islanded operation of converter fed microgrids, in: Proceedings of IEEE Power Engineering Society General Meeting, 2006, pp. 6–12.

[57] H. Karimi, H. Nikkhajoei, R. Iravani, Control of an electronically-coupled distributed resource unit subsequent to an islanding event, IEEE Trans. Power Deliv. 23 (1) (2008) 493–501.

[58] C. Sao, P.W. Lehn, Voltage balancing of converter fed microgrids with single phase loads, in: Proceedings of the IEEE Power and Energy Society General Meeting-Conversion and Delivery of Electrical Energy in the 21st Century, 2008, pp. 1–7.

[59] C. Sao, P.W. Lehn, Control and power management of converter fed microgrids, IEEE Trans. Power Syst. 23 (3) (2008) 1088–1098.

[60] S.M.A. Shabestary, M. Saeedmanesh, A. Rahimi-Kian, E. Jalalabadi, Real-time frequency and voltage control of an islanded mode microgrid, in: The 2nd Iranian Conference on Smart Grids (ICSG), 2012, pp. 1–6.

[61] S.N. Bhaskara, M. Rasheduzzaman, B.H. Chowdhury, Laboratory-based microgrid setup for validating frequency and voltage control in islanded and grid-connected modes, in: IEEE Green Technologies Conference, 2012, pp. 1–6.

[62] M. Babazadeh, H. Karimi, Robust decentralized control for islanded operation of a microgrid, in: IEEE Power and Energy Society General Meeting, 2011, pp. 1–8.

[63] R. Moradi, H. Karimi, M. Karimi-Ghartemani, Robust decentralized control for islanded operation of two radially connected DG systems, in: IEEE International Symposium on Industrial Electronics (ISIE), 2010, pp. 2272–2277.

[64] F. Katiraei, M.R. Iravani, Power management strategies for a microgrid with multiple distributed generation units, IEEE Trans. Power Syst. 21 (2006) 1821–1831.

[65] D.J. Lee, L. Wang, Small-signal stability analysis of an autonomous hybrid renewable energy power generation/energy storage system part I: time-domain simulations, IEEE Trans. Energy Convers. 23 (1) (2008) 311–320.

[66] A.M.O. Haruni, A. Gargoom, M.E. Haque, M. Negnevitsky, Dynamic operation and control of a hybrid wind-diesel stand alone power systems, in: Proceedings of the Twenty-Fifth Annual IEEE Applied Power Electronics Conference and Exposition (APEC), 2010, pp. 162–169.

[67] G.K. Kasal, B. Singh, Voltage and frequency controllers for an asynchronous generator-based isolated wind energy conversion system, IEEE Trans. Energy Convers. 26 (2) (2011) 402–416.

[68] M.H. Ashourian, M.A.A. Zin, A.S. Mokhtar, S.J. Mirazimi, Z. Muda, Controlling and modeling power-electronic interface DERs in islanding mode operation microgrid, in: Proceedings of the IEEE Symposium on Industrial Electronics and Applications (ISIEA), 2011, pp. 161–166.

[69] X. Zhang, J. Liu, T. Liu, L. Zhou, A novel power distribution strategy for parallel inverters in islanded mode microgrid, in: Proceedings of the Twenty-Fifth Annual IEEE Applied Power Electronics Conference and Exposition (APEC), 2010, pp. 2116–2120.

[70] P.K. Ray, S.R. Mohanty, N. Kishor, Dynamic modeling and control of renewable energy based hybrid system for large band wind speed variation, in: Proceedings of the IEEE PES Innovative Smart Grid Technologies Conference Europe (ISGT Europe), 2010, pp. 1–10.

[71] N.S. Jayalakshmi, D.N. Gaonkar, Performance study of isolated hybrid power system with multiple generation and energy storage units, in: Proceedings of the International Conference on Power and Energy Systems (ICPS), 2011, pp. 1–5.

[72] M. Babazadeh, H. Karimi, Robust decentralized control for islanded operation of a microgrid, in: Proceedings of the IEEE Power and Energy Society General Meeting, 2011, pp. 1–8.

[73] R. Moradi, H. Karimi, M. Karimi-Ghartemani, Robust decentralized control for islanded operation of two radially connected DG systems, in: Proceedings of the IEEE International Symposium on Industrial Electronics (ISIE), 2010, pp. 2272–2277.

[74] S.M.A. Shabestary, M. Saeedmanesh, A. Rahimi-Kian, E. Jalalabadi, Real-time frequency and voltage control of an islanded mode microgrid, in: Proceedings of the 2nd Iranian Conference on Smart Grids (ICSG), 2012, pp. 1–6.

[75] F. Katiraei, M.R. Iravani, P.W. Lehn, Microgrid autonomous operation during and subsequent to islanding process, IEEE Trans. Power Deliv. 20 (1) (2005) 248–257.

[76] H. Karimi, H. Nikkhajoei, R. Iravani, Control of an electronically-coupled distributed resource unit subsequent to an islanding event, IEEE Trans. Power Deliv. 23 (1) (2008) 493–501.

[77] F.D. Kanellos, A.I. Tsouchnikas, N.D. Hatziargyriou, Microgrid simulation during grid-connected and islanded modes of operation, in: Proceedings of the International Conference on Power Systems Transients (IPST), Canada, 2005.

[78] A.K. Saha, S. Chowdhury, S.P. Chowdhury, P.A. Crossley, Modelling and simulation of microturbine in islanded and grid-connected mode as distributed energy resource, in: Proceedings of the IEEE Power and Energy Society General Meeting, Pittsburgh, 2008, pp. 1–7.

[79] J.H. Watts, Microturbines: a new class of gas turbine engines, Gas Turbine News Brief 39 (1) (1999) 5–11.

[80] M. Nagpal, A. Moshref, G.K. Morison, et al., Experience with testing and modeling of gas turbines, in: Proceedings of the IEEE/PES 2001 Winter Meeting, January/February, Ohio, USA, 2001, pp. 652–656.

[81] J.C.H. Phang, D.S.H. Chan, J.R. Phillips, Accurate analytical method for the extraction of solar cell model parameters, Electron. Lett. 20 (10) (1984) 406–408.

[82] R.J. Wai, W.H. Wang, Grid connected photovoltaic generation system, IEEE Trans. Circuits Syst. I Regul. Pap. 55 (2008) 953–964.

[83] M.A. Eltawil, Z. Zhao, Grid-connected photovoltaic power systems: technical and potential problems—a review, Renew. Sustain. Energy Rev. 14 (2010) 112–129.

[84] R. Mukund, Wind and Solar Power Systems, CRC Press, Boca Raton, FL, 1999.

[85] N. Jenkins, R. Allan, P. Crossley, D. Kirschen, G. Strbac, Embedded Generation, The Institution of Electrical Engineers, United Kingdom, 2000.

[86] T. Ackermann, Wind Power in Power Systems, John Wiley and Sons, Royal Institute of Technology, Stockholm, Sweden, 2005.

[87] G.N. Kariniotakis, N.L. Soultanis, A.I. Tsouchnikas, S.A. Papathanasiou, N.D. Hatziargyriou, Dynamic modeling of microgrids, in: IEEE International Conference on Future Power Systems, 2005.

[88] M. Jamshidi, Control of system of systems, in: Proceedings of the 7th IEEE International Conference on Industrial Informatics, 2009, pp. 1–16.

[89] M. Jamshidi, System of systems engineering—new challenges for the 21st century, IEEE Syst. Mag. (2002).

[90] M.O. Jamshidi, System of Systems Engineering: Principles and Applications, CRC Press, Boca Raton, FL, 2008.

[91] M.O. Jamshidi, System of systems engineering—new challenges for the 21st century, IEEE Aerosp. Electron. Syst. Mag. 23 (2008) 4–19.

[92] W. Zhang, M. Branicky, S. Phillips, Stability of networked control systems, IEEE Control Syst. Mag. 21 (1) (2001) 84–99.

[93] M. Henshaw, C. Siemieniuch, M. Sinclair, S. Henson, V. Barot, M.O. Jamshidi, D. Delaurentis, C. Ncube, S.L. Lim, H. Dogan, Systems of systems engineering: a research imperative, in: Proceedings of the IEEE International Conference on System Science and Engineering, Hungary, 2013, pp. 389–394.

[94] M.W. Maier, Architecting principles for systems-of-systems, Syst. Eng. 1 (4) (1998) 267–284.

[95] A.H. Etemadi, E.J. Davison, R. Iravani, A Decentralized robust control strategy for multi-DER microgrids—part I: fundamental concepts, IEEE Trans. Power Deliv. 27 (4) (2012) 1843–1853.

[96] A.H. Etemadi, R. Iravani, Eigenvalue and robustness analysis of a decentralized voltage control scheme for an islanded multi-DER microgrid, IEEE Trans. Power Deliv. 27 (4) (2012) 1447–1455.

[97] C.L. Chen, J.S. Lai, D. Martin, Y.S. Lee, State-space modeling, analysis, and implementation of paralleled inverters for microgrid applications, in: 2010 Twenty-Fifth Annual IEEE Applied Power Electronics Conference and Exposition (APEC), 2010.

[98] M. Zhu, H. Li, X. Li, Improved state-space model and analysis of islanding inverter-based microgrid, in: IEEE International Symposium on Industrial Electronics (ISIE), 2013.

[99] J.M. Guerrero, J.C. Vasquez, J. Matas, L.G. de Vicuna, M. Castilla, Hierarchical control of droop-controlled AC and DC microgrids—a general approach toward standardization, IEEE Trans. Ind. Electron. 58 (1) (2011) 158–172.

[100] A.L. Dimeas, N.D. Hatziargyriou, Operation of a multi-agent system for microgrid control, IEEE Trans. Power Syst. 20 (3) (2005) 1447–1455.

[101] E. Planas, A. Gil-de Muro, J. Andreu, I. Kortabarria, I.M. de Algeria, General aspects, hierarchical controls and droop methods in microgrids: a review, Renew. Sustain. Energy Rev. 17 (2013) 147–159.

[102] O. Palizban, K. Kauhaniemi, J.M. Guerrero, Microgrids in active network management—part I: hierarchical control, energy storage, virtual power plants, and market participation, Renew. Sustain. Energy Rev. 36 (2014) 428–439.

[103] M. Barnes, J. Kondoh, H. Asano, J. Oyarzabal, G. Ventakaramanan, R. Lasseter, N. Hatziargyriou, T. Green, Real-world microgrids—an overview, in: Proceedings of IEEE International Conference on System of Systems Engineering, 2007, pp. 1–8.

[104] R. Zamora, A.K. Srivastava, Controls for microgrids with storage: review, challenges, and research needs, Renew. Sustain. Energy Rev. 14 (7) (2010) 2009–2018.

[105] Y. Yang, F. Blaabjerg, Synchronization in single-phase grid connected photovoltaic systems under grid faults, in: Proceedings of the IEEE International Symposium on Power Electronics for Distributed Generation Systems, 2012, pp. 476–482.

[106] E.J. Coster, J.M.A. Myrzik, B. Kruimer, W.L. Kling, Integration issues of distributed generation in distributed grids, Proc. IEEE 99 (1) (2011) 28–39.

[107] A. Nicastri, A. Nagliero, Comparison and evaluation of the PLL techniques for the design of the grid-connected inverter systems, in: Proceedings of the IEEE International Symposium on Industrial Electronics, 2010, pp. 3865–3870.

[108] A. Nagliero, R.A. Mastromauro, M. Liserre, A. Dell'Aquila, Synchronization techniques for grid connected wind turbines, in: Proceedings of the IEEE Annual Conference on Industrial Electronics, 2009, pp. 4606–4613.

[109] F. Blaabjerg, R. Teodorescu, M. Liserre, A.V. Timbus, Overview of control and grid synchronization for distributed power generation systems, IEEE Trans. Ind. Electron. 53 (5) (2006) 1398–1409.

[110] M.A. Pedrasa, T. Spooner, A Survey of techniques used to control microgrid generation and storage during island operation, in: Australian Universities Power Engineering Conference, 2006.

[111] M.S. Mahmoud, Multilevel systems: information flow in large linear problems, in: M.G. Singh, A. Titli (Eds.), Handbook of Large Scale Systems Engineering Applications, North Holland, Amsterdam, 1979, pp. 96–109.

[112] M.S. Mahmoud, Decentralized Control and Filtering in Interconnected Dynamical Systems, CRC Press, New York, 2010.

[113] A. Bidram, A. Davoudi, F.L. Lewis, J.M. Guerrero, Distributed cooperative secondary control of microgrids using feedback linearization, IEEE Trans. Power Syst. 28 (3) (2013) 3462–3470.

[114] J.A. Fax, R.M. Murray, Information flow and cooperative control of vehicle formations, IEEE Trans. Autom. Control 49 (9) (2004) 1465–1476.

[115] R. Olfati-Saber, J.A. Fax, R.M. Murray, Consensus and cooperation in networked multi-agent systems, Proc. IEEE 95 (1) (2007) 215–233.

[116] R. Olfati-Saber, R.M. Murray, Consensus protocols for networks of dynamic agents, in: Proceedings of the American Control Conference, 2003, pp. 951–956.

[117] Z. Lin, M. Broucke, B. Francis, Local control strategies for groups of mobile autonomous agents, IEEE Trans. Autom. Control 49 (4) (2004) 622–629.

[118] W. Ren, R.W. Beard, Formation feedback control for multiple spacecraft via virtual structures, Proc. IET Control Theory Appl. 151 (3) (2004) 357–368.

[119] E.S. Kazerooni, K. Khorasani, Optimal consensus seeking in a network of multiagent systems: an LMI approach, IEEE Trans. Syst. Man Cybern. B Cybern. 40 (2) (2010) 540–547.

[120] D. Wu, T. Dragicevic, J.C. Vasquez, J.M. Guerrero, Y. Guan, Secondary coordinated control of islanded microgrids based on consensus algorithms, in: Proceedings of the IEEE Energy Conversion Congress and Exposition, 2014, pp. 4290–4297.

[121] D. He, D. Shi, R. Sharma, Consensus-based distributed cooperative control for microgrid voltage regulation and reactive power sharing, in: Proceedings of IEEE/PES Innovative Smart Grid Technologies, 2014, pp. 1–6.

[122] H. Liang, B.J. Choi, W. Zhuang, X. Shen, Stability enhancement of decentralized inverter control through wireless communications in microgrids, IEEE Trans. Smart Grid 4 (1) (2013) 321–331.

[123] W.D. Zheng, J.D. Cai, A multi-agent system for distributed energy resources control in microgrid, in: Proceedings of the IEEE 5th International Conference on Critical Infrastructure (CRIS), 2010, pp. 1–5.

[124] J. Oyarzabal, J. Jimeno, J. Ruela, A. Engler, C. Hardt, Agent based micro grid management system, in: Proceedings of the IEEE International Conference on Future Power Systems, 2005, pp. 1–6.

[125] K.D. Brabandere, K. Vanthournout, J. Driesen, G. Deconinck, R. Belmans, Control of microgrids, in: Proceedings of the IEEE Power Engineering Society General Meeting, 2007, pp. 1–7.

[126] Y.A.R.I. Mohamed, A.A. Radwan, Hierarchical control system for robust microgrid operation and seamless mode transfer in active distribution systems, IEEE Trans. Smart Grid 2 (2) (2011) 352–362.

[127] A. Mehrizi-Sani, R. Iravani, Potential-function based control of a microgrid in islanded and grid-connected modes, IEEE Trans. Power Syst. 25 (4) (2010) 1883–1891.

[128] Q. Shafiee, J.M. Guerrero, J.C. Vasquez, Distributed secondary control for islanded microgrids—a novel approach, IEEE Trans. Power Electron. 29 (2) (2014) 1018–1031.

[129] G. Diaz, C. Gonzalez-Moran, J. Gomez-Aleixandre, A. Diez, Complex-valued state matrices for simple representation of large autonomous microgrids supplied by PQ and Vf generation, IEEE Trans. Power Syst. 24 (4) (2009) 1720–1730.

[130] P.H. Divshali, A. Alimardani, S.H. Hosseinian, M. Abedi, Decentralized cooperative control strategy of microsources for stabilizing autonomous VSC-based microgrids, IEEE Trans. Power Syst. 27 (4) (2012) 1949–1959.

[131] A. Yazdani, R. Iravani, A unified dynamic model and control for the voltage-sourced converter under unbalanced grid conditions, IEEE Trans. Power Deliv. 21 (3) (2006) 1620–1629.

[132] N. Pogaku, M. Prodanovic, T.C. Green, Modelling, analysis and testing of autonomous operation of an inverter based microgrid, IEEE Trans. Power Electron. 22 (2) (2007) 613–625.

[133] A.J. del Real, A. Arce, C. Bordons, Hybrid model predictive control of a two-generator power plant integrating photovoltaic panels and a fuel cell, in: Proceedings of the 46th IEEE Conference on Decision and Control, 2007, pp. 5447–5452.

[134] D.L. Peters, A.R. Mechtenberg, J.W. Whitefoot, P.Y. Papalambros, Model predictive control of a microgrid with plug-in vehicles: error modeling and the role of prediction horizon, in: Proceedings of the ASME Dynamic Systems and Controls Conference, 2011.

[135] W. Zheng, H. Ma, X. He, Modeling, analysis and implementation of real time network controlled parallel multi-inverter systems, in: Proceedings of the IEEE 7th International Conference on Power Electronics and Motion Control, 2012.

[136] Y.W. Li, C.N. Kao, An accurate power control strategy for power-electronics-interfaced distributed generation units operating in a low-voltage multi-bus micro-grid, IEEE Trans. Power Electron. 24 (12) (2009) 2977–2988.

[137] M. Nagahara, Y. Yamamoto, S. Miyazaki, T. Kudoh, N. Hayashi, H-infinity control of microgrids involving gas turbine engines and batteries, in: Proceedings of the IEEE 51st Annual Conference on Decision and Control (CDC), 2012, pp. 4241–4246.

[138] A. Nejati, A. Nobakhti, H. Karimi, Multivariable control strategy for autonomous operation of a converter-based distributed generation system, in: Proceedings of the IEEE/PES Power Systems Conference and Exposition (PSCE), 2011, pp. 1–8.

[139] J.A.P. Lopes, A.G. Madureira, C.C.L.M. Moreira, A view of microgrids, WIREs Energy Environ. 2 (1) (2012) 86–103.

[140] R.H. Lasseter, A. Akhil, C. Marnay, J. Stephens, J. Dagle, R. Guttronmson, A.S. Meliopoulous, R. Yinger, J. Eto, White paper on integration of distributed energy resources—the CERTS microgrid concept, Available at: http://certs.lbl.gov/pdf/50829-app.pdf.

[141] R.H. Lasseter, P. Paigi, Microgrids: a conceptual solution, in: 35th Annual IEEE Power Electronics Specialists Conference, Germany, 2004.

[142] A. Kim, M. Kim, E. Puchaty, M. Sevcovic, D. Delaurentis, A system-of-systems framework for the improved capability of insurgent tracking missions involving unmanned aerial vehicles, in: Proceedings of the IEEE 5th International Conference on System of Systems Engineering, 2010, pp. 1–6.

CHAPTER 2

Distributed Control Techniques in Microgrids

A. Mehrizi-Sani
Washington State University, Pullman, WA, United States

1. INTRODUCTION

The existing power system is a legacy system with components dating as far back as 40–50 years ago [1]. The next-generation power system, frequently referred to as the *smart grid*, is expected to operate under an updated philosophy with a significant increase in the level of monitoring, communication, and control and coordination [2]. A number of these paradigm shifts are necessitated by "green" initiatives that encourage and/or require the use of renewable energy resources. While even the existing power system has high levels of renewable power generation (hydropower is a prime example), the use of renewable resources in the smart grid is distinct by its distributed nature (e.g., photovoltaics (PVs) and wind) as opposed to the large centralized power plants in the current grid.

Of paramount importance is the ability to control and manage these facilities. This poses a significant challenge compared with the existing practices, and centralized schemes will not be able to operate under the significantly increased computational burden. The reasons include the following:

- unavailability of a dedicated management unit;
- computational burden due to the multitude of the controllable resources; for example, distributed generation (DG) units and loads;
- communication requirement due to the geographical expansiveness of such resources;
- frequent redesign requirements since a change in even one unit affects the central controller;
- difficulty or unwillingness to share data;
- reliability and security vulnerability of the central controller as a common point of failure.

Therefore noncentralized techniques are better suited to provide the required functionality. Noncentralized techniques, in turn, can be either decentralized or distributed. Decentralized techniques ignore the interaction between units. Such techniques are not limited to the smart grid; rather they are already employed as a natural yet simplistic way to deal with large systems. For example, in the United States, different interconnects and independent system operators (ISO) in each interconnect (for regions, states, or provinces that do have an ISO; e.g., New York Independent System Operator in New York and the Independent Electricity System Operator in Ontario) are an effort to achieve decentralized management of the power system. Supervisory control and data acquisition (SCADA) is used to provide situational awareness to such systems. Other large systems (e.g., chemical plants) also use decentralized approaches, although there are inherent and vast differences between a power system and a chemical plant in terms of the time frame and geographical extent.

As mentioned, decentralized methods assume that the interaction between subsystems is negligible. This assumption, however, is not always valid and can result in poor system-wide performance. The widespread blackout of August 2003 in North America [3] is an example of the consequences of the drawbacks of such a control strategy. Each subsystem, trying to maintain its stability, tripped and transferred the extra load to other subsystems, which in turn made the overload severer and caused a cascading event.

Unlike decentralized control techniques, distributed control techniques do consider the interactions among units. Assigning the control task to different units on the basis of operation in different time frames is what constitutes the idea of control hierarchy (primary, secondary, and tertiary controls). Within the higher control levels (secondary and tertiary) the need for distributed approaches arises because of the desire and need for higher reliability, security, and situational awareness. The promise of the smart grid and developments made in the realms of communication and computation have brought such distributed control strategies closer to reality.

The objective of this chapter is to provide a study of existing distributed control and management strategies. This chapter also identifies and proposes future directions for research in this area. A number of articles provide a survey of the state of the art of control strategies in the smart grid [4], microgrids [5], and AC converters [6]. This chapter is distinct from these articles in that it studies specifically the distributed techniques employed in the power system [7, 8].

This chapter is organized as follows. The next section defines a microgrid and discusses its characteristics that differentiate it from the existing power system. Section 3 provides perspective on control requirements and structure in a smart grid. Section 4 reviews the state of the art of distributed control techniques, and Section 5 discusses the different functionalities and applications of distributed control. Finally, Section 6 discusses research opportunities and challenges.

2. DEFINITIONS OF THE SMART GRID AND A MICROGRID

The smart grid is the vision of the future electric power delivery system. As a fairly new concept, no universally accepted definition yet exists for the smart grid; however, its core elements are the use of (1) information technology, (2) communication, and (3) power electronic devices [9–11]. The key technologies required to achieve the vision of the smart grid include:

1. advanced components (e.g., power electronics and storage systems) [12, Appendix B3];
2. advanced control technologies (e.g., distributed intelligent agents);
3. integrated communications (e.g., WiMAX and broadband over power line);
4. sensing and measurement (e.g., advanced metering infrastructure);
5. improved interfaces and decision support systems (e.g., 3D visualization systems).

The concept of a microgrid [13] is introduced as a building block of the smart grid as a solution for reliable interconnection of distributed energy resource (DER) units. Thus a microgrid is presented to the host grid as a single controllable entity that provides power and/or ancillary services. A microgrid is a collection of collocated DER units—for example, DG units, distributed storage (DS) units, and loads—that are connected through a point of common coupling to the host power system. Examples of DG units include wind turbines, PVs, fuel cells, and microturbines. Examples of DS units include flywheels, batteries, and supercapacitors. A microgrid can, in general, have any arbitrary configuration. Microgrid research, development, and demonstration (RD&D) is performed by numerous initiatives (e.g., Hydro-Québec in Canada, the Consortium for Electric Reliability Technology Solutions (CERTS) in the United States, More Microgrids in Europe, and the New Energy and Industrial Technology Development Organization (NEITDO) in Japan).

A microgrid and its various evolved forms (e.g., an active distribution system (ADS), a cognitive microgrid, and a virtual power plant (VPP) [14–18]) can be considered and exploited as the main building block of the smart grid. An ADS is a microgrid equipped with power management and supervisory control for DG units, DS units, and loads [19]. A cognitive microgrid is an intelligent microgrid that features an adaptive approach for the control of the microgrid components. In the context of a VPP [16], the cognitive microgrid is presented to the host grid at the point of common coupling as a single controllable entity with a prespecified performance. The internal mechanics of the virtual power plant is hidden from the host power system.

A microgrid is capable of operation in grid-connected mode, islanded mode, and the transition between these two. While the microgrid is in grid-connected mode, the shortfall or excess power of the microgrid is exchanged with the main grid. When the microgrid is islanded, the real and reactive power generated within the microgrid should equal the demand of local loads. IEEE Standard 1547 includes guidelines for interconnection of DER units [20]. Islanding (i.e., disconnection of the microgrid from the host grid) can be either intentional (scheduled) or unintentional [21, 22]. Intentional islanding can occur in situations such as scheduled maintenance and degraded power quality of the main grid. Unintentional islanding can occur because of faults and other unscheduled events that are unknown to the microgrid. Proper detection of such a disconnection is imperative for the safety of personnel, proper operation of the microgrid, and implementation of required controller changes. The technical literature includes a wealth of islanding detection algorithms, which may use frequency/voltage measurements or disturbance injection [21, 22].

3. OVERVIEW OF THE CONTROL STRUCTURE

Microgrids and integration of DER units introduce a number of operational challenges that arise because assumptions related to conventional power systems no longer hold. For example, integration of DER units can cause reverse power flow and lead to complications in protection coordination, power flow patterns, fault current distribution, and voltage profile control. Vast integration of generation at the distribution level also necessitates further studies on the distribution system that were traditionally performed only at the transmission level (e.g., stability and power flow). The dominance of constant-power loads, the prevalence of three-phase balanced conditions, the inductive nature of transmission lines, and the abundance of inertia of synchronous generators are among the assumptions that do not necessarily hold for microgrids. Optimal operation of a microgrid that includes a multitude of DER units with different technologies requires the solving of a large problem with parameters ranging from fuel efficiency to weather forecast and consumption trends and the associated uncertainty.

3.1. Control Requirements

The most prevailing challenges and desired traits in microgrid control include the following:

- **Communication-based control**. Distributed techniques are common in that they rely on the existence of some form of communication between a subset of units in a microgrid.
- **Time variance and topological changes**. Communication links in a microgrid can be time variant (e.g., due to fading of a wireless signal). Network topological changes (e.g., disconnection of DER units) can also affect the system performance.
- **Low inertia and stability issues due to new components**. Many DER units are interfaced with the power system through fast-acting power electronic devices. While such interfaces can enhance the system dynamic performance, they can also lead to severe excursion of voltage and frequency in islanded mode in the absence of proper controls.

3.2. Control Hierarchy

Distributed control and control hierarchy are two closely related concepts. In any system with a sufficiently high number of units, one way to provide a coordination strategy is to employ a control hierarchy [23, 24] as shown in Fig. 2.1.

3.2.1. Primary Control

Primary control, also known as *local or internal control*, is the first control level and features the fastest response. Primary control responds to system dynamics and ensures that the system variables (e.g., voltage and frequency) track their set points. Primary control is based on locally measured signals and requires no communication. Because of their speed implications, output control [25–33], power sharing, and islanding detection and the subsequent change of controller modes [21, 34] also lie in this control level.

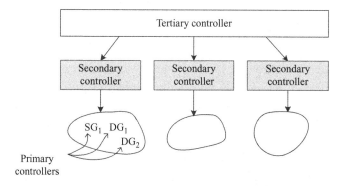

Fig. 2.1 Hierarchical control levels: primary, secondary, and tertiary.

3.2.2. Secondary Control

Secondary control [35], also known as *central control* or an *energy management system* [36], is responsible for ensuring power quality and mitigating long-term voltage and frequency deviations by determining the set points for the primary control. It operates on a slower time frame than that of the primary control to (1) decouple secondary control from primary control and (2) reduce the required communication bandwidth.

3.2.3. Tertiary Control

Tertiary control is the highest level of control and sets long-term set points depending on the requirements of the host power system on the basis of the information received about the status of the DER units, market signals, and other system requirements. It is responsible for managing multiple microgrids and is part of the host grid.

3.3. Distributed Control Within Control Hierarchy

As mentioned, primary control is performed locally and does not need communication. Higher control levels, however, can be either centralized or distributed. Distributed approaches fall mainly in the secondary and tertiary control levels.

4. OVERVIEW OF DISTRIBUTED TECHNIQUES

The common problem that distributed techniques aim to address is to solve the underlying optimization problem in a distributed manner with availability of communication. Therefore the distinction between different distributed techniques is not always clear; rather it is due to different problem formulation approaches and perspectives.

4.1. Droop-Based Techniques

Droop characteristics is a widely used method [28, 37–39]. Droop originates from the principle of power balance in synchronous generators. An imbalance between the input mechanical power and the output electric power causes a change in the rotor speed and electrical frequency. Similarly, variation in output reactive power results in voltage magnitude deviation.

Such a characteristic can be artificially created for electronically interfaced DG units. In droop, the relationships between real power and frequency and reactive power and voltage are as follows:

$$f = f^* - K_P(P - P^*),$$
$$V = V^* - K_Q(Q - Q^*).$$

Use of active power-voltage and reactive power-frequency droop, for low-voltage microgrids, is also reported [40].

The main advantage of droop control is elimination of the need for communication by use of local measurements. However, the droop control method has several disadvantages, which limit its applicability for a modern power system, such as poor transient performance, ignoring load dynamics, lack of black startup ability, poor performance in distribution networks, inability to provide accurate power sharing with output impedance uncertainties, unsuitability for nonlinear loads, and inability to impose a fixed system frequency [23, 40–42]. Recently, a washout filter-based method to improve the steady-state performance of droop control was reported [43].

4.2. Distributed Model Predictive Control-Based Techniques

Model predictive control (MPC) is a de facto industry standard for control of large process plants [44–46] and offers (1) handling of multivariable control problems, (2) ease of tuning, and (3) explicit consideration of constraints.

As shown in Fig. 2.2, MPC is a discrete-time control strategy in which the control sequence of the system is determined by minimization of a cost function associated with the system performance over a finite number of future time steps with use of the system model. The cost function is a combination of terms corresponding to minimization of the deviation of

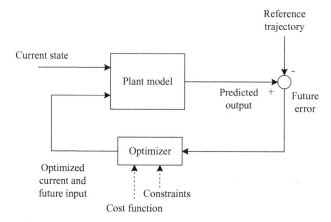

Fig. 2.2 Model predictive control.

system states and those reflecting the deviation from set points. In linear MPC, a linearized model of the discrete-time system is employed.

Each time step of MPC includes calculation of the control sequence for N future time steps, where N is the prediction horizon, to minimize the weighted sum of errors; however, only the first input is applied and the rest are updated in the next time step.

4.3. Consensus-Based Techniques

The literature on distributed computation goes back to the seminal work of Tsitsiklis et al. [47] that focuses on distributed optimization where each unit is fully aware of the global objective function. The more recent literature studies this problem under the term *consensus*. Consensus is an approach for solving distributed optimization problems and offers a flexible formulation that has promise for extendability and scalability.

The goal of consensus is to have different DER units converge to a single value. A consensus-based approach achieves global optimality using limited, possibly time-varying communication between neighboring units, without needing a dedicated unit. Study of unconstrained consensus algorithms in a distributed environment is discussed in [48–50], for example, and extension to the constrained case is discussed in [51, 52].

4.4. Agent-Based Techniques

Another distributed control strategy is the use of multiagent systems. Agents are entities that act on the environment and have communication capability, some level of autonomy based on their own goals, and limited knowledge of the environment [53]. An intelligent agent is reactive (reacts to the changes in the environment), proactive (seeks initiatives), and social (relies on communication).

A multiagent system is especially suitable for large complex systems, such as power systems, in which a large number of agents of different types interact and most of the required information is locally available.

A multiagent system can be applied to power systems in areas such as monitoring, control, modeling and simulation, and protection [54].

4.5. Decomposition-Based Techniques

Several decomposition schemes have been proposed, including the auxiliary problem principle, the predictor-corrector proximal multiplier method, and the alternating direction method [55–57].

These approaches are based on decomposition of the original optimization problem into a number of subproblems that are solved iteratively until there is convergence.

5. APPLICATIONS IN POWER SYSTEMS

5.1. Application in Primary Control

Generally, distributed control is not suitable for fast control. The control scheme proposed in [58] divides the control action of a voltage source converter (VSC) between a local controller and a central controller. While the central controller ensures that the units track their steady-state reference points and reject low-frequency disturbances, the local controllers are responsible for rejecting the high-frequency component of disturbances. Since the central controller generates only the low-frequency component of the reference current of the VSC, the signal can be transmitted to the VSC by means of a limited bandwidth communication channel.

Distributed control of a multimodule inverter system is studied in [59] to improve the reliability of the overall wind power system. Droop characteristic is another application within primary control (discussed in Section 4.1).

5.2. Application in Voltage Coordination

The objective of voltage coordination is to provide a flat voltage profile in the microgrid—ideally zero voltage regulation [60, 61]—to avoid flow of large current. Earlier work (e.g., [62]) proposed a secondary voltage regulation scheme using reactive power injection. This, however, is a steady-state operation and works only for the transmission level. Moradzadeh et al. [45] propose a voltage coordination strategy in multiarea power systems using distributed MPC and communication between neighboring controllers.

Anand et al. [63] propose a strategy based on droop characteristics for improved voltage regulation and power sharing in DC microgrids. A two-stage algorithm for voltage control is proposed in [64]. In the first stage, DER units inject the necessary reactive power into the system to mitigate voltage violations. If this does not fully mitigate the problem, in the second stage, local controllers request additional reactive power support from neighboring units. A secondary control method for improving voltage quality and rejecting harmonics is proposed in [65], where harmonics are explicitly considered within a hierarchical scheme.

5.3. Application in Economic Power Coordination

Power coordination and economic operation are important operational considerations. The optimal power flow (OPF) problem aims to solve this in a centralized manner. OPF has a long history in power systems [66] dating back to the 1960s. However, OPF is nonconvex and present solutions do not guarantee convergence to the global optimum. Moreover, generic distributed optimization strategies do not consider time variance of communication links and require extensive calculations—all of which hamper the effectiveness of distributed OPF for microgrid applications. Some methods do consider time variance of the communication links but require the constraint sets to be the same for each local generator. Lin and Lin [67] formulate distributed OPF with discrete variables, and Baldick et al. [68, 69] propose a coarse-grained OPF. A comparison of three decomposition-based methods is provided in [70].

Qi et al. [71] employ distributed MPC for coordination of wind and solar energy systems in a DC grid. Optimal charging of electric vehicles is studied in [72] by exploitation of the elasticity of the electric vehicle loads as an iterative optimal control problem. Mudumbai et al. [73] propose a consensus-based algorithm for optimal economic dispatch by adjusting the power-frequency set point of each generator using aggregate power imbalance in the network area control error (ACE) with and without losses.

5.4. Application in Frequency Coordination

In traditional power systems the inherent physics of synchronous generators forces exact coupling between mechanical rotational speed and electrical frequency. However, most renewable energy resources, including all PVs and most wind turbines, do not have this physical coupling. This is because these devices employ power electronic converters that essentially decouple the frequencies of different parts of the system. As a result, no natural inertia is present in these systems to aid frequency regulation. There are some efforts in the literature to provide virtual inertia for such resources by mimicking the operation of synchronous generators. An example is [28], which adds a term proportional to the time derivative of frequency in the power control loop. However, this is essentially a measurement-based derivative feedback term, which is subject to bandwidth limits [74]. Furthermore, this method underutilizes the smart grid (e.g., availability of phasor measurement unit (PMU) measurements and communication infrastructure [9]).

A common approach for frequency regulation of DER units is to emulate droop characteristics (see Section 4.1).

Namara et al. [44] apply distributed MPC for optimal coordination of frequency in multilink high-voltage DC (HVDC) systems. A consensus-based frequency control for multiterminal HVDC is proposed in [75].

Frequency control is relevant when (1) DER units are electronically interfaced so they can control frequency independently or (2) when a segmented power system is used (small power systems or microgrids connected together through a multiterminal HVDC link—a futuristic application, but with its merits). In frequency control the objective is to have different units converge to a common frequency. In this case, one unit has to have an independent frequency set point so that the problem does not become indeterminate. Each unit may have its own minimum and maximum values for power and terminal voltage.

6. CONCLUSIONS AND FUTURE TRENDS

6.1. Challenges Ahead

It is clear that there is a need for the development of a comprehensive framework that considers different modes of operation of a microgrid as well as the presence of a high number of DER units. In this chapter, we reviewed the main techniques employed for distributed control and discussed their applications. The main challenges ahead in perfecting and implementing these techniques include the following:

- **Conversation between control and power engineering communities**. Distributed control has been studied in the control area for much longer than in the power area. Discussion between these two communities can efficiently improve and expedite the development of meaningful approaches applicable to the power system. An example effort is joint panels at the IEEE Power and Energy Society General Meeting.

- **Stability and convergence results**. In a distributed control scheme, each system performs optimization only on its own control inputs; therefore it becomes prone to resulting in either a local optimal or a Nash equilibrium [76].
- **Technological requirements**. The communication requirement of distributed control, although minimal and slow, is in contrast to the existing practices in the power system to avoid communication because of its inherent uncertainties. Future advances and innovations are expected to make communication more abundant in the grid. The small geographical span of a microgrid makes possible the use of affordable and standard communication protocols.
- **Utility adaptation**. The ultimate goal of this RD&D effort is to implement the proposed approaches in the power system. Utilities and other stakeholders in the power area are (understandably) conservative. Extensive real-time simulations and pilot projects need to be conducted to overcome this barrier for eventual implementation.
- **Regulatory considerations**. Distributed control may require exchange of data and models that are not part of the standard protocols and practices in the power system.
- **Non-model-based approaches**. Control approaches that do not require a detailed model of the system and are relatively robust to changes in system parameters are especially effective for microgrids, in which the on/off status of apparatus can change frequently and affect the system performance, necessitating retuning of controller parameters [77, 78]. This may pose problems for the effective and fast mitigation of transients, especially when the operating points are close to their limits, in utility-based microgrids as well as other microgrids such as those in naval ships, military systems, and aircraft, in which spinning reserve may not exist, inertia is not significant, and the notion of slack bus is not necessarily valid.

6.2. High-Voltage DC

Offshore wind power extraction is a promising area; it can yield up to 50% more energy than a comparable wind power extraction onshore. However, challenges include the large capacitance of undersea cables

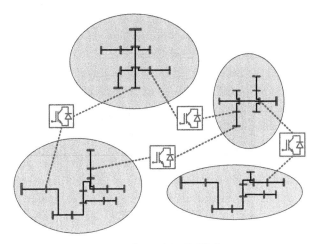

Fig. 2.3 Power system segmentation through HVDC links.

and their erosion. HVDC transmission can address the problem of high capacitance but it is commercially competitive only at distances of more than 100 km and power levels between 200 and 900 MW. The benefits of HVDC include allowing frequencies at the two ends to be different (and less power conditioning requirement at the source end) and lower losses. HVDC can also be used in a scenario called *segmentation* [79] (see Fig. 2.3). In segmentation, a large AC system, which is difficult to manage, is divided into smaller loosely coupled sections. Each segment is then effectively a microgrid. DC connection of different segments ensures independent operations of these microgrids.

6.3. Nonmodel-Based Control

Traditional control design methods such as model-based design (e.g., [80, 81]), model-based automatic tuning (e.g., [82–85]), and simulation-based optimization (e.g., [86, 87]) are not always adequate for microgrid applications, in which the on/off status of apparatus can change frequently and affect the performance of the rest of the system and necessitate retuning of controller parameters. To redesign and reimplement new controllers, these methods require (1) updated models, (2) a computational infrastructure, and (3) access to the controller parameters. However, these facilities are not always available. Consequently, design methods that are relatively insensitive to system parameters and are nonintrusive can provide more desirable outcomes. Methods such as online adjustment of controller gains [88] and application of game theory for controller design [89] have been reported. However, they are limited in performance (dependent on the initial choice of gains) and scope respectively.

In such scenarios, methods that can autonomously and without requiring a detailed model of the system ensure reference set point tracking can show great promise. A strategy called *set point automatic adjustment with correction enabled* (SPAACE) to improve set point tracking performance of DG units in a microgrid was proposed in [77, 78, 90, 91]. Fig. 2.4 shows the basics of this method. SPAACE augments the controllers that are already implemented and monitors the controlled variable $x(t)$ and on the basis of its variations and deviation from the set point $u^*(t)$ temporarily modulates the set point $u(t)$ so that $x(t)$ closely tracks the set point. In a typical implementation of SPAACE, the variation in the set point is introduced as step changes from u^* to $(1 - m)u^*$, where m is a design parameter. The salient features of SPAACE are (1) robustness to changes in the system, (2) not requiring the system model, and (3) scalability and reliance merely on local signals. Fig. 2.5 shows how SPAACE is related to the primary and secondary control levels.

6.4. Model Predictive Control

As mentioned, in many plants ignoring the interaction between subsystems leads to significant performance deterioration. This necessitates the use of a complete system model that includes the interaction between subsystems. Nevertheless, the control strategy still has to be distributed. One achieves this in MPC by composing an objective function that includes all

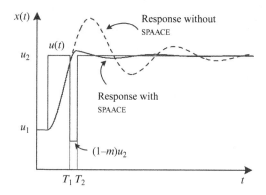

Fig. 2.4 Example of application of SPAACE to improve reference set point tracking.

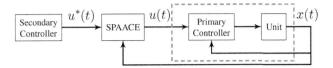

Fig. 2.5 Relationship between SPAACE and primary controller.

system-wide performance parameters. However, the optimization at each subsystem is performed only on its own control inputs, and it is assumed the input to other subsystems is constant. Note that this results in a Nash equilibrium instead of a Pareto-optimal set of control inputs [76]. To refine the results, the optimization is run for a number of iterations until convergence of the results is achieved or the allowable optimization time is reached. Note that as computations are done in real time, determination of the input sequence must be completed within the time step. This requires the generation of intermediate solutions that are feasible for the system. This is explicitly enforced through constraints of the optimization problem.

REFERENCES

[1] G. Heydt, R. Ayyanar, K.W. Hedman, V. Vittal, Electric power and energy engineering: the first century, Proc. IEEE 100 (5) (2012) 1315–1328.

[2] G.W. Arnold, Challenges and opportunities in smart grid: a position article, Proc. IEEE 99 (6) (2011) 922–927.

[3] Final report on the August 14, 2003 Blackout in the United States and Canada: Causes and Recommendations, Technical report, U.S.-Canada Power System Outage Task Force, 2004, https://reports.energy.gov/BlackoutFinal-Web.pdf.

[4] J. Guerrero, P. Loh, T.-L. Leeand, M. Chandorkar, Advanced control architectures for intelligent microgrids—Part I, IEEE Trans. Ind. Electron. 60 (4) (2012) 1254–1262.

[5] A. Bidram, A. Davoudi, Hierarchical structure of microgrids control system, IEEE Trans. Smart Grid 3 (4) (2012) 1963–1976.

[6] J. Rocabert, A. Luna, F. Blaabjerg, P. Rodríguez, Control of power converters in AC microgrids, IEEE Trans. Power Electron. 27 (11) (2012) 4734–4749.

[7] M. Yazdanian, A. Mehrizi-Sani, Distributed control techniques in microgrids, IEEE Trans. Smart Grid 5 (6) (2014) 2901–2909.

[8] D. Olivares, A. Mehrizi-Sani, A. Etemadi, C. Canizares, R. Iravani, M. Kazerani, A. Hajimiragha, O. Gomis-Bellmunt, M. Saeedifard, R. Palma-Behnke, G. Jimenez-Estevez, N. Hatziargyriou, Trends in microgrid control, IEEE Trans. Smart Grid 5 (4) (2014) 1905–1919.

[9] NETL Modern Grid Strategy—Powering Our 21st-Century Economy: A Compendium of Smart Grid Technologies, White Paper, National Energy Technology Laboratory (NETL) for the U.S. Department of Energy Office of Electricity Delivery and Energy Reliability, 2009.

[10] CERTS Berkeley 2005 Symposium on Microgrids: Participant Questionnaires, Collection of Responses, Consortium for Electric Reliability Technology Solutions, 2005, http://der.lbl.gov/certsmicrogrids/questionnaires_list.html.

[11] IEEE Guide for Design, Operation, and Integration of Distributed Resource Island Systems with Electric Power Systems, IEEE Std. 1547.4–2011, 2011.

[12] A systems view of the modern grid, White Paper, National Energy Technology Laboratory for the U.S. Department of Energy Office of Electricity Delivery and Energy Reliability, 2007, http://www.netl.doe.gov/moderngrid/docs/ASystemsViewoftheModernGrid_Final_v2_0.pdf.

[13] N. Hatziargyriou, H. Asona, R. Iravani, C. Marnay, Microgrids, IEEE Power Energy Mag. 5 (4) (2007) 78–94.

[14] N. Ruiz, I. Cobelo, J. Oyarzabal, A direct load control model for virtual power plant management, IEEE Trans. Power Syst. 24 (2) (2009) 959–966.

[15] H. Morais, P. Kádár, M. Cardoso, Z.A. Vale, H. Khodr, VPP operating in the isolated grid, in: Power and Energy Society General Meeting—Conversion and Delivery of Electrical Energy in the 21st Century, July 2008.

[16] D. Pudjianto, C. Ramsay, G. Strbac, Virtual power plant and system integration of distributed energy resources, IET Renew. Power Gener. 1 (1) (2007) 10–16.

[17] A. Molderink, V. Bakker, M.G.C. Bosman, J.L. Hurink, G.J.M. Smit, Management and control of domestic smart grid technology, IEEE Trans. Smart Grid 1 (2) (2010) 109–119, http://dx.doi.org/10.1109/TSG.2010.2055904.

[18] D. Pudjianto, C. Ramsay, G. Starbac, Microgrids and virtual power plants: concepts to support the integration of distributed energy resources, Proc. Inst. Mech. Eng. A: J. Power Energy 222 (7) (2008) 731–741, http://dx.doi.org/10.1243/09576509JPE556.

[19] C. D'Adamo, P. Taylor, S. Jupe, B. Buchholz, F. Pilo, C. Abbey, J. Marti, Active distribution networks: general features, present status of implementation and operation practices (update of WG C6.11 activities), Electra (246) (2009) 22–29.

[20] IEEE Standard for Interconnecting Distributed Resources with Electric Power Systems, IEEE Std 1547–2003, 2003.

[21] H. Karimi, A. Yazdani, M.R. Iravani, Negative-sequence current injection for fast islanding detection of a distributed resource unit, IEEE Trans. Power Electron. 23 (1) (2008) 298–307.

[22] T. Funabashi, K. Koyanagi, R. Yokoyama, A review of islanding detection methods for distributed resources, in: IEEE PowerTech Conference, June 2003, Bologna, Italy, 2003.

[23] J.M. Guerrero, J.C. Vasquez, J. Matas, L.G. de Vicuña, M. Castilla, Hierarchical control of droop-controlled AC and DC microgrids—a general approach towards standardization, IEEE Trans. Ind. Electron. 58 (1) (2011) 158–172.

[24] A. Mehrizi-Sani, R. Iravani, Potential-function based control of a microgrid in islanded and grid-connected modes, IEEE Trans. Power Syst. 25 (4) (2010) 1883–1891, http://dx.doi.org/10.1109/TPWRS.2010.2045773.

[25] J.A.P. Lopes, C.L. Moreira, A.G. Madureira, Defining control strategies for microgrids islanded operation, IEEE Trans. Power Syst. 21 (2) (2006) 916–924.

[26] F. Blaabjerg, R. Teodorescu, M. Liserre, A.V. Timbus, Overview of control and grid synchronization for distributed power generation systems, IEEE Trans. Ind. Electron. 53 (5) (2006) 1398–1409.

[27] F. Katiraei, M.R. Iravani, P.W. Lehn, Micro-grid autonomous operation during and subsequent to islanding process, IEEE Trans. Power Del. 20 (1) (2005) 248–257.

[28] F. Gao, M.R. Iravani, A control strategy for a distributed generation unit in grid-connected and autonomous modes of operation, IEEE Trans. Power Del. 23 (2) (2008) 850–859.

[29] M. Yazdanian, A. Mehrizi-Sani, Internal model-based current control of the RL filter-based voltage-sourced converter, IEEE Trans. Energy Convers. 29 (4) (2014) 873–881.

[30] C. Stone, A. Mehrizi-Sani, Set point adjustment strategy for mitigating transients in a microgrid, in: IEEE Industrial Electronics Society Annual Conference (IECON 13), November 2013, Vienna, Austria, 2013.

[31] A. Mehrizi-Sani, A strategy to improve reference tracking of distributed energy resources, in: IEEE Innovative Smart Grid Technologies (ISGT), February 2013, Washington, DC, 2013.

[32] C. Stone, A. Mehrizi-Sani, Improved dynamic response for LCC-based HVDC systems, in: EPRI HVDC/FACTS Conference, August 2013, Palo Alto, CA, 2013.

[33] M. Yazdaniane, A. Mehrizi-Sani, Case studies on cascade voltage control of islanded microgrids based on the internal model control, in: 9th IFAC Symposium Control of Power and Energy Systems (CPES), December 2015, New Delhi, India, 2015.

[34] H. Karimi, H. Nikkhajoei, M.R. Iravani, Control of an electronically-coupled distributed resource unit subsequent to an islanding event, IEEE Trans. Power Del. 23 (1) (2008) 493–501.

[35] A. Mehrizi-Sani, R. Iravani, Secondary control of microgrids: application of potential functions, in: CIGRÉ Session 2010, August 2010, Paris, France, 2010.

[36] Master Controller Requirements Specification for Perfect Power Systems (as outlined in the Galvin Electricity Initiative), White Paper, EPRI, Palo Alto, CA and The Galvin Project, Inc., Chicago, IL, 2007.

[37] K.D. Brabandere, B. Bolsens, J.V. den Keybus, A. Woyte, J. Driesen, R. Belmans, A voltage and frequency droop control method for parallel inverters, IEEE Trans. Power Electron. 22 (4) (2007) 1107–1115, http://dx.doi.org/10.1109/TPEL.2007.900456.

[38] M.C. Chandorkar, D.M. Divan, R. Adapa, Control of parallel connected inverters in standalone AC supply systems, IEEE Trans. Ind. Appl. 29 (1) (1993) 136–143.

[39] C. Sao, P. Lehn, Autonomous load sharing of voltage source converters, IEEE Trans. Power Del. 20 (2) (2005) 1009–1016, http://dx.doi.org/10.1109/TPWRD.2004.838638.

[40] X. Yu, A.M. Khambadkone, H. Wang, S.T.S. Terence, Control of parallel-connected power converters for low-voltage microgrid—Part I: a hybrid control architecture, IEEE Trans. Power Electron. 25 (12) (2010) 2962–2970, http://dx.doi.org/10.1109/TPEL.2010.2087393.

[41] J. Kim, J.M. Guerrero, P. Rodriguez, R. Teodorescu, K. Nam, Mode adaptive droop control with virtual output impedances for an inverter-based flexible AC microgrid, IEEE Trans. Power Electron. 26 (3) (2011) 689–701, http://dx.doi.org/10.1109/TPEL.2010.2091685.

[42] M.B. Delghavi, A. Yazdani, An adaptive feedforward compensation for stability enhancement in droop-controlled inverter-based microgrids, IEEE Trans. Power Del. 26 (3) (2011) 1764–1773.

[43] M. Yazdanian, A. Mehrizi-Sani, Washout filter-based power sharing, IEEE Trans. Smart Grid 7 (2) (2015) 967–968.

[44] P.M. Namara, R.R. Negenborn, B.D. Schutter, G. Lightbody, Optimal coordination of a multiple HVDC link system using centralized and distributed control, IEEE Trans. Control Syst. Technol. 21 (2) (2013) 302–314.

[45] M. Moradzadeh, R. Boel, L. Vandevelde, Voltage coordination in multi-area power systems via distributed model predictive control, IEEE Trans. Power Syst. 28 (1) (2013) 513–521.

[46] S. Roshany-Yamchi, M. Cychowski, R. Negenborn, B. Schutter, K. Delaney, J. Connell, Kalman filter-based distributed predictive control of large-scale multi-rate

systems: application to power networks, IEEE Trans. Control Syst. Technol. 21 (1) (2013) 27–39.

[47] J.N. Tsitsiklis, D.P. Bertsekas, M. Athans, Distributed asynchronous deterministic and stochastic gradient optimization algorithms, IEEE Trans. Autom. Control AC-31 (9) (1986) 803–812.

[48] T. Keviczky, F. Borrelli, K. Fregene, D. Godbole, G.J. Balas, Decentralized receding horizon control and coordination of autonomous vehicle formations, IEEE Trans. Control Syst. Technol. 16 (1) (2008) 19–33.

[49] E. Frazzoli, F. Bullo, Decentralized algorithms for vehicle routing in a stochastic time-varying environment, in: 43rd IEEE Conference on Decision and Control, December 2004, Atlantis, Paradise Island, Bahamas, 2004, pp. 3357–3363.

[50] R. Olfati-Saber, A.A. Fax, R.M. Murray, Consensus and cooperation in networked multi-agent systems, Proc. IEEE 95 (1) (2007) 215–233.

[51] A. Nedić, A. Ozdaglar, Distributed subgradient methods for multi-agent optimization, IEEE Trans. Autom. Control 54 (1) (2009) 48–60.

[52] A. Nedić, A. Ozdaglar, P.A. Parrilo, Constrained consensus and optimization in multi-agent networks, IEEE Trans. Autom. Control 55 (4) (2010) 922–938.

[53] S. Russell, P. Norvig, Artificial Intelligence: A Modern Approach, 2nd ed., Prentice Hall, Upper Saddle River, NJ, 2003.

[54] S. McArthur, E. Davidson, V. Catterson, A. Dimeas, N. Hatziagyriou, F. Ponci, T. Funabashi, Multi-agent systems for power engineering applications—Part I: Concepts, approaches, and technical challenges, IEEE Trans. Power Syst. 22 (4) (2007) 1743–1752.

[55] A.G. Beccuti, T. Demiray, G. Andersson, M. Morari, A Lagrangian decomposition algorithm for optimal emergency voltage control, IEEE Trans. Power Syst. 23 (4) (2010) 1769–1779.

[56] A. Ravindran, K. Ragsdell, G. Reklaitis, Engineering Optimization: Methods and Applications, 2nd ed., John Wiley, Hoboken, NJ, 2006, 260–330.

[57] A.J. Conejo, E. Castillo, R. Mínguez, R. García-Bertrand, Decomposition Techniques in Mathematical Programming: Engineering and Science Applications, Springer-Verlag, The Netherlands, 2006.

[58] M. Prodanović, T.C. Green, High-quality power generation through distributed control of a power park microgrid, IEEE Trans. Ind. Electron. 53 (5) (2006) 1471–1482.

[59] M. Parker, L. Ran, S. Finney, Distributed control of a fault-tolerant modular multilevel inverter for direct-drive wind turbine grid interfacing, IEEE Trans. Ind. Electron. 60 (2) (2013) 509–522.

[60] P. Vovos, A. Kiprakis, R. Wallace, G. Harrison, Centralized and distributed voltage control: impact on distributed generation penetration, IEEE Trans. Power Syst. 22 (1) (2007) 476–483.

[61] P. Ferreira, P. Carvalho, L.A. Ferreira, M. Ilic, Distributed energy resources integration challenges in low-voltage networks: voltage control limitations and risk of cascading, IEEE Trans. Sustain. Energy 4 (1) (2012) 82–88.

[62] M. Ilic-Spong, J. Christensen, K.L. Eichorn, Secondary voltage control using pilot point information, IEEE Trans. Power Syst. 3 (2) (1988) 660–668.

[63] S. Anand, B.G. Fernandes, J. Guerrero, Distributed control to ensure proportional load sharing and improve voltage regulation in low-voltage DC microgrids, IEEE Trans. Power Electron. 28 (4) (2013) 1900–1913.

[64] B.A. Robbins, C.N. Hadjicostis, A.D. Domínguez-García, A two-stage distributed architecture for voltage control in power distribution systems, IEEE Trans. Power Syst. 28 (2) (2012) 1470–1482.

[65] M. Savaghebi, A. Jalilian, J. Vasquez, J. Guerrero, Secondary control for voltage quality enhancement in microgrids, IEEE Trans. Smart Grid 3 (4) (2012) 1893–1902.

[66] H.W. Dommel, W.F. Tinney, Optimal power flow solutions, IEEE Trans. Power App. Syst. PAS-87 (10) (1968) 1866–1874.

[67] C.-H. Lin, S.-Y. Lin, Distributed optimal power flow with discrete control variables of large distributed power systems, IEEE Trans. Power Syst. 23 (2) (2008) 1383–1392.

[68] R. Baldick, B.H. Kim, C. Chase, Y. Luo, A fast distributed implementation of optimal power flow, IEEE Trans. Power Syst. 14 (3) (1999) 858–864.

[69] B. Kim, R. Baldick, Coarse-grained distributed optimal power flow, IEEE Trans. Power Syst. 12 (2) (1997) 932–939.

[70] B. Kim, R. Baldick, A comparison of distributed optimal power flow algorithms, IEEE Trans. Power Syst. 15 (2) (2000) 599–604.

[71] W. Qi, J. Liu, P.D. Christodes, Distributed supervisory predictive control of distributed wind and solar energy systems, IEEE Trans. Control Syst. Technol. 21 (2) (2012) 504–512.

[72] L. Gan, U. Topcu, S.H. Low, Optimal decentralized protocol for electric vehicle charging, IEEE Trans. Power Syst. 28 (2) (2012) 940–951.

[73] R. Mudumbai, S. Dasgupta, B. Cho, Distributed control for optimal economic dispatch of a network of heterogeneous power generators, IEEE Trans. Power Syst. 27 (4) (2012) 1750–1760.

[74] C.A. Baone, C.L. DeMarco, From each according to its ability: distributed grid regulation with bandwidth and saturation limits in wind generation and battery storage, IEEE Trans. Control Syst. Technol. 21 (2) (2012) 384–394.

[75] J. Dai, Y. Phulpin, A. Sarlette, D. Ernst, Coordinated primary frequency control among non-synchronous systems connected by a multi-terminal high-voltage direct current grid, IEE Gener. Transm. Distrib. 6 (2) (2012) 99–108.

[76] X. Du, Y. Xi, S. Li, Distributed model predictive control for large-scale systems, in: American Control Conference, June, 2001, pp. 3142–3143.

[77] A. Mehrizi-Sani, R. Iravani, Online set point adjustment for trajectory shaping in microgrid applications, IEEE Trans. Power Syst. 27 (1) (2012) 216–223, http://dx.doi.org/10.1109/TPWRS.2011.2160100.

[78] A. Mehrizi-Sani, R. Iravani, Online set point modulation to enhance microgrid dynamic response: theoretical foundation, IEEE Trans. Power Syst. 27 (4) (2012) 2167–2174.

[79] H. Clark, A.-A. Edris, M. El-Gasseir, K. Epp, A. Isaacs, D. Woodford, Softening the Blow of disturbances: segmentation with grid shock absorbers for reliability of large transmission interconnections, IEEE Power Energy Mag. 8 (1) (2008) 30–41, http://dx.doi.org/10.1109/MPAE.2008.4412938.

[80] F. Katiraei, M.R. Iravani, P.W. Lehn, Small-signal dynamic model of a micro-grid including conventional and electronically interfaced distributed resources, IET Gener. Transm. Distrib. 1 (3) (2007) 369–378.

[81] D.E. Davison, E.J. Davison, Optimal transient response shaping of the servomechanism problem, J. Optim. Theory Appl. 115 (3) (2002) 491–515.

[82] K.J. Åstrøm, Auto-tuning, adaptation and expert control, in: American Control Conference, June 1985, Boston, MA, 1985, pp. 1514–1519.

[83] K.J. Åstrøm, T. Hägglund, C.C. Hang, W.K. Ho, Automatic tuning and adaptation for PID controllers—a survey, Contr. Eng. Pract. 1 (4) (1993) 699–714.

[84] K.J. Åstrøm, T. Hägglund, PID Controllers: Theory, Design and Tuning, 2nd ed., Instrument Society of America, Research Triangle Park, NC, 1995.

[85] K.J. Åstrøm, T. Hägglund, Automatic tuning of PID controllers, in: W. Levin (Ed.), The Control Handbook, vol. 53, CRC Press and IEEE Press, 1996.

[86] A.M. Gole, S. Filizadeh, R.W. Menzies, P.L. Wilson, Optimization-enabled electromagnetic transient simulation, IEEE Trans. Power Del. 20 (1) (2005) 512–518, http://dx.doi.org/10.1109/TPWRD.2004.835385.

[87] M. Heidari, S. Filizadeh, A.M. Gole, Support tools for simulation-based optimal design of power networks with embedded power electronics, IEEE Trans. Power Del. 23 (3) (2008) 1561–1570, http://dx.doi.org/10.1109/TPWRD.2007.916177.

[88] H. Li, F. Li, Y. Xu, D.T. Rizy, J.D. Kueck, Adaptive voltage control with distributed energy resources: algorithm, theoretical analysis, simulation, and field test verification, IEEE Trans. Power Syst. 25 (3) (2010) 1638–1647.

[89] W.W. Weaver, P.T. Krein, Game-theoretic control of small-scale power systems, IEEE Trans. Power Del. 24 (3) (2009) 1560–1567.

[90] A. Mehrizi-Sani, R. Iravani, Performance evaluation of a distributed control scheme for overvoltage mitigation, in: CIGRÉ International Symposium Electric Power System of Future: Integrating Supergrids and Microgrid, September 2011, Bologna, Italy, 2011.

[91] H. Ghaffarzadeh, C. Stone, A. Mehrizi-Sani, Predictive set point modulation to mitigate transients in lightly damped balanced and unbalanced systems, IEEE Trans. Power Syst., accepted for publication (TPWRS-01262-2015.R1), 2016.

CHAPTER 3

Hierarchical Power Sharing Control in DC Microgrids

S. Peyghami*, H. Mokhtari*, F. Blaabjerg†
*Sharif University of Technology, Tehran, Iran
†Aalborg University, Aalborg, Denmark

1. INTRODUCTION

The first electrical power distribution system was developed by Edison to lighten a street in lower Manhattan with a direct current (DC) in 1882. On the other hand, with the development of alternating current (AC) arc lightening systems, transformers, and induction machines in 1880s, it became possible to lighten long distances by cheaper wires with a high AC voltage [1, 2]. Edison's DC power system suffered from the transmission capability for long distances, and thus was applicable only for high-density customers [3]. The competition between AC and DC distribution systems in that century was called the *war of currents* [4], which was finally won by AC.

However, today the important concerns such as limited energy sources and environmental issues require governments and engineers to find more reliable, efficient, cost-effective, and environmentally friendly solutions for the energy resources and technologies at the generation, transmission, and consumption levels. Nowadays, most countries, especially in Europe, are investing in renewable energy sources such as wind farms and photovoltaic parks. Integrating renewable energy sources into power systems reduces the dependency on the limited fossil fuels and decreases the pollution and heating of the environment as well. On the other hand, installing renewable resources in the vicinity of loads as distributed generators (DGs) has some advantages, including reducing the transmission losses, increasing the reliability, no need to develop transmission lines because of increasing energy consumption, and reducing the costs. In recent years, microgrid technology has been employed to add controllability to DGs and to increase the reliability by supporting critical loads in the absence of a utility grid [5–9]. Because of advances in power electronic converters, the major problem of DC systems is solved and the *"war of currents"* has been reborn [4]. Because of the main concerns about the sources and environment as well as high efficiency and reliability, DC power systems are competing with AC systems at the generation, transmission, and distribution levels.

1.1. Advantages of DC Microgrids

DC microgrids have major advantages over AC microgrids, including higher efficiency, reliability, and stability. Many energy resources directly produce DC power or AC power with variable frequency, and hence they require power electronic converters to connect to a grid. DC-based energy sources and storage units such as photovoltaic arrays, fuel cell modules, and batteries need single-phase or three-phase inverters to connect to AC microgrids. Furthermore, variable-frequency sources such as wind turbines and microturbines require a double-stage AC-DC-AC converter (full converter) to connect to an AC microgrid. On the other hand, because of advances in power electronic technologies, most of the loads in domestic and industrial applications are naturally DC or need a frequency other than the fundamental 50/60 Hz. Domestic loads such as laptops, PCs, cell phones, TVs, and lighting are supplied through a rectifier from the AC grid. Some domestic loads such as resistive loads for heating and lighting as well as universal motor-based loads such as vacuum cleaners and mixers can work with both AC and DC power. The significant part of industrial loads (i.e., motors) is connected to the grid through a double-stage variable-speed drive. Moreover, data centers are also DC-type loads. Therefore integrating both DC-based loads and sources into a DC-based power system reduces the power conversion stages, and hence increases the efficiency and reliability.

Furthermore, another advantage of a DC power system over an AC power system is the lack of reactive power. Reactive power causes power loss in the lines as well as oversizing of the inverters and DC link capacitors and reduction of the power transmission capability of the lines, which affect the efficiency and reliability of AC power systems. However, in DC microgrids, only the active power is transmitted in the lines, which reduces the wire sizing and DC link capacitors. Therefore the efficiency and reliability of DC microgrids are higher than those of AC microgrids. Meanwhile, eliminating the power conversion stages in full converter-based sources and variable-speed drives will further reduces the cost.

Harmonic currents of nonlinear loads increase the power loss in the lines, transformers, and converters, and consequently reduce the reliability of the transformers and converters in AC microgrids. Furthermore, nonlinear loads make the control system more complicated and may even lead to instability in some cases. Considering the high penetration of nonlinear loads in power distribution systems, from the efficiency, reliability, and stability standpoints, DC microgrids are again preferred to AC microgrids.

In comparison with AC microgrids, the control of DC microgrids is very simple, since the power is controlled by the DC link voltage, and the control complexity due to the angular and frequency stability does not appear in DC microgrids. Therefore a DC microgrid seems to be an efficient, reliable, and cost-effective option in some applications, which are explained in the next subsection.

1.2. Applications of DC Microgrids

DC microgrid technology has been used in spacecraft, space stations [10, 11], aircraft [12], electric vehicles [56], and ships [13, 14], which have no access to a utility grid. In these applications, the reliability, efficiency, weight, and volume of the system are important factors for designers. For example, the newest onboard DC distribution grid designed by ABB for a marine power and propulsion system is illustrated in Fig. 3.1. The benefits of this onboard DC grid include [15]:

- more functional vessel layout through more flexible placement of electrical components;
- reduced maintenance of engines by more efficient operation;
- improved dynamic response and maneuverability;
- increased space for payload through the lower electrical footprint and more flexible placement of electrical components;
- a system platform that allows plug-and-play retrofitting possibilities to adapt to future energy sources;
- up to 20% fuel savings due to higher efficiency.

Hybrid AC-DC-type microgrids are also employed in ships and airplanes as well [12, 14, 16, 17]. DC-type power systems are also reliable options for space applications [10, 11]. Data centers are the other application area of DC power systems [5, 18, 19].

DC power distribution systems are also efficient options for home systems [5, 20] to integrate renewable resources and storage units to make zero-net-energy buildings [5]. Furthermore, DC-based power systems are used in traction systems as well as high-voltage DC transmission power systems.

On the basis of the advantages of DC systems over AC systems and the expected increase in the application of DC systems in the future power grids, Edison's power system may be the winner of the war of currents [4].

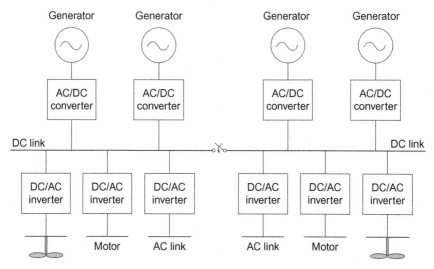

Fig. 3.1 Onboard DC grid for marine power designed by ABB.

However, it seems that the future power grids will be made up of multiterminal AC-DC power systems (i.e., hybrid AC-DC grids) [10, 21–25]. DC power systems are connected to the utility grid through one or multiple power electronic converters as *interlinking converters* (ICs) [25]; hence, the DC power system is dynamically decoupled from the utility grid [6]. Therefore, DC power systems always operate in microgrid mode (i.e., islanded mode) since there is no stiff DC grid.

In this chapter, different control levels of DC power systems (i.e., DC microgrids) are studied and discussed. Different energy units (EUs) in DC microgrids are classified. The necessity and responsibility of a power management system for the control of these EUs is discussed in Section 2. Furthermore, power sharing control in conventional AC power systems is briefly described to make an analogy between the power management systems of AC and DC grids.

The primary droop-based controller is explained in Section 3. A cost-based droop scheme to optimize the operational costs is also discussed in this section. In Section 4, centralized and decentralized secondary controllers are explained. Different control structures of the tertiary controller to control the power flow among different EUs and/or the utility grid as well as the cluster of DC microgrids are presented in Section 5. In Section 6, different autonomous droop-based control approaches are described for proper power dispatching among dispatchable and non-dispatchable EUs. Finally, Section 7 summarizes the power sharing issues in DC microgrids.

2. POWER MANAGEMENT ISSUES

An appropriate power management system and a control structure are required for stable and reliable operation of a microgrid. There are different kinds of EUs in DC microgrids, the control systems of which depend on the nature of their prime energy resources. Hence a classification of different EUs is mandatory before power management systems in DC microgrids are discussed. This classification is explained in this section. The necessity for the power management system and its responsibilities in DC grids are also discussed in this section.

2.1. Classification of Energy Units in DC Microgrids

EUs in DC microgrids include dispatchable and non-dispatchable DGs as well as distributed storage (DS) units. The output power of dispatchable units and DS units can be controlled to supply the required load power. However, non-dispatchable units usually operate in a *maximum power point tracking* (MPPT) mode. Furthermore, ICs can be controlled to inject or absorb the demanded power and hence, they behave as a dispatchable source or a controlled load. Different EUs are classified in Table 3.1.

A single-line block diagram of a typical DC microgrid is shown in Fig. 3.2. All EUs need a power electronic converter to connect to the DC microgrid. Different converters can be employed for each energy unit depending on the nature of the prime energy resource as well as the voltage and power levels. Furthermore, microturbines and wind turbines require only a single-stage converter to connect to the microgrid. DS units and utility ICs also employ bidirectional converters. An inner current and/or voltage regulator is required to control the current and voltage

Table 3.1 Classification of different energy units in DC microgrids

Source type	Specification	Example
Generators	Dispatchable	Fuel cell, microturbine
	Non-dispatchable	Photovoltaic, wind turbine
Storage units	Dispatchable/fast	Battery
	Dispatchable/slow	Fly wheel, regenerative fuel cell
Utility grid	Dispatchable	Interlinking converter

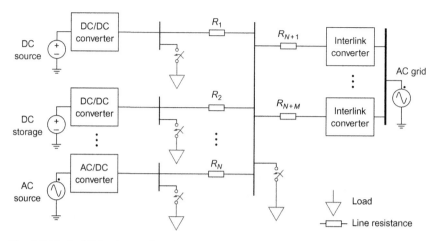

Fig. 3.2 Typical DC microgrid with AC and DC sources, storage units, and distributed loads.

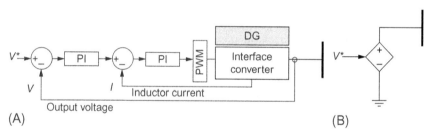

Fig. 3.3 Grid-forming converter: (A) control block diagram; (B) simplified electrical model.

of the microgrid to support the load demand and control the DC link voltage.

Dispatchable units are responsible for forming the voltage of the microgrid. These converters, which can be controlled as constant voltage sources, are also called *grid-forming converters*. A control block diagram of a grid-forming converter is shown in Fig. 3.3. The control system causes the output voltage (V) to settle at a reference value (V^*). An inner current controller can be used to improve the dynamics of the control system by regulating the current of the DC inductor of the converter. In the case of multiple grid-forming converters in a microgrid, the output current of the DGs will be shared on the basis of the line resistances.

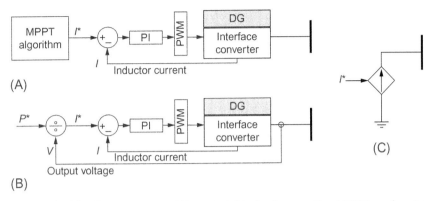

Fig. 3.4 Grid-feeding converter: (A) control block diagram for *MPPT*-based units; (B), control block diagram for dispatchable units in constant power mode; and (C) simplified electrical model.

Non-dispatchable units usually operate in an MPPT mode. Hence they only inject MPPT power or current (I^*) into the grid. A control block diagram of an MPPT-based unit is shown in Fig. 3.4A. Other dispatchable units can also work in a constant power mode to inject the set point power (P^*) into the grid. A control block diagram of a dispatchable unit in constant power mode is shown in Fig. 3.4B. In these cases, the converters, also called *grid-feeding converters*, are controlled as a constant current source as shown in Fig. 3.4C. For proper operation of a grid-feeding converter, at least one voltage former is required in the microgrid to form the DC link voltage.

2.2. Hierarchical Power Sharing in DC Microgrids

As already discussed, all converters can be modeled either as current sources or as voltage sources on the basis of the input energy resources and the control strategy. The voltage of the DC microgrid is regulated by the voltage source converters. On the other hand, the current flow in the lines requires different voltages at both ends of the line. Therefore both power flow control and voltage regulation need to be controlled by the grid-forming converters. However, the maximum supplied power limited by the rated value of these converters has to be taken into account to prevent overstressing of the converters. Furthermore, to decrease the operational costs, the load/demand of the microgrid should be economically dispatched among the different sources. Consequently, some complementary controllers are required to control the power flow and voltage regulation in addition to the converters' internal voltage and/or current controllers. The main objectives of the complementary controllers as a power management system include:

- current/power sharing among converters to prevent overstressing of components;
- improving voltage stability;
- preventing circulating current among the converters;
- regulating the system voltage within acceptable intervals;
- power flow control between the DC microgrid and the AC utility grid or the neighboring AC microgrid;
- economic power dispatch among the converters and/or between the DC microgrid and the AC utility grid, the neighboring AC microgrid, and DC microgrids.

The importance and timescale of these objectives are different, and hence a hierarchical control system is required to provide the controllers with different dynamic responses. The hierarchical control structure consists of three levels of control as shown in Fig. 3.5. The primary controller (level I) is responsible for proper power/current sharing among the units, improving the voltage stability, and preventing the circulating current among the converters. Keeping the microgrid voltage within acceptable values is performed by the secondary controller at level II. A *wide-area monitoring* system (i.e., a communication system) is required to collect the voltage at different buses in the microgrid (i.e., V_{MG}). The secondary

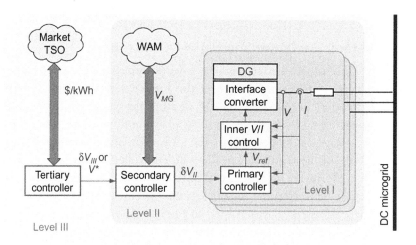

Fig. 3.5 Hierarchical control (three levels) structure of DC microgrids.

controller regulates the microgrid voltages (V_{MG}) at a desired value (V^*) determined by the tertiary controller. The correction term (δV_{II}) is sent to all the EUs to regulate V_{MG}.

The tertiary controller is responsible for performing:

- optimal power flow among different DGs in the microgrid dictated by the distributed system operator (DSO) or the transmission system operator (TSO) regarding the electricity market;
- optimal power flow among the DC microgrid and the stiff DC grid or AC grid dictated by the TSO/DSO;
- power flow control among the cluster of DC microgrids;
- power flow control among the DC microgrids to decrease the conversion losses and operational constants.

This controller determines the reference voltage (V^*) for the secondary controller or the correction term (δV_{III}) for the primary controller to regulate the output power of the related energy unit.

As the hierarchical control structure in DC microgrids mimics the control and operation of conventional AC power systems, the traditional power sharing concept is briefly explained in the next subsection, and in the following sections the power management controller for DC microgrids is discussed in more detail.

2.3. Hierarchical Power Sharing in Conventional Power Systems

In conventional AC power systems, to have a stable and reliable power system, the frequency, and voltages at all the buses must remain at acceptable values, which are determined by the standard recommendations. Therefore it is important that the generators in the system participate in voltage regulating and frequency control. In conventional power systems, either load or line reactive power is supplied by the synchronous generators and reactive power compensators. The active power of the load is provided by the synchronous generators. The portion of each generator in the load sharing is determined by the electricity market. Some generators operate in constant power mode and supply only the power planned by the grid. However, the load of the grid is a stochastic variable and cannot be exactly forecasted to plan the output power of the generators. Therefore some generators are responsible for adjusting the differences between the planned load and the actual load. This is performed by the droop controller of the governor system. This droop characteristic allows different generators to work in parallel to supply the loads. Since the frequency is the same

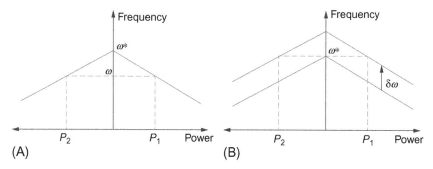

Fig. 3.6 Droop characteristics of two synchronous generators: (A) primary controller effect; (B) secondary controller effect. The secondary controller shifts up the droop characteristic to settle the frequency at ω^* and $\delta\omega = \omega^* - \omega$.

all over the grid, different generators can be coordinated with the droop characteristics [26, 27]. For example, the droop characteristics of two synchronous generators are shown in Fig. 3.6, where ω^* is the reference value of the frequency and P_1 and P_2 are the output power of the first and the second generator. The steady-state value of the frequency settles at ω. The frequency drop from the reference value can be compensated by the secondary controller. Therefore the secondary controller shifts up the droop characteristics where the frequency becomes equal to the reference value as shown in Fig. 3.6. The reference value is determined by the tertiary controller, which is responsible for the power exchange among the different generators and areas on the basis of the electricity market. These concepts have been applied in AC microgrids [28, 29]. In the case of a microgrid, the tertiary level controls the power exchange between the microgrid and the main grid. On the basis of the commands from the dispatcher, the frequency reference (ω^*) can be adjusted to control the power flow between the microgrid and the main grid.

These control levels have been applied to DC microgrids [29]. Instead of the frequency-power droop, a voltage-power droop as shown in Fig. 3.7A has been employed for power sharing in DC microgrids. Increasing the load causes a voltage drop in the microgrid. The droop controller senses the voltage drop and increases the output power. Meanwhile, a secondary controller regulates the voltage drop due to the droop controller. The effect of the secondary controller is shown in Fig. 3.7B as well.

Simplified AC and DC power systems are shown in Fig. 3.8. In the AC grid (see Fig. 3.8A), the active power flow is proportional to the relative

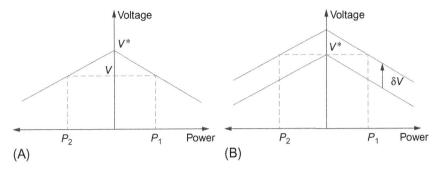

Fig. 3.7 Droop characteristics of two grid-forming converters: (A) primary controller effect; (B) secondary controller effect. The secondary controller shifts up the droop characteristic to settle the voltage at V^* and $\delta V = V^* - V$.

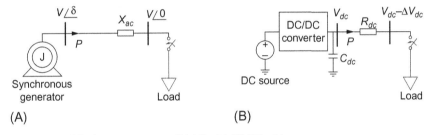

Fig. 3.8 Simplified power systems: (A) AC grid; (B) DC grid.

voltage angle (δ), which is also proportional to the frequency. Therefore the active power can be controlled by the frequency. The active power (P) can be calculated as

$$P = \frac{V \cdot V}{X_{ac}} \sin(\delta), \tag{3.1}$$

where V is the AC voltage and X_{ac} is the line inductance.

In the DC grid (see Fig. 3.8B), the active power flow is proportional to the DC voltage (V_{dc}). Therefore the active power can be controlled by the DC link voltage as calculated in Eq. (3.2):

$$P = V_{dc} \frac{\Delta V_{dc}}{R_{dc}}, \tag{3.2}$$

where ΔV_{dc} is the voltage drop over the line resistance (R_{dc}).

In AC systems, load changes are responded to by the energy of rotating inertia (J) of the synchronous generators. A governor system senses the

Table 3.2 Correspondence between AC and DC parameters [30]

Characteristics	AC grid	DC grid
Control parameter	Frequency (ω)	DC link voltage (V_{dc})
Line impedance	X_{ac}	R_{dc}
Active power (P)	$V^2 \sin \delta / X_{ac}$	$V_{dc} V_{dc} / R_{dc}$
System inertia	J	C_{dc}
System energy	$\frac{1}{2} J \omega^2$	$\frac{1}{2} C V_{dc}^2$

frequency drop and increases the mechanical input power of the generator. In a DC grid, increasing the load at first decreases the voltage of the DC link capacitor (C_{dc}). A droop controller senses the voltage drop and increases the output power to supply the load. The analogies between an AC grid and a DC grid are summarized in Table 3.2 [30].

In the next sections, three levels of a power sharing system in DC microgrids including a primary droop controller, a secondary regulator, and a tertiary controller are explained.

3. PRIMARY CONTROL: LEVEL I

As discussed in the previous section, a primary controller, as the first level in the power management hierarchy, is required to control the load sharing among the parallel dispatchable units. The primary controller is also responsible for improving the voltage stability and preventing circulating current among converters [29, 31–39]. A voltage-power or voltage-current droop scheme, the same as a frequency-power droop control of a synchronous generator, can be employed as the primary controller. Load changes in the microgrid cause DC link voltage variation. Therefore the droop controller can change the output power of the converter by voltage variation. This concept is analogous to the control system of the governor in conventional power plants. In this section, the droop scheme is explained in more detail.

3.1. Droop Control

Droop-based approaches for the control of power/current sharing among parallel converters are reliable and resilient methods. Droop controllers can automatically set the reference of the input voltage regulator. Increasing/decreasing the load power decreases/increases the DC link voltage of the microgrid. The droop controller determines the output current/power

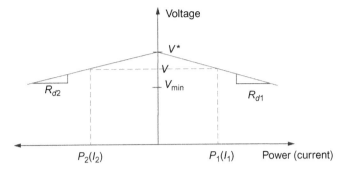

Fig. 3.9 Voltage-power droop characteristics of two converters used for the primary control.

of each converter. This concept is shown graphically in Fig. 3.9 for two converters. As can be seen, the load current/power is shared between the converters on the basis the slope of their droop characteristics. At the operating point, the DC voltage is equal to V, and the load power is equal to $P_1 + P_2$, where P_1 and P_2 are the output power of the first and the second converter.

The droop slope can be determined according to the rated current (I_n) or rated power (P_n) of the converter. Considering V_{min}/V^* as the minimum/reference voltage of the microgrid, we can calculate the droop slope (R_d) as

$$R_d = \frac{V^* - V_{min}}{I_n} \quad \text{or} \quad R_d = \frac{V^* - V_{min}}{P_n}. \tag{3.3}$$

In DC microgrids the power is proportional to the current, and hence both voltage and current can be used to implement the droop controller. However, the relation between the current and the voltage as well as the power and the square of the voltage is linear, and it would be better to use the droop control of $V - I$ or $V^2 - P$ to have a linear controller. Therefore the droop slope in the second case can be determined as [3]

$$R_d = \frac{(V^*)^2 - V_{min}^2}{P_n}. \tag{3.4}$$

This equation indicates that the output current/power of the converters is proportional to their rated current/power, and hence it prevents overstressing of the converters in heavy loading conditions. The droop slope or the droop controller gain must be determined such that it guarantees the overall stability of the system. Therefore the maximum allowable voltage variations as well as the stability of the closed loop system are the factors that must be taken into account in the droop gain design procedure. The output of the primary droop controller is the reference voltage (V_{ref}) for the inner voltage regulator and can be calculated as

$$V_{ref} = V_{max} - R_d I \quad \text{or} \quad V_{ref} = V_{max} - R_d P = V_{max} - (R_d V)I. \quad (3.5)$$

Eq. (3.5) shows that a droop-controlled converter behaves as an ideal voltage source connected in series with a virtual resistor. The droop controller can be implemented in an inverse or forward form. As shown in Fig. 3.10A, an inverse droop method contains two inner voltage and current controllers. The forward droop method, as shown in Fig. 3.10B, has only one current controller, and hence is faster than an inverse droop method. However, in the forward droop method the droop gains must be carefully chosen since they can affect the phase margin of the current controller. Without loss of generalization, in this chapter, the inverse droop scheme is used to discuss the other levels of the power management system.

A simplified electrical model of the converter is shown in Fig. 3.10C. These converters can form the voltage of the microgrid as well as control the current flow among parallel converters. Hence they are called *grid-supporting converters*. Therefore a grid-forming converter or grid-supporting

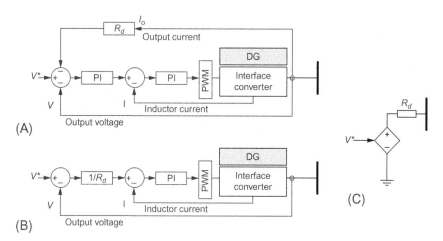

Fig. 3.10 Grid-supporting converter: (A) control block diagram of the inverse droop; (B) control block diagram of the forward droop; (C) simplified electrical model.

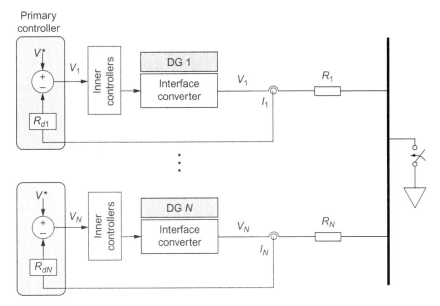

Fig. 3.11 Primary droop control of different converters in the DC microgrid.

converter is an essential unit in the microgrid to form the voltage of the microgrid.

A simplified DC microgrid with droop controllers is shown in Fig. 3.11. Its steady-state model is represented in Fig. 3.12. In this microgrid, only grid-supporting units are considered. Considering $R_1, R_2, \ldots, R_N$ as the line resistors and $R_{d1}, R_{d2}, \ldots, R_{dN}$ as the droop gains of the converters, we can determine the relation among the output currents of the units as

$$I_1 : I_2 : \ldots : I_N \equiv \frac{1}{(R_{d1} + R_1)} : \frac{1}{(R_{d2} + R_2)} : \ldots : \frac{1}{(R_{dN} + R_N)}. \quad (3.6)$$

If $R_{d1}, R_{d2}, \ldots, R_{dN} \gg R_1, R_2, \ldots, R_N$, as shown in Eq. (3.7), the output currents are inversely proportional to the droop gains; that is,

$$I_1 : I_2 : \ldots : I_N \cong \frac{1}{R_{d1}} : \frac{1}{R_{d2}} : \ldots : \frac{1}{R_{dN}}. \quad (3.7)$$

Higher droop gains can reduce the effect of line resistances on the current sharing accuracy. Lower droop gains cause mismatches among the output currents. However, higher droop gains cause large voltage drops in the output terminal of the converters. A microgrid is considered with

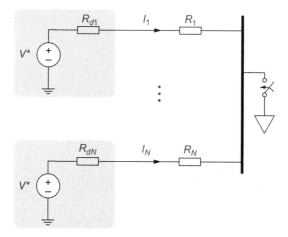

Fig. 3.12 Simplified electrical model of droop-controlled converters of the microgrid shown in Fig. 3.11.

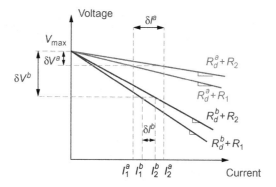

Fig. 3.13 Effect of the droop gain on the current sharing.

two converters and $R_{d1} = R_{d2} = R_d$ and $R_1 > R_2$. The effect of the line resistances on the droop characteristics is shown in Fig. 3.13 for two different droop gains ($R_d^a < R_d^b$). Since the line resistors are not equal, the effective droop slopes are not the same. As can be seen in Fig. 3.13, with a small droop gain, the output current of the converters has a large mismatch (δI^a); however, the voltage drop (δV^a) has a small value. On the other hand, increasing the droop gain (R_d^b) reduces the mismatches between the currents (δI^b) but increases the voltage drop (δV^b). To compensate for the current mismatch and voltage drop due to the droop controller, a high–level secondary regulator is required.

3.2. Cost-Based Droop Control

To decrease/optimize the operational costs of a microgrid, a cost-based droop controller for AC and DC microgrids is proposed in [40–42]. Typical cost-based droop characteristics for three DGs in a DC microgrid are shown in Fig. 3.14, where DG 1 has the highest operating cost and DG 3 has the lowest. Therefore the load power can be economically dispatched among DGs to achieve the lowest operational costs.

The cost of different type of DGs can be represented by Eq. (3.8):

$$C_i(P_i) = \alpha + \beta \left(\frac{P_i}{P_{n,i}} \right) + \xi \left(\frac{P_i}{P_{n,i}} \right)^2, \tag{3.8}$$

where C_i is the normalized cost per rated power, $P_{n,i}$, of the ith converter and coefficients α, β, and ξ model the cost of the prime energy source and the power loss in the converters [41].

Therefore, Eq. (3.5) can be modified to Eq. (3.9) to consider the operational costs:

$$V_{ref,i} = V_{max} - \chi \left(C_i(P_i) - C_{NL,i} \right),$$
$$\chi = \frac{V_{max} - V_{min}}{\max_{i=1:N} \left\{ C_i(P_{n,i}) - C_{NL,i} \right\}}, \tag{3.9}$$

where $C_{NL,i}$ is the no-load cost of the ith DG and N is the total number of DGs [41].

As already mentioned, the primary controller in DC microgrids causes current mismatches and voltage drop. To control the current/power sharing

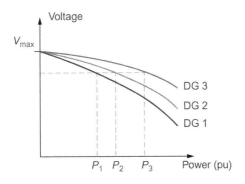

Fig. 3.14 Typical cost-based droop characteristics for three distributed generators.

and voltage regulation, a secondary controller is required. This control level will be explained in the next section.

4. SECONDARY CONTROL: LEVEL II

After load sharing among different the converters in a microgrid, the secondary controller compensates for the voltage drop caused by the primary controller and controls the accurate current sharing as well. The main objective of the secondary controller is to regulate the voltage in the microgrid [29]. Meanwhile, a supervisory controller can be considered at this level to compensate for the current mismatches [43]. However, in the case of small microgrids with short feeders, the effect of line resistances on the current sharing is negligible, and hence the current regulator is not required.

A control block diagram of the voltage and average current regulator (ACR) is shown in Fig. 3.15A. The voltage regulator controls the DC link voltage at the rated value (V^*). The ACR controls the output current of the converters at the weighted average current of the active converters. The normalized per rated current (I_{rated}) of the converters needs to be shared through a communication network. The effects of both controllers are shown graphically in Fig. 3.15B. The DC link voltage drops to V_{MG} as the load increases. The secondary controller measures the DC link voltage and calculates the correction term of δV. It then sends this term to all converters to increase their output voltage. The ACR locally adjusts the slope of the droop characteristic by calculating the correction term δV_c

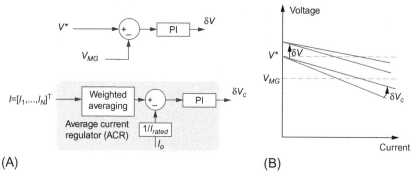

Fig. 3.15 Secondary controller: (A) voltage and ACR; (B) effect of the secondary controller on the droop characteristics of the converter.

to regulate the output current at the weighted average value. The voltage regulator can be implemented with a centralized or decentralized control policy. Furthermore, from the control theory point of view, the secondary voltage regulator should be slower than the ACR, and both of them need to be slower than the primary controller.

The secondary controller in DC microgrids requires some data transmission to regulate the voltage and current. This information includes the output current of the converters, $I = [I_1, I_2, \ldots, I_N]^T$, the DC link voltage ($V$), or the output voltage of the converters $V = [V_1, V_2, \ldots, V_N]^T$, and the number of active converters (N). Therefore, to have effective control of the DC microgrid, a communication layer needs to be considered. However, it can affect the reliability and stability of the system since any link failure results in the failure of the whole microgrid [44]. To get rid of the communication among the converters, some decentralized (distributed) controllers have been employed.

4.1. Centralized Controller

In the centralized control approach, a central control unit regulates the voltage of the microgrid. In [22, 29], the voltage of the localized loads at one bus or the coupling point into the utility grid (i.e., the *point of common coupling*, PCC) is considered to be regulated by the secondary controller. A control block diagram of a centralized secondary controller without an ACR is shown in Fig. 3.16. The voltage of the PCC (V_{MG}) needs to

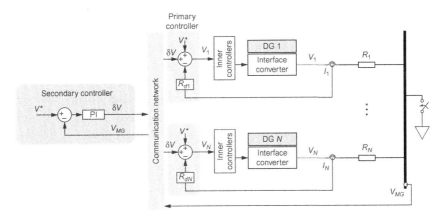

Fig. 3.16 Centralized secondary control without an ACR in a hierarchical control scheme for a DC microgrid.

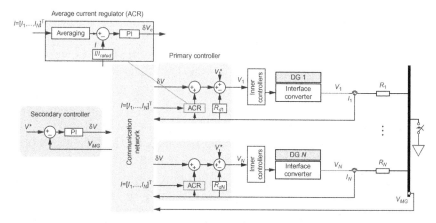

Fig. 3.17 Centralized secondary control with an ACR in a hierarchical control scheme for a DC microgrid.

be regulated at its rated voltage ($V^\star$), which can be determined by the higher control level. The voltage of the PCC and the correction term of the secondary level (δV) should be communicated between the central unit and the local converters.

The centralized secondary controller with an ACR is also shown in Fig. 3.17, where the per-unit output current of the converters, $I = [I_1, I_2, \ldots, I_N]^T$, is also communicated among the converters. The accuracy of the current sharing in this approach is higher than that of the previous one (see Fig. 3.16); however, more data transmission is required.

In this control approach, the voltage of only one bus (i.e., PCC) is regulated. However, in practice, the loads are not localized in a common bus, and hence the control of the voltage of one bus may not guarantee desirable voltage regulation in the microgrid. A simple approach to control the voltage of a microgrid is to regulate the average voltage of the desired buses (V_{avg}) at the secondary level. As shown in Fig. 3.18, in this approach, the average voltage of the loads and converters, $V = [V_1, V_2, \ldots, V_N, V_{L1}, \ldots, V_{LM}]^T$, is regulated by the secondary controller. However, the regulation of the voltage of the loads is mandatory. In this control approach, a point-to-point communication is required to regulate the voltages and currents of the converters. Disconnection of the central unit may lead to the failure of the whole system. To overcome this issue, some decentralized approaches are explained in the next subsection.

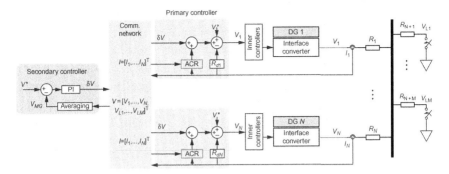

Fig. 3.18 Centralized secondary control considering an average voltage control of the desired buses with an ACR in a hierarchical control scheme for a DC microgrid.

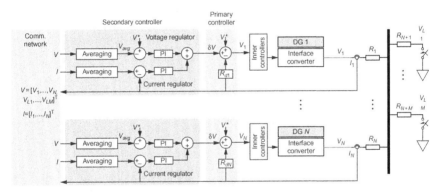

Fig. 3.19 Decentralized (distributed) secondary controller.

4.2. Decentralized Controller

In this approach, secondary controllers can locally regulate the voltage in the microgrid. A general decentralized (distributed) secondary controller is shown in Fig. 3.19. A communication network collects the information on the voltage of the desired buses, $V = [V_1, V_2, \ldots, V_N]^T$, and the output current of the converters, $I = [I_1, I_2, \ldots, I_N]^T$. The secondary controller of each converter calculates the average voltage (V_{avg}) and the weighted average current of the converters, and then regulates its output voltage and current accordingly. Implementation of the secondary controller in the distributed policy improves the reliability of the system since the central controller is replaced by decentralized regulators.

However, regulation of the average of the voltages and currents requires more data transmission through the microgrid. To overcome this problem, consensus algorithms are employed to regulate the global variables in distributed systems [38, 44]. In consensus algorithms, each converter requires to communicate with the neighboring converters. An updating protocol as shown in Fig. 3.20, also called a *dynamic consensus protocol*, carries out both the estimated global average voltage by the neighboring converter $V_{avg(j)}$ and the local voltage V_i in order to estimate the global average voltage $V_{avg(i)}$ [32, 44–47]. The average voltage of the jth converter, $V_{avg(j)}$, can be used to calculate the average voltage of the ith converter, $V_{avg(i)}$, with the dynamic consensus estimator shown in Fig. 3.20, where the coefficient a_{ij} is the weight of information exchanged between converters i and j. After some iterations, the estimated average voltage of the converters converges to a value that is equal to the average voltage of the converters (V_{avg}). As shown in Fig. 3.21, this average voltage is regulated by the secondary controller to settle at the reference value.

On the other hand, the per–unit current of the neighboring converter can be used to compensate for the mismatch of the currents of all

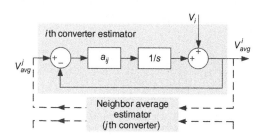

Fig. 3.20 Consensus global average voltage estimator.

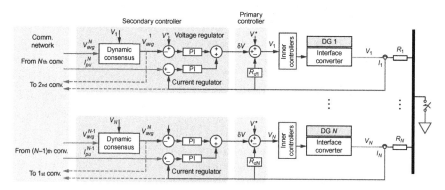

Fig. 3.21 Decentralized (distributed) secondary controller.

converters. This approach is analogous to the circular chain control in parallel inverters in AC microgrids, which is used to improve the current mismatches among the inverters [9, 28, 48–50].

In this approach, each converter shares the average voltage and per-unit current with the neighboring converter. Therefore sparse communication among the converters with low volume of transmitted information improves the reliability and stability of the system [44]. However, in consensus algorithms, only the average voltage of the grid-supporting buses is regulated. In practice, the loads may not be connected to the grid-supporting converters and might be distributed over the microgrid. Therefore regulation of the voltage of the grid-supporting buses may not guarantee the voltage regulation at load buses.

5. TERTIARY CONTROL: LEVEL III

The highest level of the power management system is the tertiary control, which controls the power flow:
1. between the microgrid and the utility grid;
2. among the cluster of microgrids, and between the cluster of microgrids and the utility grid;
3. among the DGs in the microgrid.

The main objectives of this controller include optimal and stable power flow in different topologies to increase the efficiency and/or decrease the operational costs as well as increase the reliability.

5.1. Power Flow Control Between the Microgrid and the Utility Grid

The exchanged power between the microgrid and the main grid is determined by the TSO/DSO. The TSO/DSO calculates the power reference according to the optimal power flow analysis based on the load and generation forecasting. If the microgrid is directly connected to a stiff DC grid, the tertiary controller can be implemented like the one shown in Figs. 3.22 and 3.23. After connection to the DC grid, a secondary controller can be replaced by the tertiary controller [29] (see Fig. 3.22) or the output of the tertiary controller can be used as a reference voltage (V_{ref}) for the secondary controller (see Fig. 3.23).

In practice, DC microgrids or DC distribution systems need to be connected to DC or AC power systems through a power electronic converter. As shown in Fig. 3.24, this converter behaves like a power

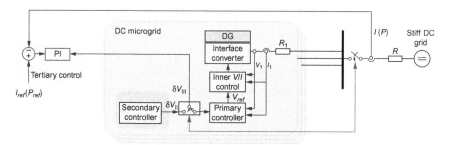

Fig. 3.22 Tertiary controller for power flow control between the microgrid and the stiff DC grid.

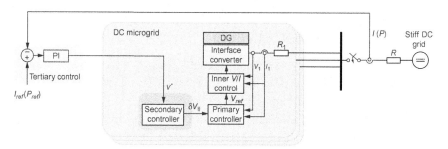

Fig. 3.23 Tertiary controller for power flow control between the microgrid and the stiff DC grid.

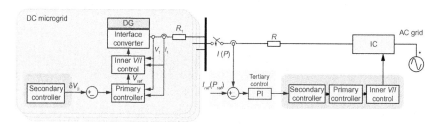

Fig. 3.24 Tertiary controller for power flow control between the microgrid and AC grid.

source or load for the microgrid. Therefore the tertiary controller can be implemented by control of this interfaced converter. Without the tertiary controller the interfaced converter operates in a droop method to support the microgrid. This approach is discussed in the next section.

5.2. Power Flow Control Among a Cluster of Microgrids

Another DC power system topology including a cluster of different DC microgrids is depicted in Fig. 3.25. The tertiary control is required to control the power flow among the different microgrids. A general tertiary control is shown in Fig. 3.25. This tertiary control is similar to the primary droop controller and is called a *cluster primary controller*. A cluster secondary controller is required to regulate the voltage drop due to the cluster primary controller. The same as for level II, the cluster secondary controller can be implemented with a centralized or a decentralized control policy. A consensus cluster control can also be employed for the tertiary control level [32, 47]. If the cluster of microgrids is connected to a stiff DC grid, a cluster tertiary controller like the one in Fig. 3.26 can be considered to control the power flow among the microgrids and the main grid [29]. Furthermore, the per-unit current of the clusters can be considered in the cluster secondary controller to improve the accuracy of the power flow among the clusters [32, 47].

5.3. Power Flow Control Among Distributed Generators in a DC Microgrid

The tertiary control can be employed to optimize the converter losses in the microgrid as well [36, 45, 46]. In this approach, one can control the power

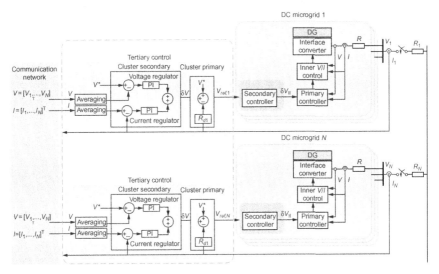

Fig. 3.25 Tertiary controller (including cluster secondary and cluster primary controller) for the power flow control among the cluster microgrids.

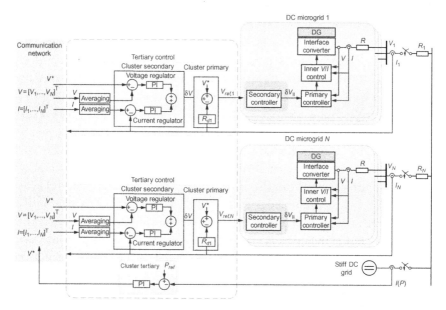

Fig. 3.26 Centralized tertiary controller for the converter power loss optimization.

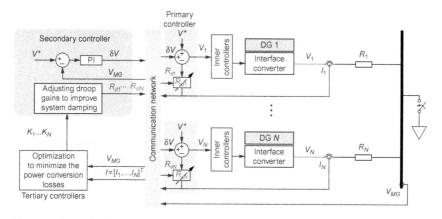

Fig. 3.27 Centralized tertiary controller for the converter power loss optimization.

flow among the different DGs in the microgrid by taking into considering the efficiency curve of the converters. As shown in Fig. 3.27, a central tertiary controller is considered to optimize the efficiency of the system on the basis of the efficiency curve of the converters. An optimization algorithm such as a genetic algorithm or partial swarm optimization calculates the total demand, microgrid voltage, and the efficiency curves of the converters to

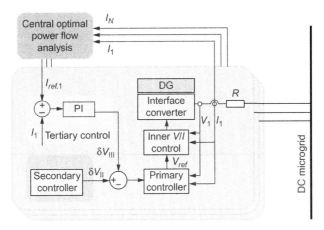

Fig. 3.28 Centralized optimal power flow controller to minimize the operational costs.

optimize the output power of the converters. This controller determines the droop coefficients of $K_1, K_2, \ldots, K_N$. The centralized secondary controller can be used in the state space model of the system to adjust the droop gains $(R_{d1}, \ldots, R_{dN})$ determined by the tertiary controller to improve the system damping. The voltage of the microgrid can be regulated by the secondary controller as well [36, 45, 46].

Furthermore, a centralized optimal power flow controller like the one shown in Fig. 3.28 can be employed to minimize the operational costs of the microgrid. An optimization algorithm determines the optimal current references of the converters (i.e., $I_{ref1}, \ldots, I_{refN}$) and then a local tertiary controller regulates the output current of the converters, $I_1, \ldots, I_N$.

As already discussed, the operational costs can be considered at the primary level by modification of the droop gains. Therefore the droop gains are not constant in this approach and may affect the damping and stability of the system. On the other hand, use of the tertiary level to optimize the operational costs requires a central control unit, in which case one encounters reliability issues.

6. AUTONOMOUS OPERATION OF DC MICROGRIDS

Since there is no stiff DC grid in practice, all DC distribution systems can be considered as a microgrid. Therefore one can employ a droop-based approach to control the power and energy flow in the microgrid by defining suitable droop characteristics for different kinds of EUs to operate in an autonomous mode. A simplified DC microgrid is shown in Fig. 3.29.

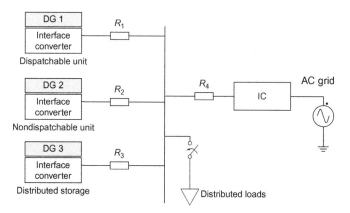

Fig. 3.29 Typical DC microgrid with dispatchable and non-dispatchable units, distributed storage, an interlinking converter, and distributed loads.

In the following subsections, possible droop schemes for different kinds of EUs are explained.

6.1. Autonomous Droop Approach

Typical droop characteristics for an IC, a DS unit, and dispatchable and non-dispatchable units are shown in Fig. 3.30 [51]. For an IC, a bidirectional droop like the one shown in Fig. 3.30A can be considered. The maximum power injected into the grid must be determined by the TSO/DSO. However, the required power that can be supported by the grid can be determined by the droop characteristic shown in Fig. 3.30A. For the storage droop control shown in Fig. 3.30B, such as a battery, a regenerative fuel cell, and a flywheel, the maximum and minimum limits can be determined by the energy level or *state of charge* (SoC) level of the battery. Dispatchable EUs such as microturbines and fuel cells can be controlled like the one shown in Fig. 3.30C. Finally, as shown in Fig. 3.30D, non-dispatchable units such as photovoltaic arrays and wind turbines can operate in a droop mode from zero to MPPT power. Therefore MPPT-based units can support the microgrid under the MPPT power. In this approach only the maximum and minimum limits of the droop characteristics are updated on the basis of the SoC or MPPT limits. Furthermore, in droop mode operation, converters work as grid-supporting converters, and in constant power mode, they work as grid-feeding converters.

Adjustment of power limits may cause voltage instability because of the lack of a voltage-forming or voltage-supporting converter in some loading and sourcing conditions. For example, if the power demand of an IC is zero,

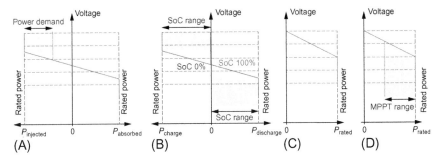

Fig. 3.30 Droop characteristics for different energy units in a DC microgrid: (A) interlinking converter; (B) storage converter; (C) dispatchable units; (D) non-dispatchable units.

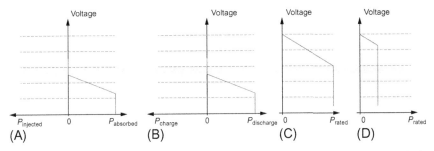

Fig. 3.31 Droop characteristics for different energy units in a DC microgrid: (A) interlinking converter; (B) storage converter; (C) dispatchable units; (D) non-dispatchable units in the highlighted region, there is no voltage forming converter and the system is unstable.

the storage unit is fully charged, and the non–dispatchable unit operates in an MPPT mode; hence, their droop characteristics must be modified like the one shown in Fig. 3.31A, B, and D respectively. If the load is equal to the sum of the rated powers of the dispatchable unit and the MPPT power of a non–dispatchable DG, the voltage can have any value in the highlighted region in Fig. 3.31. In this area, there is no voltage-forming converter, and the voltage cannot converge to a steady-state value.

To avoid operating all converters in constant power mode at the same time, which may occur in some loading and sourcing conditions, in addition to adjustment of the power limits, the slope of the droop controls should be adjusted. This approach is illustrated in Fig. 3.32. However, a control system needs to be designed to guarantee the stability of the system in different loading and sourcing conditions since the droop gains affect the stability of the system.

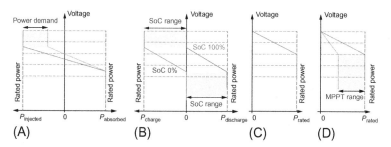

Fig. 3.32 Droop characteristics for different energy units: (A) interlinking converter; (B) storage converter; (C) dispatchable units; (D) non-dispatchable units.

The droop characteristics shown in Figs. 3.30 and 3.32 guarantee the parallel operation of the ICs in the DC microgrid. However, in the case of islanded hybrid AC-DC microgrids as well as in the absence of the tertiary control DSO/TSO, it would be better to support the microgrid demand by its internal sources. Hence the excess power or required power can be supported by the ICs. This approach is capable of controlling the energy flow between AC-DC microgrids [52]. The droop characteristics of two parallel ICs are shown in Fig. 3.33B and C. If the voltage lies between V_H and V_L, the power of the ICs is zero; otherwise the output power of the ICs can be determined by the droop characteristics as shown in Fig. 3.33B. Furthermore, the power sharing between ICs can be properly done by this droop control approach [52].

Different types of droop characteristics may be defined for various energy sources in the DC microgrid. To operate the microgrid in an autonomous mode, it is important to determine a voltage range for the

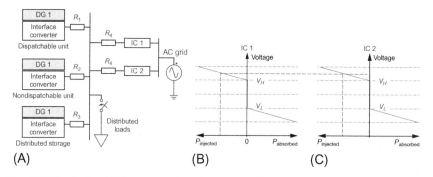

Fig. 3.33 Typical AC-DC microgrid with two interlinking converters: (A) single-line diagram of the power system; (B), (C) droop characteristics of the interlinking converters.

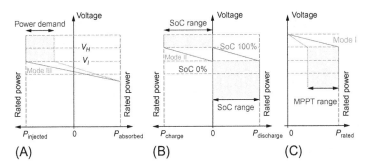

Fig. 3.34 Autonomous droop-based power sharing in a DC microgrid: (A) interlinking converter; (B) battery converter; (C) photovoltaic array.

droop gain of each unit to have at least one voltage-forming converter in the microgrid. For example, an autonomous control system based on the droop is shown in Fig. 3.34 [53], where it can be seen that in mode I, the voltage is controlled by the photovoltaic array. In mode II, a battery is responsible for forming the voltage, and in mode III, the IC controls the DC link voltage.

6.2. Control of the Converters in Autonomous DC Microgrids

The control systems of grid-forming, grid-feeding, and grid-supporting converters were explained in previous sections. It was mentioned that the dispatchable units operate in grid-forming or grid-supporting mode. Non-dispatchable units work in grid-feeding mode. However, as discussed in the previous subsection, non–dispatchable units can also operate in grid-supporting mode (droop mode) under the MPPT power. Four possible control approaches are shown in Fig. 3.35. The mode transition schemes shown in Fig. 3.36A [35] and Fig. 3.36B [32] include two modes; that is, a constant power mode (upper pass) and a droop mode (lower pass). In these approaches a supervisory controller is required to change the mode of operation. However, seamless control can also be implemented like the one shown in Fig. 3.36C [53] and Fig. 3.36D [33]. In Fig. 3.36C a current limiter is considered to limit the output current at the MPPT current (I_{MPPT}), which can be determined by the MPPT controller. In all of the approaches, the droop gain can be modified on the basis of maximum and minimum available currents and Eq. (3.3).

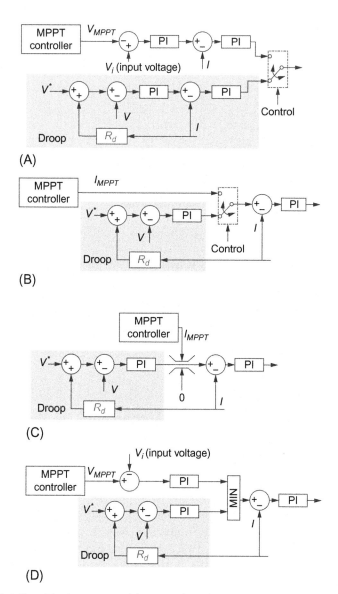

Fig. 3.35 Adjustable droop control for MPPT-based units: (A), (B) mode transition approach; (C), (D) seamless approach.

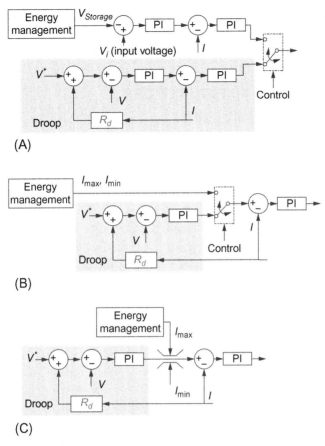

Fig. 3.36 Adjustable droop control for storage units: (A), (B) mode transition approach; (C) seamless approach.

To control the energy level of the storage units, like non–dispatchable units, two seamless and mode transition approaches are possible as shown in Fig. 3.36. In the control approach shown in Fig. 3.36A [35] and Fig. 3.36B [32], when the storage unit is empty, the control mode needs to be switched to the charging mode by a supervisory controller. However, in the control method shown in Fig. 3.36C [53], the controller is seamlessly switched between the constant power mode and droop mode. As discussed in Section 6.1, charging and discharging of the storage units can be controlled by the droop method. Therefore the droop gain and maximum and minimum limits can be modified on the basis of the energy level of the storage units, which can be locally estimated by the energy management system [54, 55].

7. CONCLUSION AND FUTURE WORK

In this section, the chapter is briefly summarized and some challenges and suggestions for future work are presented.

DC power systems are more reliable, robust, efficient, and environmentally friendly than AC systems. Because of advances in power electronics, DC power systems are competing with AC systems at different levels of a power system (i.e., generation, transmission, and distribution) as well as in islanded applications. In this chapter, power management issues in DC microgrids were discussed. The main objectives of a power management system at different levels of a hierarchy ere explained. Three levels of power sharing system were considered for robust, resilient, stable, and optimal power flow control in DC microgrids.

The droop-based primary controller is employed to control the current/power sharing and improve the voltage stability as well as prevent the circulating current among the converters. Droop methods can be implemented in an inverse or a forward control form. The cost-based droop method to optimize the operational costs in the microgrid was also explained. The secondary controller is used to compensate for the voltage drop due to the droop gains and regulate the voltage of the microgrid. Centralized and decentralized methods can be used to implement the secondary controller.

Different tertiary control policies for efficient and optimal power flow in DC microgrids were explained. The main objectives of the tertiary controller include power flow control between the microgrid and the main grid, power flow control among the clusters of microgrids, power flow control between the clusters and the main grid, and optimal power flow among the DGs in the microgrid to increase the efficiency or decrease the operational costs.

Models of different converters and control systems were presented. Different EUs were categorized as dispatchable and non–dispatchable units, where dispatchable units are modeled as grid-forming or grid-supporting converters and non-dispatchable units are modeled as grid-feeding converters. In the microgrid, at least one grid-forming or grid-supporting converter should form the DC link voltage. Autonomous operation of

DC microgrids is possible with appropriate droop characteristics of the converters. In an autonomous operation, all EUs are considered to work in a droop mode. Therefore non-dispatchable units can operate with the droop mode control under MPPT power to support the microgrid. The droop gain of the EUs for non-dispatchable DGs and DS units must be modified on the basis of the MPPT power of non-dispatchable DGs and the energy level of the storage units. The control systems of different kinds of EUs were presented as well.

Employing the communication network to improve the current sharing accuracy and voltage regulation decreases the reliability of the power sharing system in DC microgrids. Voltage-based droop methods suffer from a poor voltage regulation and current sharing since the voltage is a local variable. Therefore a global variable needs to be defined to improve the power sharing accuracy.

The secondary controller is employed to compensate for the voltage drop due to the droop gains. However, in the literature there is no control scheme to regulate the load voltage. In the centralized controller the voltage of the localized loads at one bus is regulated by the secondary controller. However, in real power systems the loads are distributed at different locations. Hence it is mandatory to regulate the load voltages.

Optimal and efficient power flow control among DGs requires the whole system model in a central controller to improve the system damping and stability. In practice, modeling the whole system and updating the system model considering the on/off state of the converters and lines requires a point-to-point communication network. This affects the reliability and stability. Therefore decentralized approaches must be presented for optimal, reliable, and stable operation of DC microgrids.

REFERENCES

[1] L. de Andrade, T.P. de Leao, A brief history of direct current in electrical power systems, in: IEEE HISTELCON, 2012, pp. 1–6.
[2] C. Sulzberger, Pearl street in miniature: models of the electric generating station [History], IEEE Power Energy Mag. 11 (2) (2013) 76–85.
[3] N.R. Chaudhuri, B. Chaudhuri, R. Mujumder, A. Yazdani, Multi-Terminal Direct-Current Grids: Modeling, Analysis and Control, John Wiley & Sons, New York, 2014.
[4] P. Fairley, DC versus AC: the second war of currents has already begun [In My View], IEEE Power Energy Mag. 10 (6) (2012) 104–103.
[5] B.T. Patterson, DC, come home: DC microgrids and the birth of the "Enernet", IEEE Power Energy Mag. 10 (6) (2012) 60–69.

[6] D. Boroyevich, I. Cvetkovic, R. Burgos, D. Dong, Intergrid: a future electronic energy network?, IEEE J. Emerg. Sel. Top. Power Electron. 1 (3) (2013) 127–138.

[7] F. Katiraei, R. Iravani, N. Hatziargyriou, A. Dimeas, Microgrids management, IEEE Power Energy Mag. 6 (3) (2008) 54–65.

[8] B. Kroposki, R. Lasseter, T. Ise, S. Morozumi, S. Papathanassiou, N. Hatziargyriou, Making microgrids work, IEEE Power Energy Mag. 6 (3) (2008) 40–53.

[9] N. Hatziargyriou, H. Asano, R. Iravani, C. Marnay, Microgrids, IEEE Power Energy Mag. 5 (4) (2007) 78–94.

[10] S.P. Barave, B.H. Chowdhury, Hybrid AC/DC power distribution solution for future space applications, in: Power Engineering Society General Meeting, vol. 65, 2007, p. 401.

[11] M.R. Patel, Spacecraft Power Systems, vol. 29, CRC Press, Boca Raton, 2004.

[12] H. Zhang, F. Mollet, C. Saudemont, B. Robyns, Experimental validation of energy storage system management strategies for a local DC distribution system of more electric aircraft, IEEE Trans. Ind. Electron. 57 (12) (2010) 3905–3916.

[13] J.G. Ciezki, R.W. Ashton, Selection and stability issues associated with a navy shipboard DC zonal electric distribution system, IEEE Trans. Power Del. 15 (2) (2000) 665–669.

[14] B. Zahedi, L.E. Norum, Modelling and simulation of hybrid electric ships with DC distribution systems, Proc. IEEE EPE 28 (10) (2013) 1–10.

[15] J. Hansen, Onboard DC grid for enhanced DP operation in ships, in: Dynamic Positioning Conference, 2011.

[16] G.F. Reed, B.M. Grainger, A.R. Sparacino, Z.H. Mao, Ship to grid: medium-voltage DC concepts in theory and practice, IEEE Power Energy Mag. 10 (6) (2012) 70–79.

[17] A. Mohamed, O. Mohammed, A study of electric power distribution architectures on shipboard power systems, 2012.

[18] D. Salomonsson, L. Soder, A. Sannino, An adaptive control system for a DC microgrid for data centers, IEEE Trans. Ind. Appl. 44 (6) (2008) 1910–1917.

[19] E. Alliance, 380 Vdc Architectures for the Modern Data Center, 2013, http://www.emergealliance.org/.

[20] E. Rodriguez-Diaz, J.C. Vasquez, J.M. Guerrero, Intelligent DC homes in future sustainable energy systems: when efficiency and intelligence work together, IEEE Consum. Electron. Mag. 5 (1) (2016) 74–80.

[21] X. Liu, P. Wang, P.C. Loh, A hybrid AC/DC microgrid and its coordination control, IEEE Trans. Smart Grid 2 (2) (2011) 278–286.

[22] D. Bo, Y. Li, Z. Zheng, Energy management of hybrid DC and AC bus linked microgrid, in: Proceedings of IEEE PEDG, 2010, pp. 713–716.

[23] K. Kurohane, S. Member, T. Senjyu, S. Member, A hybrid smart AC/DC power system, IEEE Trans. Smart Grid 1 (2) (2010) 199–204.

[24] R. Majumder, A hybrid microgrid with DC connection at back to back converters, IEEE Trans. Smart Grid 5 (1) (2014) 251–259.

[25] P.C. Loh, D. Li, Y.K. Chai, F. Blaabjerg, Autonomous control of interlinking converter with energy storage in hybrid AC-DC microgrid, IEEE Trans. Ind. Appl. 49 (3) (2013) 1374–1382.

[26] H. Bevrani, Robust Power System Frequency Control, Springer International Publishing, Cham, 2014.

[27] P. Kundur, N. Balu, M. Lauby, Power System Stability and Control, McGraw-Hill, New York, 1994.

[28] J.M. Guerrero, L. Hang, J. Uceda, Control of distributed uninterruptible power supply systems, IEEE Trans. Ind. Electron. 55 (8) (2008) 2845–2859.

[29] J.M. Guerrero, J.C. Vasquez, J. Matas, L.G. De Vicuna, M. Castilla, Hierarchical control of droop-controlled AC and DC microgrids—a general approach toward standardization, IEEE Trans. Ind. Electron. 58 (1) (2011) 158–172.

[30] K. Rouzbehi, C. Gavriluta, J.I. Candela, A. Luna, P. Rodriguez, Comprehensive analogy between conventional AC grids and DC grids characteristics, in: Proceedings of IEEE IECON, 2013, pp. 2004–2010.

[31] V. Nasirian, A. Davoudi, F.L. Lewis, Distributed adaptive droop control for DC microgrids, IEEE Trans. Energy Convers. 29 (4) (2014) 1147–1152.

[32] Q. Shafiee, T. Dragicevic, J.C. Vasquez, J.M. Guerrero, Hierarchical control for multiple DC-microgrids clusters, IEEE Trans. Energy Convers. 29 (4) (2014) 922–933.

[33] A. Khorsandi, M. Ashourloo, H. Mokhtari, A decentralized control method for a low-voltage DC microgrid, IEEE Trans. Energy Convers. 29 (4) (2014) 793–801.

[34] C. Dierckxsens, K. Srivastava, M. Reza, S. Cole, J. Beerten, R. Belmans, A distributed DC voltage control method for VSC MTDC systems, Electr. Power Syst. Res. 82 (1) (2012) 54–58.

[35] T. Dragicevic, J.M. Guerrero, J.C. Vasquez, D. Skrlec, Supervisory control of an adaptive-droop regulated DC microgrid with battery management capability, IEEE Trans. Power Electron. 29 (2) (2014) 695–706.

[36] L. Meng, T. Dragicevic, J.M. Guerrero, J.C. Vasquez, Optimization with system damping restoration for droop controlled DC-DC converters, in: 2013 IEEE Energy Conversion Congress and Exposition, 2013, pp. 65–72.

[37] K. Strunz, E. Abbasi, D.N. Hui, DC microgrid for wind and solar power integration, IEEE J. Emerg. Sel. Top. Power Electron. 2 (1) (2014) 115–126.

[38] X. Lu, J.M. Guerrero, K. Sun, J.C. Vasquez, An improved droop control method for DC microgrids based on low bandwidth communication with DC bus voltage restoration and enhanced current sharing accuracy, IEEE Trans. Power Electron. 29 (4) (2014) 1800–1812.

[39] Y.A.R.I. Mohamed, E.F. El-Saadany, Adaptive decentralized droop controller to preserve power sharing stability of paralleled inverters in distributed generation microgrids, IEEE Trans. Power Electron. 23 (6) (2008) 2806–2816.

[40] I.U. Nutkani, P.C. Loh, F. Blaabjerg, Cost-based droop scheme with lower generation costs for microgrids, in: 2013 IEEE ECCE Asia Downunder (ECCE Asia), 2013, pp. 339–343.

[41] I.U. Nutkani, W. Peng, P.C. Loh, F. Blaabjerg, Cost-based droop scheme for DC microgrid, in: 2014 IEEE Energy Conversion Congress and Exposition (ECCE), 2014, pp. 765–769.

[42] I.U. Nutkani, P.C. Loh, F. Blaabjerg, Droop scheme with consideration of operating costs, IEEE Trans. Power Electron. 29 (3) (2014) 1047–1052.

[43] S. Anand, B.G. Fernandes, J.M. Guerrero, Distributed control to ensure proportional load sharing and improve voltage regulation in low-voltage DC microgrids, IEEE Trans. Power Electron. 28 (4) (2013) 1900–1913.

[44] V. Nasirian, A. Davoudi, F.L. Lewis, Distributed adaptive droop control for DC microgrids, IEEE Trans. Energy Convers. 29 (4) (2014) 1147–1152.

[45] L. Meng, T. Dragicevic, J.C. Vasquez, J.M. Guerrero, Tertiary and secondary control levels for efficiency optimization and system damping in droop controlled DC-DC converters, IEEE Trans. Smart Grid 6 (6) (2015) 2615–2626.

[46] L. Meng, T. Dragicevic, J.M. Guerrero, J.C. Vasquez, Dynamic consensus algorithm based distributed global efficiency optimization of a droop controlled DC microgrid, in: Proceedings of ENERGYCON, 2014, pp. 1276–1283.

[47] S. Moayedi, A. Davoudi, Distributed tertiary control of DC microgrid clusters, IEEE Trans. Power Electron. 30 (9) (2015) 1–10.

[48] T.-F. Wu, Y.-K. Chen, Y.-H. Huang, 3C strategy for inverters in parallel operation achieving an equal current distribution, IEEE Trans. Ind. Electron. 47 (2) (2000) 273–281.

[49] G. Ding, F. Gao, S. Zhang, P.C. Loh, F. Blaabjerg, Control of hybrid AC/DC microgrid under islanding operational conditions, J. Mod. Power Syst. Clean Energy 2 (2014) 1–22.

[50] X. Lu, J.M. Guerrero, K. Sun, J.C. Vasquez, R. Teodorescu, L. Huang, Hierarchical control of parallel AC-DC converter interfaces for hybrid microgrids, IEEE Trans. Smart Grid 5 (2) (2014) 683–692.

[51] D. Boroyevich, I. Cvetkovic, D. Dong, R. Burgos, F. Wang, F. Lee, Future electronic power distribution systems—a contemplative view, in: Proceedings of IEEE OPTIM, 2010, pp. 1369–1380.

[52] P.C. Loh, D. Li, Y.K. Chai, F. Blaabjerg, Autonomous operation of AC-DC microgrids with minimized interlinking energy flow, IET Power Electron. 6 (8) (2013) 1650–1657.

[53] Y. Gu, X. Xiang, W. Li, X. He, Model-adaptive decentralized control for renewable DC microgrid with enhanced reliability and flexibility, IEEE Trans. Power Electron. 29 (9) (2014) 5072–5080.

[54] P.C. Loh, Y.K. Chia, D. Li, F. Blaabjerg, Autonomous operation of distributed storages in microgrids, IET Power Electron. 7 (1) (2014) 23–30.

[55] P.C. Loh, F. Blaabjerg, Autonomous control of distributed storages in microgrids, in: Proceedings of IEEE ICPE-ECCE Asia, 2011, pp. 536–542.

[56] C.C. Chan, An overview of electric vehicle technology, Proc. IEEE 81 (9) (1993) 1202–1213.

CHAPTER 4

Master/Slave Power-Based Control of Low-Voltage Microgrids

P. Tenti, T. Caldognetto
University of Padova, Padova, Italy

1. INTRODUCTION

This chapter describes a master-slave architecture integrating a particular control algorithm, called *power-based control*. The aim of the master-slave architecture is to enable low-voltage grids to efficiently support the functionalities of smart microgrids, such as high distribution efficiency, demand response, islanded operation, and flexibility in hosting different energy resources. To that purpose, distributed energy resources (DERs) interface with the microgrid by means of conventional current-driven inverters (energy gateways [EGs], slave units), and a voltage-driven grid-interactive inverter (utility interface [UI], master unit) governs the interaction between the microgrid and the utility at their point of common coupling (PCC). In the architecture the operation of distributed energy resources interfaced with the grid by means of EGs is coordinated by power-based control. Power-based control is a *model-free* control algorithm in which the master controller drives all distributed units to pursue the goals of secondary and tertiary control. In particular, the algorithm aims at regulating the power injection of controllable DERs (i.e., UI and EGs) so that the following objectives are simultaneously fulfilled:

- The power flow at the PCC of the microgrid follows a preassigned profile.
- The voltage magnitudes at the point of connection of controllable DERs are below a given threshold.

The former is an extremely valuable feature for a microgrid clustering renewable sources because it enables dispatchability. The latter allows one to manage distribution network congestion during periods of peak production so as to limit the stress on electrical infrastructures.

Microgrid
http://dx.doi.org/10.1016/B978-0-08-101753-1.00004-8

To theses purposes, each active unit is committed to contribute to microgrid power needs in proportion to its *power capability*, leading to a uniform utilization of DERs and thermal stress in the power electronic interfacing units.

It is shown that, when it is integrated into the master–slave architecture, the final microgrid has favorable features in terms of stability, robustness to grid parameter variations, dynamic response, scalability, and implementation requirements.

2. MASTER-SLAVE ARCHITECTURE

The microgrid scenario considered herein concerns low-voltage networks with high penetration of DERs. The microgrid structure referred to in this chapter to efficiently use the available resources is outlined in Fig. 4.1. In this structure, DERs are interfaced with the distribution network by means of so-called EGs, and the microgrid is interfaced with the utility by means of a UI; an information and communication technology (ICT)

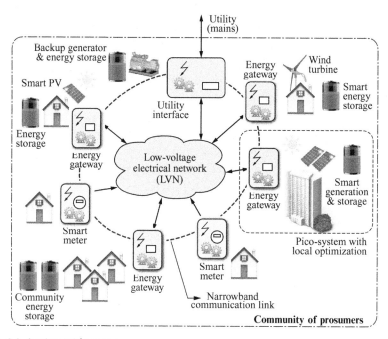

Fig. 4.1 A microgrid scenario.

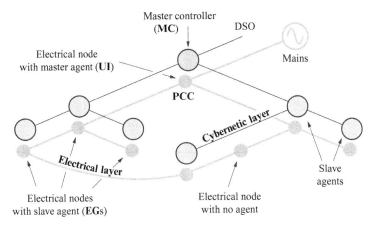

Fig. 4.2 Master-slave microgrid architecture.

infrastructure that links the resources is assumed to be available for low-speed data collection and data exchange.

The microgrid architecture is depicted in Fig. 4.2, which highlights two fundamental layers: the electrical layer and the cybernetic layer. The electrical layer represents the electrical infrastructure, which comprises, in particular, the mains, the DERs, and the electrical distribution network. The cybernetic layer represents the ICT infrastructure needed for the monitoring and control of the electrical layer, and comprises the sensors, the computation units, and the communication modules and links.

The main devices constituting the architecture are described in detail in the following.

- *Utility interface (UI).* The UI is an electronic power processor (EPP) equipped with energy storage and connected at the PCC of the microgrid with the utility. In grid-connected mode, the UI performs as a voltage-driven voltage support source synchronized with the mains, while in islanded mode it becomes the voltage-forming device for the whole microgrid [1, 2].[1] The UI constitutes by itself a controllable energy resource that can be employed to improve the microgrid behavior seen by the mains. To that purpose, the local control unit of the UI may interact with other intelligent devices via the cybernetic layer.

---

[1]A detailed discussion on electronic power converter control and behavior in AC microgrids can be found in [3].

- *Energy gateway (EG).* An EG is a DER that can be controlled to contribute to microgrid power needs. In addition to the energy resource, which can be a combination of renewable sources and storage devices, an EG is equipped with a local control unit and an EPP. The local control unit collects all the quantities needed to determine the state of the local resources and generates the reference set point of the power to be injected into the grid. The references calculated by the local control unit are then actuated by the EPP, which electrically interfaces the energy resources with the grid. The EG interacts through the EPP with the electrical layer, and through the local control unit with the cybernetic layer. A node connecting an EG to the grid is referred to as an *active node*.
- *Passive nodes.* The remaining nodes that host passive devices (i.e., link-up loads only) are referred to as *passive nodes*. Although not necessarily endowed with any particular kind of intelligent measurement or control device, passive nodes may be equipped with smart meters performing local measurements and handling one-way communication to a centralized microgrid controller.

In the cybernetic layer displayed in Fig. 4.2, a master-slave control is considered to supervise the operation of the controllable devices introduced previously: namely, the UI and the EGs. It is assumed that the microgrid's *master controller* is deployed in the UI, whereas the EGs, which are geographically distributed, play the role of slave agents. The master unit (i.e., the master controller) can communicate with the slave units (i.e., the EGs) via a communication channel (e.g., via power-line communication).[2]

2.1. Control Principle

In the proposed microgrid architecture it is assumed that the UI permanently performs as a voltage source, while EGs are driven as current sources. In grid-connected operation the UI behaves as a grid-supporting voltage source and can implement ancillary control functions (e.g., management of the UI's energy storage, compensation of residual load unbalance, and distortion). Since the power balance is ensured by the mains, the local control needs may prevail, and each active node makes available only

---

[2]A detailed investigation of the feasibility of such approaches in terms of the performance required for the communication channel is provided by Angioni et al. [4], referring to, specifically, long-term evolution technology.

its residual power and energy capacity for microgrid control. In spite of this limitation, the power flow from EGs can be adjusted by the master controller to meet global needs (e.g., grid voltage stabilization, power loss minimization, peak power shaving, demand response, day-ahead planning, low-voltage ride through).

A different scenario occurs in islanded operation, during transitions from on-grid to off-grid, and under black start. In these cases the UI acts as a grid-forming voltage source, and the master controller manages the entire energy reserve of the microgrid to ensure power balance. The EGs keep behaving as current sources, but the whole energy generated and stored locally is made available to sustain microgrid operation. The EGs can also be driven to a controlled overload condition to meet temporary energy constraints.

In all cases the distributed units cooperate to fulfill the microgrid's needs. However, while in grid-connected operation local requirements may prevail, in any other operating condition the microgrid's needs are given higher priority. This change of priority does not require modification of the control algorithms; it is simply determined by the master controller, which knows the overall power capacity of the DERs, by properly assigning power commands to EGs.

2.2. Control Strategy

A generic strategy of the control algorithm can be defined as follows. At the beginning of each control period T (lasting a few line cycles) the master controller in the UI polls all the nodes of the microgrid. The active nodes return the values of active and reactive power that are available for microgrid control, while passive nodes may return their active and reactive power consumption. In more detail, the data packet sent by the nth EG to the master controller includes:

1. the power rating of the grid-connected EPP (a_n);
2. the active and reactive power (p_n, q_n) exchanged with the grid;
3. the active and reactive power (p_l, q_l) absorbed by local loads;
4. the estimated active power $\hat{p}_n$ generated by local power sources;
5. the estimated maximum additional energy that can be stored in the energy storage unit ($\hat{e}_{S,n}^{in}$) and the maximum energy that can be extracted from the energy storage unit ($\hat{e}_{S,n}^{out}$);
6. the estimated upper and lower limits ($\hat{p}_n^{min}$, $\hat{p}_n^{max}$) of the active power deliverable by the EG, including energy storage power limits ($\hat{p}_{Sn}^{in}$, $\hat{p}_{Sn}^{out}$),

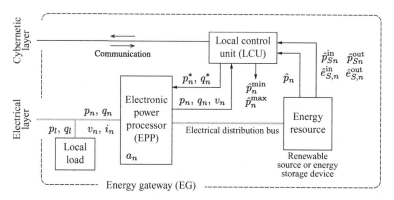

Fig. 4.3 Generic structure of an energy gateway.

local constraints on admissible power injection at the node, and other needs that are specific to the particular node.

The same data structure can be used for passive nodes if we disregard the fields related to power generation. Fig. 4.3 presents schematically the generic structure of an EG and reports all the relevant parameters; management algorithms may use all, or a subset of, these parameters for control purposes.

After collecting data from all the nodes, the master controller can compute a set of quantities defining the power and energy state of the microgrid. The master controller then executes a control algorithm that depends on the operating mode (grid-connected or islanded mode) and the relative amount of generated and absorbed power. To that purpose, the master controller computes the total power consumed ($p_{l,tot}$) and the range of power that can be generated ($\hat{p}_{l,tot}, \hat{p}_{l,tot}^{min}, \hat{p}_{l,tot}^{max}$) within the microgrid as

$$p_{l,tot} = \sum_{l=1}^{L} p_l, \quad \hat{p}_{n,tot} = \sum_{n=1}^{N} \hat{p}_n, \quad \hat{p}_{n,tot}^{min} = \sum_{n=1}^{N} \hat{p}_n^{min}, \quad \hat{p}_{n,tot}^{max} = \sum_{n=1}^{N} \hat{p}_n^{max},$$

$$(4.1)$$

and identifies the most appropriate control action to be demanded of EGs. The following cases are identified.

2.2.1. Grid-Connected Operation

In this case, power control is noncritical since the mains ensures the power balance. Thus the local control units of EGs can choose the power to deliver, their actual choice being dependent on the kind of local power source and energy needs. In fact, renewable energy sources (e.g., wind or photovoltaics) should be fully exploited, while cost-benefit issues can drive the choice for other types of sources (e.g., small hydro, fuel cells, gas turbines). In any case, EGs can feed reactive power to support load demand so as to reduce distribution losses, improve node voltage stability, and increase the power factor at the PCC. By request of the master controller, EGs can also adjust their active power flow. This can be done to meet special needs of the microgrid (e.g., voltage support, thermal limitation in feeders, intentional islanding conditions) or to respond to requests from the utility (e.g., demand-response).

2.2.2. Islanded Operation

In this case the power balance must be ensured within the microgrid. Two situations can be distinguished:

- *Overgeneration* ($\hat{p}_{n,tot} > p_{l,tot}$). In this situation, the total power generated by distributed sources exceeds the load consumption. Under steady-state conditions, the extra power is stored in distributed energy storage devices according to their state of charge, and the EGs are driven accordingly.

    Under transient conditions the dynamic power unbalance is temporarily faced at the expense of the energy stored in the UI since the UI acts as a voltage source and automatically fulfills every dynamic power request. However, within a few line cycles the EGs power commands are adapted to the new situation and the load power demand is shared among DERs. The state of charge of the UI's energy storage is promptly restored to ensure the capability to face new transients.

    If overgeneration lasts too long, the power generated by renewable sources is scaled down to meet the actual power consumption according to a suitable sharing criterion. Within their kilovolt-ampere ratings, the EGs can also feed reactive power to meet load demand and stabilize node voltages.

- *Undergeneration* ($\hat{p}_{n,tot} < p_{l,tot}$). In this situation the power generated within the microgrid is not enough to fulfill the demand from loads. The power balance must therefore be ensured by use of the distributed energy storage units according to their energy availability. Clearly, this kind of operation can be maintained for a limited time. Then nonpriority loads must be disconnected to prevent full discharge of energy storage units and, in particular, of the UI's energy storage. Also in this case, EGs

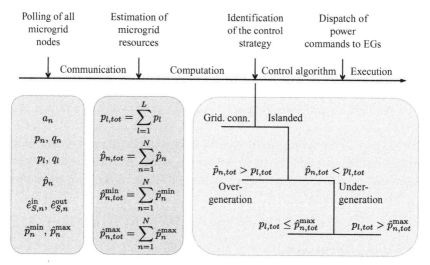

Fig. 4.4 Master-slave control principle.

can feed reactive power, within their power capability, to stabilize node voltages and reduce distribution losses.

The operations composing the control scheme described previously are represented in Fig. 4.4.

3. POWER-BASED CONTROL

The master-slave microgrid architecture introduced in the previous section (shown in Figs. 4.1 and 4.2) comprises a UI, N EGs, and a set of passive nodes. As explained earlier, the UI permanently performs as a voltage source, and behaves as a grid-supporting unit in grid–connected operation and as a grid-forming unit in islanded operation. The EGs are always controlled as current sources. An ICT infrastructure provides the communication link between the UI located at the PCC and hosting the microgrid's master controller and the EGs spread over the grid and performing as slave units.

To regulate the PCC power flow, power-based control uses the principles described in Sections 2.1 and 2.2. Specifically, in power-based control the interaction among the master controller and the EGs takes place in two phases. In the first phase the master controller *gathers* from each EG a data packet that conveys the information on its local energy availability; in the second phase the master controller *broadcasts* to *all* the EGs a

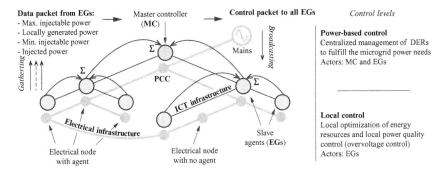

Fig. 4.5 Master-slave microgrid architecture with power-based control.

common control packet that is finally translated by each EG into a particular power reference.[3] Power references are derived by EGs taking into account constraints on power availability and the maximum voltage magnitude (V^{max}) at the point of connection. Fig. 4.5 highlights the two phases.

The following sections describe in detail the operation of the power-based control algorithm.

4. DATA COLLECTION AND PREPROCESSING

As a first step in the control of distributed EGs, the master controller collects the status of the EGs and preprocesses the acquired information to estimate the overall microgrid's power needs, which take into account the load's power absorption (together with the associated losses) and the power requested at the PCC. The following sections describe how data collection and preprocessing are performed.

4.1. Data Collection

At the beginning of the $(\ell + 1)$th cycle (i.e., at time instant ℓT) the master controller determines the total active power $p_{MG}(\ell)$ and reactive power $q_{MG}(\ell)$ absorbed by the microgrid at the PCC.[4] This power is equal to

---

[3]It is worth mentioning that broadcasting a single, unique reference to *all* the controllable units is an advantageous feature of the approach because it limits the workload of the communication infrastructure.

[4]For convenience, we denote quantities simply by indicating the relevant control cycle; for example, by using $p_{MG}(\ell)$ in place of $p_{MG}(\ell T)$.

the sum of the power drawn from the mains (i.e., p_G, q_G) and the power delivered by the UI (i.e., p_{UI}, q_{UI}). Moreover, for $n \in \{1, \dots, N\}$, the *local controller* of the nth EG (i.e., EG$_n$) sends the following data to the master controller:

- The active power $p_n(\ell)$ and the reactive power $q_n(\ell)$ generated during the ℓth control cycle.
- The estimated active power $\hat{p}_n(\ell + 1)$ that will be generated by the local renewable source in the current control cycle—namely, during the time interval $(\ell T, (\ell + 1) T)$.[5]
- The estimated minimum active power $\hat{p}_n^{\min}(\ell + 1)$ and maximum active power $\hat{p}_n^{\max}(\ell+1)$ that the EG can inject during the current control cycle by taking into account all the local constraints, including the maximum power that can be delivered ($\hat{p}_{Sn}^{\text{out}}$) or absorbed ($\hat{p}_{Sn}^{\text{in}}$) by the local energy storage unit. In particular
  - If $|\underline{v}_n(\ell)| < V^{\max}$ (i.e., no voltage violations), then

$$\begin{cases} \hat{p}_n^{\min}(\ell + 1) = \hat{p}_n(\ell + 1) - \hat{p}_{Sn}^{\text{in}}(\ell + 1), \\ \hat{p}_n^{\max}(\ell + 1) = \hat{p}_n(\ell + 1) + \hat{p}_{Sn}^{\text{out}}(\ell + 1), \end{cases} \quad (4.2)$$

during *grid-connected* operation, and

$$\begin{cases} \hat{p}_n^{\min}(\ell + 1) = -\hat{p}_{Sn}^{\text{in}}(\ell + 1), \\ \hat{p}_n^{\max}(\ell + 1) = \hat{p}_n(\ell + 1) + \hat{p}_{Sn}^{\text{out}}(\ell + 1), \end{cases} \quad (4.3)$$

during *islanded* operation.
  - If $|\underline{v}_n(\ell)| \geq V^{\max}$, which corresponds to an overvoltage condition at the point of connection of the nth EG, then $\hat{p}_n^{\min}$, $\hat{p}_n^{\max}$, and $\hat{p}_n$ are set equal to $p_n(\ell)$, that is,

$$\hat{p}_n^{\min}(\ell + 1) = \hat{p}_n^{\max}(\ell + 1) = \hat{p}_n(\ell + 1) = p_n(\ell). \quad (4.4)$$

- The rated apparent power $\hat{a}_n(\ell + 1)$ of the EPP of the EG inverter and its temporary overloading capability $\hat{a}_n^{\text{over}}(\ell + 1)$.

In a basic implementation the estimated quantities for cycle $\ell + 1$ are simply considered equal to the values at control cycle ℓ. In more advanced implementations, during grid-connected operation it is possible to take advantage of additional information (e.g., node voltage statistics, weather

---

[5] This estimate can be done on the basis of the status of the renewable source adopted (e.g., irradiation measurements for photovoltaic modules).

forecasts) to learn how to conveniently define, on a long-term basis, the parameters $\hat{p}_{Sn}^{out}$ and $\hat{p}_{Sn}^{in}$, for example, to maximize the local energy production.

It is worth remarking how the definitions in Eqs. (4.2), (4.3) given for the estimated minimum active power $\hat{p}_n^{min}$ reflect different control priorities in grid-connected and islanded operation (see Section 2.2). Indeed, during grid-connected operation it is more advantageous to extract all the power available from renewables (e.g., by operating photovoltaic sources at their maximum power point), whereas during islanded operation it is of paramount importance to guarantee the active power balance for the islanded system. In this light, $\hat{p}_n^{min}$ is set equal to $\hat{p}_n - \hat{p}_{Sn}^{in}$ in grid-connected mode so that each EG will produce at least the power available from the local source, independently of the state of charge of the local energy storage, whereas it is set equal to $-\hat{p}_{Sn}^{in}$ during islanded mode to allow EGs to provide nonpositive active power injection when local generation exceeds absorption.

4.2. Preprocessing

Concurrently with the operation of EGs, on the basis of the data collected the *master controller* determines:

- the total active and reactive power delivered by EGs during cycle ℓ:

$$p_{n,tot}(\ell) = \sum_{n=1}^{N} p_n(\ell), \tag{4.5}$$

$$q_{n,tot}(\ell) = \sum_{n=1}^{N} q_n(\ell); \tag{4.6}$$

- the total active and reactive power absorbed within the microgrid during cycle ℓ by noncontrollable units (i.e., noncontrollable loads and DERs),

$$p_{l,tot}(\ell) = p_{MG}(\ell) + p_{n,tot}(\ell), \tag{4.7}$$

$$q_{l,tot}(\ell) = q_{MG}(\ell) + q_{n,tot}(\ell), \tag{4.8}$$

which takes into account comprehensively the overall electrical load and the losses in the power lines.

- the estimated active power $\hat{p}_{l,tot}(\ell+1)$ and reactive power $\hat{q}_{l,tot}(\ell+1)$ that will be absorbed by microgrid loads in the next control cycle $\ell+1$ and the reference for the total power $p_{n,tot}^*(\ell+1)$, $q_{n,tot}^*(\ell+1)$ to be delivered by EGs:

$$\hat{p}_{l,tot}(\ell+1) = p_{l,tot}(\ell),$$
$$p_{n,tot}^*(\ell+1) = \hat{p}_{l,tot}(\ell+1) - p_{MG}^*(\ell+1), \tag{4.9}$$

$$\hat{q}_{l,tot}(\ell+1) = q_{l,tot}(\ell),$$
$$q_{n,tot}^*(\ell+1) = \hat{q}_{l,tot}(\ell+1) - q_{MG}^*(\ell+1), \tag{4.10}$$

where $p_{MG}^*(\ell+1)$ and $q_{MG}^*(\ell+1)$ represent the assigned reference power flows for the microgrid for the next control cycle;

• the estimated total active power generated by EGs in cycle $\ell+1$ and the corresponding upper and lower limits:

$$\hat{p}_{n,tot}(\ell+1) = \sum_{n=1}^{N} \hat{p}_n(\ell+1), \tag{4.11}$$

$$\hat{p}_{n,tot}^{\min}(\ell+1) = \sum_{n=1}^{N} \hat{p}_n^{\min}(\ell+1), \tag{4.12}$$

$$\hat{p}_{n,tot}^{\max}(\ell+1) = \sum_{n=1}^{N} \hat{p}_n^{\max}(\ell+1); \tag{4.13}$$

• the estimated maximum reactive power that the active nodes can deliver in normal operation or in an overloading condition in cycle $\ell+1$:

$$\hat{q}_n^{\max}(\ell+1) = \sqrt{\hat{a}_n^2(\ell+1) - \hat{p}_n^2(\ell+1)},$$
$$\hat{q}_n^{\max}(\ell+1) = \sum_{n=1}^{N} \hat{q}_n^{\max}(\ell+1), \tag{4.14}$$

$$\hat{q}_n^{\text{over}}(\ell+1) = \sqrt{\hat{a}_n^{\text{over}2}(\ell+1) - \hat{p}_n^2(\ell+1)},$$
$$\hat{q}_n^{\text{over}}(\ell+1) = \sum_{n=1}^{N} \hat{q}_n^{\text{over}}(\ell+1). \tag{4.15}$$

On the basis of the global status of controllable DERs (i.e., EGs) obtained above, the master controller regulates the power flow at the PCC to track the references p_{MG}^*, q_{MG}^*, given a preassigned power absorption profile from the main grid p_G^*, q_G^*. Accordingly, the power exchange at the terminals of the UI are

$$\hat{p}_{UI}(\ell + 1) = p^*_{MG}(\ell + 1) - p^*_G(\ell + 1), \qquad (4.16)$$

$$\hat{q}_{UI}(\ell + 1) = q^*_{MG}(\ell + 1) - q^*_G(\ell + 1). \qquad (4.17)$$

While references p^*_G, q^*_G, which are actuated by the UI [5], are either set according to the negotiation on energy exchange with the distributed system operator (taking place in the tertiary control layer [6]) or set to zero during the islanded operating mode, references p^*_{MG}, q^*_{MG} are locally adjusted by the master controller according to the energy state of the UI, as, for example, in [7].

5. SET POINT COMPUTATION

The estimated quantities represented by Eqs. (4.9)–(4.15) are the input data for the control algorithm that drives the distributed EGs. To activate it, the master controller generates two control variables α_p and α_q (both ranging in the interval $[0, 2]$) that are finally *broadcasted* to all the EGs (i.e., *applied to the whole microgrid*). Once new control coefficients are available, each EG autonomously computes its own set point of power to be exchanged with the grid (p^*_n, q^*_n).

The following sections describe how control variables α_p and α_q and set points p^*_n and q^*_n are calculated by the master controller and the EGs, respectively.

5.1. Active Power Control

The active power is controlled by variable α_p, which is set by the master controller depending on the operation mode. Four operating modes can be distinguished:

1. $p^*_{n,tot} (\ell + 1) < \hat{p}^{min}_{n,tot} (\ell + 1)$. In this case the loads are expected to absorb a total active power lower than the minimum power the active nodes can deliver. As a result, the master controller sets

$$\alpha_p = 0, \qquad (4.18)$$

and each EG sets its active power reference $p^*_n (\ell + 1)$ at the minimum allowed value:

$$p^*_n (\ell + 1) = \hat{p}^{min}_n (\ell + 1). \qquad (4.19)$$

The power balance can temporarily be ensured by diversion of the power in excess to the UI, which stores it in its energy storage device, as described in [8]. Of course, this situation can be sustained for a limited time, then loads and/or generators must be readjusted (e.g., maximum power point trackers (MPPTs) must be de-tuned so as to extract less power) to restore equilibrium.

2. $\hat{p}_{n,tot}^{\min}(\ell+1) \leq p_{n,tot}^{*}(\ell+1) < \hat{p}_{n,tot}(\ell+1)$. The expected load power is lower than the generated power but the excess power can be temporarily diverted into distributed storage units. In this case the UI does not contribute to power balance, and the master controller sets

$$\alpha_p = \frac{p_{n,tot}^{*}(\ell+1) - \hat{p}_{n,tot}^{\min}(\ell+1)}{\hat{p}_{n,tot}(\ell+1) - \hat{p}_{n,tot}^{\min}(\ell+1)}, \quad 0 \leq \alpha_p < 1. \tag{4.20}$$

Correspondingly, each active node sets its active power reference as

$$p_n^{*}(\ell+1) = \hat{p}_n^{\min}(\ell+1) + \alpha_p \left(\hat{p}_n(\ell+1) - \hat{p}_n^{\min}(\ell+1) \right). \tag{4.21}$$

3. $\hat{p}_{n,tot}(\ell+1) \leq p_{n,tot}^{*}(\ell+1) < \hat{p}_{n,tot}^{\max}(\ell+1)$. The expected load power is higher than the generated power but the difference can be supported, temporarily, by distributed energy storage. In this case the UI does not contribute to power balance, and the master controller sets

$$\alpha_p = 1 + \frac{p_{n,tot}^{*}(\ell+1) - \hat{p}_{n,tot}(\ell+1)}{\hat{p}_{n,tot}^{\max}(\ell+1) - \hat{p}_{n,tot}(\ell+1)}, \quad 1 \leq \alpha_p < 2. \tag{4.22}$$

Correspondingly, each active node sets its active power reference as

$$p_n^{*}(\ell+1) = \hat{p}_n(\ell+1) + (\alpha_p - 1) \left(\hat{p}_n^{\max}(\ell+1) - \hat{p}_n(\ell+1) \right). \tag{4.23}$$

4. $p_{n,tot}^{*}(\ell+1) \geq \hat{p}_{n,tot}^{\max}(\ell+1)$. The loads are expected to absorb a total power that is greater than the maximum power the active nodes can deliver. In this case the master controller sets

$$\alpha_p = 2. \tag{4.24}$$

Correspondingly, each active node sets its active power reference as

$$p_n^{*}(\ell+1) = \hat{p}_n^{\max}(\ell+1). \tag{4.25}$$

The power balance can temporarily be ensured at the expense of the energy stored in the UI. After some time, of course, some of the loads and/or generators will have to be readjusted to restore the equilibrium.

By considering Eqs. (4.19), (4.21), (4.23), (4.25), we can write the control law for EGs in a compact form as

$$p_n^* = \hat{p}_n^{\min}(\ell + 1)$$
$$+ \left(\hat{p}_n(\ell + 1) - \hat{p}_n^{\min}(\ell + 1)\right) \cdot \min\left(\alpha_p, 1\right)$$
$$+ \left(\hat{p}_n^{\max}(\ell + 1) - \hat{p}_n(\ell + 1)\right) \cdot \max\left(\alpha_p - 1, 0\right). \qquad (4.26)$$

The control principle behind Eq. (4.26) is to make EGs contribute to the power needs of the microgrid—measured at the microgrid's PCC—in proportion to their *capability* of deliver or absorb power, thus allowing a uniform exploitation of available resources to be obtained without preventing EGs from pursue local objectives.

Finally, to maintain the voltage magnitudes at active nodes below a given threshold $V^{\max}$, for $n \in \{1, \ldots, N\}$, EG_n applies a voltage control that is based only on local measurements of the voltage magnitude. Precisely, EG_n continuously measures $|\underline{v}_n|$ and, if an overvoltage occurs at some time instant $\bar{t}$, it adjusts the power injection according to the following rule:

$$\dot{p}_n(t) = -k_I \left(|\underline{v}_n(t)| - V^{\max}\right) \qquad (4.27)$$

for $t \geq \bar{t}$, where $p_n(\bar{t}) = p_n^*$. In this case, EG_n keeps applying the *purely local voltage control* described in Eq. (4.27) as long as the injected power p_n is lower than the power reference calculated as per Eq. (4.26) by using the received α_p and the *actual* (i.e., those indicated in Eq. 4.2 or 4.3) parameters $\hat{p}_n$, $\hat{p}_n^{\min}$, and $\hat{p}_n^{\max}$.

The rationale behind this local control law as follows. If the node at which EG_n is connected is experiencing an overvoltage, this means that node n cannot accept the power EG_n is injecting; accordingly, EG_n starts to decrease p_n as described in Eq. (4.27), relying on the fact that in mainly resistive networks, by decreasing the active power injection at grid nodes, the corresponding voltage magnitudes decrease as well. The effectiveness of this choice is discussed further in Section 6, and a more formal analysis of the control scheme is provided in Section 7.

Observe also that if there exists $\tilde{t}$ such that $|\underline{v}(\tilde{t})| = V^{\max}$, then the power $p_n(\tilde{t})$ represents the power that the node where EG_n is connected can receive without experiencing overvoltages. In general, it might happen that, during the overvoltage condition, $\hat{p}_n(\ell + 1) < \hat{p}_n^{\min}(\ell + 1)$; in this case the overproduction $\hat{p}_n^{\min}(\ell + 1) - \hat{p}_n(\ell + 1)$ is assumed to be curtailed.

5.2. Reactive Power Control

The reactive power is controlled by variable α_q, which is set by the master controller depending on the operation mode. There are two operation modes:

1. $q^*_{n,tot}(\ell + 1) \leq \hat{q}^{max}_{n,tot}(\ell + 1)$. Load requirements can be met by distributed EGs. In this case the master controller sets

$$\alpha_q = \frac{q^*_{n,tot}(\ell + 1)}{\hat{q}^{max}_{n,tot}(\ell + 1)}, \quad 0 \leq \alpha_q \leq 1. \tag{4.28}$$

Correspondingly, each active node sets its reactive power reference as

$$q^*_n(\ell + 1) = \alpha_q \cdot \hat{q}^{max}_n(\ell + 1). \tag{4.29}$$

2. $q^*_{n,tot}(\ell + 1) > \hat{q}^{max}_{n,tot}(\ell + 1)$. Load requirements can be met only by overloading the EGs. In this case the master controller sets

$$\alpha_q = 1 + \frac{q^*_{n,tot}(\ell + 1) - \hat{q}^{max}_{n,tot}(\ell + 1)}{\hat{q}^{over}_{n,tot}(\ell + 1) - \hat{q}^{max}_{n,tot}(\ell + 1)}, \quad 1 < \alpha_q \leq 2. \tag{4.30}$$

Correspondingly, each active node sets its reactive power reference as

$$q^*_n(\ell + 1) = \hat{q}^{max}_n(\ell + 1) + (\alpha_q - 1)(\hat{q}^{over}_n(\ell + 1) - \hat{q}^{max}_n(\ell + 1)). \tag{4.31}$$

5.3. Grid-Connected Mode: Active and Reactive Power Control

The strategy described previously allows the microgrid's power absorption p_{MG} to be controlled in both grid-connected and islanded mode of operation. The different definitions given in Eqs. (4.2), (4.3) are meant to guarantee the regulation of the power absorbed by the microgrid while operating islanded and the complete extraction of the power potentially available from renewables while operating connected to the grid.

It is worth remarking that while the microgrid is operating connected to the mains, the above control strategy can also be used to obtain a conventional grid-connected operation, where DERs simply inject the locally generated power in compliance with grid standards. This is achieved by the master controller by setting $\alpha_p = 1$, so that the total power generated by DERs is injected into the grid. Local power needs (e.g., to restore the state of charge of the energy storage at the nominal value) and constraints (e.g., to limit voltage magnitudes) continue to be taken into account by

EGs, by controlling the locally generated power. In any case, the power balance is ensured by the utility grid.

As far as reactive power compensation is concerned, the UI first decides its contribution q_{UI} for the next control cycle. Then it adjusts the total reactive power requested from EGs according to the equation

$$q^*_{n,tot}(\ell + 1) = \hat{q}_{l,tot}(\ell) - q_{UI}(\ell + 1). \tag{4.32}$$

For both active power and reactive power, the master controller in the UI can also distribute the references differently in the three phases to compensate for load unbalance, as described in [9].

6. REMARKS ON VOLTAGE CONTROL

To the end of controlling voltage profiles, the use of reactive power capabilities of DERs, which does not virtually involve additional costs, has been shown to be an effective and advantageous solution in medium-voltage networks [10, 11], where interconnection impedances are mainly inductive (i.e., characterized by low R/X ratios). However, this approach is not adequate in low-voltage networks, where, instead, interconnection impedances are typically resistive (i.e., R/X ratios are high) [12, 13]. Indeed, in networks with high R/X ratios, the reactive power injection that would be needed to counteract voltage rises caused by excessive active power injections may be so intense as to lead to detrimental effects on the electrical infrastructure (e.g., overload of medium–voltage/low-voltage transformer and distribution cables) and affect EPPs' reliability [14]. Therefore approaches based on active power control, such as the one in Eq. (4.27), fit better in high R/X networks, with the main drawback of the potential reduction in the overall power production, which, though, can be alleviated or even eliminated with small local accumulation.

With regard to the effectiveness of the approach to counteracting over-voltages and undervoltages at grid nodes described, it is worth observing the following facts. One of the advantages of distributed generation (with devices installed, ideally, at the consumers' premises) is to have active power generation closer to loads (i.e., closer to the point where the power is consumed). This helps to compensate for the voltage drops that would be naturally present due to the active power absorption by loads [15]. In addition, the control proposed herein coordinates the power injection by DERs (specifically by EGs) so that it can adapt to the needs of the microgrid's loads and share the load among the generators in a fair way (specifically according to their power availability). This contributes to sustaining the grid voltage to avoid *undervoltage* conditions. On the other hand, *overvoltages* are an intrinsic issue of distributed generation in low-voltage grids. Indeed, overvoltages may occur due to congestion of distribution lines during periods of peak production from renewables, or can arise from an excessive power injection by one or more distributed resources (e.g., in response to a remote control signal, such as the one referred to here as α_p). These particular aspects, instead, are dealt with by the local overvoltage control technique (see Eq. 4.27).

7. CONTROL ANALYSIS

In the control structure devised, an EG in overvoltage mode behaves as a noncontrollable source, which does not respond to coefficient α_p; such an EG can be considered as a load absorbing negative power. In addition, because of Eq. (4.4) and how the coefficient is calculated in Eqs. (4.18), (4.20), (4.22), (4.24), EGs in overvoltage mode do not affect the value of α_p. Therefore the stability of the control system can be studied in two phases. The first considers the situation with no overvoltage occurrences; this allows one to analyze the main operations of power-based control, also taking into account the nonidealities of realistic application cases. The second phase considers a complete model of the system, including, specifically, the description of the distribution grid topology; this allows one to analyze how the overvoltage control affects the dynamics of the system.

7.1. Power-Based Principle Analysis

A simplified model of power-based control is presented in Figs. 4.6 and 4.7, where a star represents one of Eqs. (4.18), (4.20), (4.22), (4.24) and a dagger represents one of Eqs. (4.19), (4.21), (4.23), (4.25).

A simplified block diagram representing the main operations of power-based control concerning active power balance is shown in Fig. 4.7.[6] Gain errors and offset errors are included to take into account the main nonidealities of a realistic application case. In general, gain errors affect the loop gain of the feedback system and have to be considered to assess system stability, whereas offset errors have to be taken into account to analyze its

---

[6] A corresponding scheme can be derived for reactive power control.

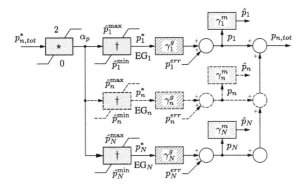

Fig. 4.6 Generation of power commands.

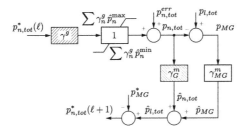

Fig. 4.7 Calculation of power references.

steady-state accuracy in regulating the controlled quantities. In Fig. 4.6, variables γ_n^g and p_n^{err} represent the gain and offset errors made by EG_n in producing the assigned power reference; p_n^*, γ_n^m, and γ_{MG}^m represent the gain errors of measurement instruments, and offset errors in measurements are ignored for now. On the basis of Fig. 4.6, the simplified model in Fig. 4.7 can be drawn, and may be used for stability analysis.

By use of the block diagram in Fig. 4.7 it is possible to derive the discrete-time transfer function between the total absorbed power $p_{l,tot}$ and the reference $p_{n,tot}^*$. If we ignore the reference input p_{MG}^*, since it varies very slowly, as mentioned in the last paragraph of Section 4.2, we have

$$p_{n,tot}^*(z) = \frac{\gamma_{MG}^m}{z + \gamma^g\left(\gamma_G^m - \gamma_{MG}^m\right)} p_{l,tot}(z). \qquad (4.33)$$

Eq. (4.33) shows that if the system is ideal (i.e., γ_n^g, γ_n^m, and γ_{MG}^m are equal to 1 and p_n^{err} is equal to 0), then the power reference EGs are committed to tracking the total absorbed power in the microgrid, with one control cycle delay; secondly, the stability condition for active power reference generation, in general, can be expressed as $\left|\gamma^g\left(\gamma_G^m - \gamma_{MG}^m\right)\right| < 1$, which can easily be met by any commercial power meter. This proves a stable control operation for operating modes 2 and 3 referred to in Section 5.1.

The diagrams in Fig. 4.7 also highlight that if the power requested by the load exceeds the total power capability of EGs, the coefficient α_p seamlessly saturates to its upper limit, so each EG continuously delivers the maximum power that is locally available ($\hat{p}_n^{\max}$). When the opposite situation occurs (i.e., the minimum injectable power from EGs is higher than the load power), the coefficient α_p is automatically saturated at its lower limit and each EG continuously delivers the minimum power ($\hat{p}_n^{\min}$). Because the control system operates on a cycle-by-cycle basis, with no memory of the grid state during previous cycles, a stable control operation is guaranteed for operating modes 1 and 4 referred to in Section 5.1.

For the regulation accuracy of the power flow at the PCC, let us first observe that

$$p_{MG}(\ell) = p_{l,tot}(\ell) - \gamma^g \hat{p}_{n,tot}^*(\ell) - p_{n,tot}^{err}(\ell), \qquad (4.34)$$

where p_{MG} is shared among the UI (p_{UI}) and the mains (p_{GRID}) (Eqs. 4.16, 4.17) according to the negotiation on energy exchange with the distributed system operator taking place in the tertiary control layer. From Eq. (4.34), the power flow at the PCC is equal to the power reference p_{MG} minus the error introduced by EGs. This error can be canceled by the master controller, for example, by employing a local integrative regulator to properly modulate the power term p_{MG} [16]. Similarly, the fluctuations in local power production can be modeled as exogenous inputs, which affect only the limits $\hat{p}_n^{\min}$ and $\hat{p}_n^{\max}$ and do not impair the stability of the system. The limits are acquired and processed by the master controller in each control cycle, allowing the control commands to EGs to be updated accordingly, so as to account for the actual generation profile. Finally, although temporary mismatches (i.e., lasting a few line cycles) among the effectively generated power and its estimate can have an effect on the injected power, this can be limited by a proper design of EG hardware. In any case, DC-link voltage deviations caused by abrupt changes in operating conditions, which may affect primarily voltage-driven inverters [17], are attenuated in the EG structure considered thanks to the current-driven approach adopted [18].

7.2. System Analysis

For the purpose of investigating the stability of the control scheme, let us consider a power system where power injections from distributed EGs are driven by power-based control with the automatic overvoltage limitation of Eq. (4.27). At the generic mth node an EG and a load, absorbing a constant active power (P_{Lm}), can be connected. The load may represent also a noncontrollable source, in this case $P_{Lm} < 0$. The voltage phasor at the mth node is indicated as $\underline{v}_m$.

In this situation, microgrid nodes can be grouped as follows:

- $\mathcal{L}$ is the set of L passive nodes injecting or absorbing the total power that is produced or consumed locally. These can be modeled as constant power sources injecting

$$p_m^{\mathcal{L}} = -P_{Lm}. \tag{4.35}$$

- $\mathcal{K}$ is the set of K active nodes, not affected by overvoltage conditions, injecting or absorbing power under the guidance of power-based control. These can be modeled as controlled power sources injecting

$$p_m^{\mathcal{K}} - P_{Lm} = P_m^{\mathcal{K}}(\alpha_p, \hat{p}_m^{\min}, \hat{p}_m, \hat{p}_m^{\max}) - p_{Lm}. \tag{4.36}$$

- $\mathcal{H}$ is the set of H active nodes where an overvoltage condition persists. For these nodes the active power injection is locally calculated to control the maximum measured node voltage amplitude. These nodes behave as controlled power sources with the control law

$$p_m^{\mathcal{H}} - P_{Lm} = -\frac{k_I}{s}\left(\left|\underline{v}_m\right| - V^{\max}\right) - P_{Lm}. \tag{4.37}$$

- Node 0, at the PCC, is characterized by a fixed nominal voltage $\underline{v}_0 = V_N\,e^{j0}$. The power absorption at this node guarantees the power balance of the system.

For the stability analysis it is convenient to linearize the relation between voltage magnitudes and power injections at grid nodes. By employing the results obtained in [19], one may model voltage magnitudes at grid nodes with the following approximated and linearized expression:

$$\left|\underline{v}\right| = V_0\mathbf{1} + \frac{1}{V_0}\Re\left(e^{j\theta}\mathbf{X}\mathbf{a}^T\right) + \frac{d(V_0)}{V_0^2}. \tag{4.38}$$

In Eq. (4.38), $\underline{v}$ is the vector of node voltages, V_0 is the voltage amplitude around which the model is linearized, $\mathbf{X}$ is a symmetric and positive semidefinite matrix, $\mathbf{1}$ is a column vector with all 1's, $\mathbf{a}^T$ is the transpose of the complex power vector $\mathbf{a} = \mathbf{p} + j\mathbf{q}$ of node power injections, θ is the phase of the grid impedance per unit length, and $d(V_0)/V_0^2$ is a bounded function when $V_0 \to \infty$. $\mathbf{X}$ carries the information on the grid topology and the interconnecting impedances. To ease the exploitation of symmetries, it is assumed in the following that nodes are arranged as $\mathbf{x} = \left(x_0, x_1^{\mathcal{L}}, x_2^{\mathcal{L}}, \ldots, x_l^{\mathcal{K}}, x_1^{\mathcal{K}}, x_2^{\mathcal{K}}, \ldots, x_k^{\mathcal{K}}, x_1^{\mathcal{H}}, x_2^{\mathcal{H}}, \ldots, x_h^{\mathcal{H}}\right)^T = \left(x_0, \mathbf{x}^{\mathcal{L}}, \mathbf{x}^{\mathcal{K}}, \mathbf{x}^{\mathcal{H}}\right)^T$, where x is a generic electrical quantity.

Focusing on active power flows, ignoring the term $d(V_0)/V_0^2$, and assuming $\theta = 0$ (exploiting the low-voltage network's property of having high R/X ratios), we may rewrite Eq. (4.38) as

$$|\underline{v}| = V_0 \mathbf{1} + \frac{1}{V_0} \mathbf{X} \mathbf{p}. \tag{4.39}$$

For the elements of vector $\mathbf{p}$ in the situation described previously, the power absorption at node 0 can be modeled as the difference between the total absorbed power and the generated power within the microgrid:

$$p_0 = \sum_{n=1}^{l} p_n^{\mathcal{L}} - \sum_{n=1}^{k} p_n^{\mathcal{K}} - \sum_{n=1}^{h} p_n^{\mathcal{H}}. \tag{4.40}$$

The power generation of nodes in $\mathcal{K}$ is proportional to the microgrid's power needs, considering the local power generation constant, because it depends on relatively slowly varying phenomena (e.g., weather conditions and the state of charge of local energy storage). Indicating with values $m_1, \ldots, m_k \in \mathbb{R}^+$, $m_n < 1$, the proportionality factors of the power contribution from the nodes in $\mathcal{K}$, we can write

$$p_n^{\mathcal{K}} = m_n \left(p_0 + \sum_{n=1}^{k} p_n^{\mathcal{K}} \right). \tag{4.41}$$

Finally, power generation of nodes in $\mathcal{H}$ behaves with the control law expressed by Eq. (4.37):

$$p_n^{\mathcal{H}} = -\frac{k_I}{s} \left(|\underline{v}_n| - V_0 \right), \tag{4.42}$$

where V_0 should assume in this case the value of the maximum voltage $V^{\max}$.

Therefore, for a specific configuration of absorption and generation, the total injected power at grid nodes can be represented as in Eq. (4.43):

$$
p =
\begin{bmatrix}
0 & 1 & \cdots & 1 & 1 & \cdots & 1 & 1 & \cdots & 1 \\
0 & 0 & \cdots & 0 & 0 & \cdots & 0 & 0 & \cdots & 0 \\
\vdots & \vdots & \ddots & \vdots & \vdots & \ddots & \vdots & \vdots & \ddots & \vdots \\
0 & 0 & \cdots & 0 & 0 & \cdots & 0 & 0 & \cdots & 0 \\
m_1 & 0 & \cdots & 0 & m_1 & \cdots & m_1 & 0 & \cdots & 0 \\
\vdots & \vdots & \ddots & \vdots & \vdots & \ddots & \vdots & \vdots & \ddots & \vdots \\
m_k & 0 & \cdots & 0 & m_k & \cdots & m_k & 0 & \cdots & 0 \\
0 & 0 & \cdots & 0 & 0 & \cdots & 0 & 0 & \cdots & 0 \\
\vdots & \vdots & \ddots & \vdots & \vdots & \ddots & \vdots & \vdots & \ddots & \vdots \\
0 & 0 & \cdots & 0 & 0 & \cdots & 0 & 0 & \cdots & 0
\end{bmatrix}
\begin{bmatrix}
p_0 \\ p_1^{\mathcal{L}} \\ \vdots \\ p_l^{\mathcal{L}} \\ p_1^{\mathcal{K}} \\ \vdots \\ p_k^{\mathcal{K}} \\ p_1^{\mathcal{H}} \\ \vdots \\ p_h^{\mathcal{H}}
\end{bmatrix}
$$

$$
-
\begin{bmatrix}
0 & 0 & \cdots & 0 & 0 & \cdots & 0 & 0 & \cdots & 0 \\
0 & 0 & \cdots & 0 & 0 & \cdots & 0 & 0 & \cdots & 0 \\
\vdots & \vdots & \ddots & \vdots & \vdots & \ddots & \vdots & \vdots & \ddots & \vdots \\
0 & 0 & \cdots & 0 & 0 & \cdots & 0 & 0 & \cdots & 0 \\
0 & 0 & \cdots & 0 & 0 & \cdots & 0 & 0 & \cdots & 0 \\
\vdots & \vdots & \ddots & \vdots & \vdots & \ddots & \vdots & \vdots & \ddots & \vdots \\
0 & 0 & \cdots & 0 & 0 & \cdots & 0 & 0 & \cdots & 0 \\
0 & 0 & \cdots & 0 & 0 & \cdots & 0 & \frac{k_I}{s} & \cdots & 0 \\
\vdots & \vdots & \ddots & \vdots & \vdots & \ddots & \vdots & \vdots & \ddots & \vdots \\
0 & 0 & \cdots & 0 & 0 & \cdots & 0 & 0 & \cdots & \frac{k_I}{s}
\end{bmatrix}
\begin{bmatrix}
0 \\ |v_1^{\mathcal{L}}| - V_0 \\ \vdots \\ |v_l^{\mathcal{L}}| - V_0 \\ |v_1^{\mathcal{K}}| - V_0 \\ \vdots \\ |v_k^{\mathcal{K}}| - V_0 \\ |v_1^{\mathcal{H}}| - V_0 \\ \vdots \\ |v_h^{\mathcal{H}}| - V_0
\end{bmatrix}
-
\begin{bmatrix}
0 \\ P_{L1} \\ \vdots \\ P_{Ll} \\ P_{Ll+1} \\ \vdots \\ P_{Ll+k+1} \\ P_{Ll+k+2} \\ \vdots \\ P_{Ln}
\end{bmatrix}.
$$

$$\tag{4.43}$$

7.3. Stability Analysis

The analysis is now focused on power absorption at nonpassive nodes. In the following, vectors referring to this set of nodes are denoted with a tilde (e.g., $\tilde{x}$). Considering, $\tilde{p} = (p_0, p^{\mathcal{K}}, p^{\mathcal{H}})^T$, we can observe that

$$
\tilde{p} = A\tilde{p} + Bp^{\mathcal{L}} + \frac{1}{s}C\left(|\tilde{\underline{v}}| - |\tilde{v}_0|\right), \tag{4.44}
$$

where $p^{\mathcal{L}}$ is the total load in the microgrid and

$$
A =
\begin{bmatrix}
0 & -\mathbf{1}_k^T & -\mathbf{1}_h^T \\
m & m\mathbf{1}_k^T & \mathbf{O}_{kh} \\
\mathbf{0}_h & \mathbf{O}_{hk} & \mathbf{O}_{hh}
\end{bmatrix}, \quad
B =
\begin{bmatrix}
1 \\ \mathbf{0}_k \\ \mathbf{0}_h
\end{bmatrix} \quad
C =
\begin{bmatrix}
0 & \mathbf{0}_k & \mathbf{0}_h \\
\mathbf{0}_k & \mathbf{O}_{hk} & \mathbf{O}_{hk} \\
\mathbf{0}_h & \mathbf{O}_{hk} & -k_I I_h
\end{bmatrix}.
\tag{4.45}
$$

In Eq. (4.45), $\mathbf{1}_i$ denotes the $i \times 1$ column vector with all 1's, $m \in \mathbb{R}^{k \times 1}$, and $\mathbf{O}_{ij} \in \mathbb{R}^{i \times j}$ denotes the matrix with all 0's.

From Eq. (4.44) it is possible to write $\tilde{p}$ as

$$\tilde{p} = (I - A)^{-1} Bp^{\mathcal{L}} + \frac{1}{s} (I - A)^{-1} C \left(\left| \underline{\tilde{v}} \right| - \left| \tilde{v}_0 \right| \right), \qquad (4.46)$$

and, finally, using Eq. (4.46) in Eq. (4.39) for nodes $\{0, \mathcal{K}, \mathcal{H}\}$, we obtain

$$
\begin{aligned}
s \left(\left| \underline{\tilde{v}} \right| - \left| \tilde{v}_0 \right| \right) &= \frac{1}{V_0} \tilde{X} (I - A)^{-1} C \left(\left| \underline{\tilde{v}} \right| - \left| \tilde{v}_0 \right| \right) \\
&+ \frac{s}{V_0} \tilde{X} (I - A)^{-1} Bp^{\mathcal{L}},
\end{aligned}
\qquad (4.47)
$$

which represents the state equation of the linearized system with state matrix

$$1/V_0 \tilde{X} (I - A)^{-1} C. \qquad (4.48)$$

These fundamental relations can be employed to evaluate numerically the eigenvalues of the system.

8. APPLICATION EXAMPLE

To clearly illustrate the control features, let us consider the simple test case shown in Fig. 4.8. It comprises two EGs, one load, and the UI. The power system considered is low voltage and the parameters of the power electronic interfaces, photovoltaic sources, and storage units adopted are those of commercial devices suited for residential applications. Distribution grid parameters are reported in Table 4.1, while the parameters of EG 1 and EG 2 are reported in Tables 4.2 and 4.3, respectively. A narrowband communication link provides the required information exchange between the UI and the two EGs.

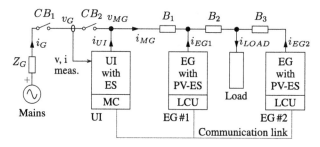

Fig. 4.8 Low-voltage power system considered.

Table 4.1 Distribution grid parameters

Parameters	Symbol	Value
Grid voltage	V_G	230 V
Grid frequency	f_G	50 Hz
Maximum voltage deviation	$\Delta v_\%^{max}$	4.0%
B_1 impedance	Z_{B1}	$0.17 + j0.04\,\Omega$
B_2 impedance	Z_{B2}	$0.26 + j0.06\,\Omega$
B_3 impedance	Z_{B3}	$0.70 + j0.16\,\Omega$
Load power factor	PF	0.95

Table 4.2 First energy gateway parameters

Parameters	Symbol	Value
EG power rating	a_{EG1}	4.2 kVA
EG overload power rating	a_{EG1}^{over}	4.6 kVA
EG nominal efficiency	η_{EG1}	0.95
PV nominal power rating	P_{PV1}	4.0 kW
ES capacity	E_{ES1}	3.6 kWh
ES maximum discharging power	$p_{S1}^{out(max)}$	2.0 kW
ES maximum recharging power	$p_{S1}^{in(max)}$	1.0 kW
ES charging efficiency	$\eta_{ES1,Rec}$	0.92
ES discharging efficiency	$\eta_{ES,Disc}$	0.92

Table 4.3 Second energy gateway parameters

Parameters	Symbol	Value
EG power rating	a_{EG2}	5.0 kVA
EG overload power rating	a_{EG2}^{over}	5.4 kVA
EG nominal efficiency	η_{EG2}	0.95
PV nominal power rating	P_{PV2}	4.0 kW
ES capacity	E_{ES2}	5.4 kWh
ES max discharging power	$p_{S2}^{out(max)}$	3.0 kW
ES max recharging power	$p_{S2}^{in(max)}$	1.5 kW
ES charging efficiency	$\eta_{ES2,Rec}$	0.92
ES discharging efficiency	$\eta_{ES,Disc}$	0.92

The results obtained from the simulation of the low-voltage power system in Fig. 4.8 in response to typical absorption and generation profiles are discussed in the following. To highlight the effect of the control approach on microgrid performance, the three specific cases of operation reported below are considered.

- *Case A—no power-based control*: In this case the EGs are not equipped with local energy storage and operate independently, injecting into the grid the total active power extracted from the local photovoltaic source. No communication and reactive power compensation is implemented.
- *Case B—power-based control*: In this case EGs are not equipped with local energy storage and operate under the supervision of the master controller. The EGs inject into the grid the active power extracted from the local photovoltaic source and the reactive power that corresponds to the received coefficient α_q. The local *active power* generation is automatically curtailed in the case of overvoltage detection.
- *Case C—power-based control with distributed energy storage*: In this case EGs are equipped with local energy storage and operate under the supervision of the master controller. On the basis of the received coefficients α_p and α_q, the EGs deliver the requested active and reactive power that correspond to the local power availability. In the case of overvoltage, the EGs limit their active power injection. The excess power production is automatically curtailed or, in the case of energy storage availability, the excess power produced is stored locally.

8.1. Active Power Profiles

Figs. 4.9–4.11 show the behavior of the measured active power flows for the considered cases in response to given generation and absorption profiles.

In case A (see Fig. 4.9) EG 1 and EG 2 exchange with the grid only the active power produced by the photovoltaic sources, without any overvoltage constraint at grid nodes being taken into account. Then the power drawn from the PCC is equal to the total power absorbed by the load (plus losses)

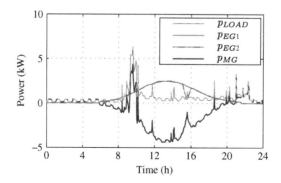

Fig. 4.9 Active power profiles without power-based control (case A).

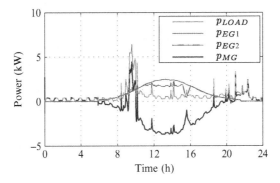

Fig. 4.10 Active power profiles with power-based control (case B).

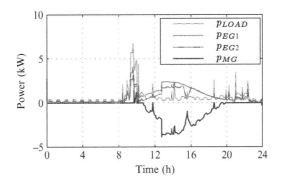

Fig. 4.11 Active power profiles with power-based control and energy storage (case C). Note how the peak in power demand, at 9:00, is eliminated.

minus the total power generated by the photovoltaic sources. Consequently, the power flow at the PCC shows the same variability of generation and absorption profiles.

In case B (see Fig. 4.10) power-based control is active. For the active power injection, when the voltage magnitudes of active nodes are within the nominal values, the active power flow behaviors in cases A and B are identical. A different situation is established for reactive power. Power-based control instructs the EGs to completely compensate for the net reactive power produced within the microgrid, thus causing a constant zero reactive power exchange at the PCC. Further details are given with the discussion of Table 4.4.

Table 4.4 Performance indexes computed at microgrid PCC

Parameters	Case A (no control)	Case B (PB control)	Case C (PB control + ES)
Produced energy (kWh)	36.5	34.1	34.2
Distribution loss (kWh)	0.83	0.65	0.47
v_{EG1} max overvoltage (%)	1.4	1.2	1.2
v_{LOAD} max overvoltage (%)	2.4	1.8	1.8
v_{EG2} max overvoltage (%)	5.5	4.0	4.0

In case C (see Fig. 4.11) the effect on microgrid operation of the integration of energy storage is shown. Energy storage enables efficient control of the active power injection from EGs. The active power injection from EGs is now driven by the needs of the loads through the supervision of the master controller. This reflects on the active power exchanged at the PCC, which appears smoother than in cases A and B thanks to the inherent *peak shaving* capability of the microgrid.

Finally, a comparison of cases B and C shows the effect of the overvoltage limitation by dynamic active power control, which causes the reduction in EG 2 power generation, needed to fulfill the imposed grid voltage magnitude constraint (see Table 4.2, parameter $\Delta v\%^{(\text{max})}$).

8.2. Power Flow at the PCC

Fig. 4.12 shows the behavior of the active power flow through the PCC for the cases considered. As discussed in the previous section, the lower variance of the power flow at the PCC happens when power-based control

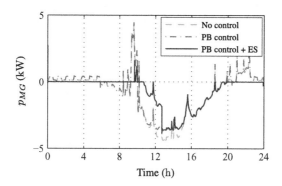

Fig. 4.12 Active power at the point of common coupling.

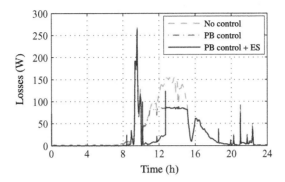

Fig. 4.13 Distribution power losses over the grid.

is active and EGs are equipped with energy storage units. In Fig. 4.12, the fluctuations in load power absorption during the 0–8-h and 20–24-h intervals are completely absorbed by EGs thanks to the available energy storage. Similarly, the peak of absorption occurring at 9 h is properly redistributed among EGs (see Fig. 4.11) and effectively reduced at the PCC.

Distributed energy storage also contributes to partially absorbing the overproduction from photovoltaic sources in the 10–13-h interval until the completion of the recharge cycle.

8.3. Distribution Losses

Fig. 4.13 shows the distribution losses obtained for the cases considered. Power-based control inherently compensates for unwanted reactive power flows within the grid in a distributed fashion. This is known to be beneficial in terms of distribution losses [20, 21]. Besides, the distributed energy storage, relevant for case C, enables the active power control, and thus further improves the loss count by reducing circulating currents. The resulting effect on distribution losses can be seen in 10–12-h interval in Fig. 4.13, where the distribution losses in case C are significantly lower than those measured in cases A and B.

8.4. Voltage Deviations at Grid Nodes

As discussed in Section 6, low-voltage distribution lines are mainly resistive; therefore the active power flow significantly affects voltage amplitudes at grid nodes. During periods of peak production from renewables, undesirable voltage deviations from the nominal values can be registered because of abnormal active power injection. In the simulation setup considered,

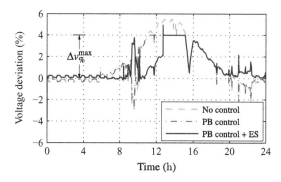

Fig. 4.14 Voltage deviation at the second energy gateway node.

overvoltage conditions are automatically detected and managed locally by the active nodes by their regulating dynamically the active power injected into the grid as per Eq. (4.27).

Fig. 4.14 shows the voltage deviation at the point of connection of EG 2. This node is more affected by these phenomena because it is the farthest from the PCC. In particular, Fig. 4.14 shows how the overvoltage control feature integrated into the control scheme allows an accurate and precise limitation of voltage magnitude at critical nodes. In the case considered, the power that cannot be injected into the grid because of overvoltage limitations is stored in the local accumulators if this is compatible with the corresponding state of charge; otherwise the generation is curtailed by modifying the way the local maximum power point tracker algorithm (MPPT) operates–relevant techniques may be, for example, those described in [5, 22].

To meet the $\Delta v_{\%}^{max}$ constraint can necessarily lead to reduced power production from renewables in grids where the distribution lines have high R/X ratios. Fig. 4.15 reports the profile of the total maximum power that can be ideally extracted from photovoltaic sources and the actual total power production obtained in cases B and C. Since the power injection is limited during an overvoltage condition, the excess power is totally curtailed in case C, whereas in case B it is curtailed for only the portion that cannot be stored in the local energy storage. Fig. 4.16 shows the corresponding state of charge of the energy storage of the two EGs; because of the proportional contribution from the EGs, the stage of charge behaviors are similar.

The behavior of coefficients α_p and α_q during the simulation scenario is reported in Fig. 4.17 just for case C. In case B, similar behaviors are obtained for the coefficient α_q, while the coefficient α_p assumes only the

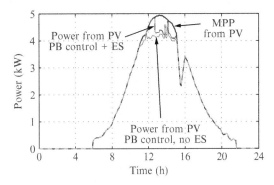

Fig. 4.15 Total power production from photovoltaic (PV) sources.

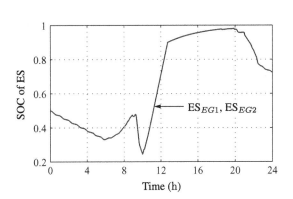

Fig. 4.16 State of charge (SOC) of energy storage (ES) devices (case C).

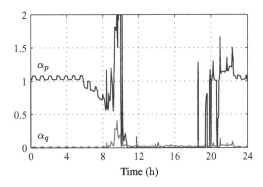

Fig. 4.17 Behavior of coefficients α_p and α_q (case C).

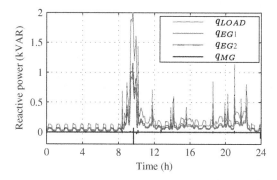

Fig. 4.18 Reactive power contributions from energy gateway (EG) 1, EG 2, and the point of common coupling, together with the absorbed load power.

values corresponding to operating modes 1 and 4 in Section 5.1 because of the absence of storage devices. In this last case, the power output from EGs is equal to the maximum power that can be generated locally while the overvoltage constraint $\Delta v_\%^{\max}$ is complied with.

8.5. Performance Indexes

To emphasize the main results illustrated so far, Table 4.4 reports some performance indexes applied to the application example considered. In particular, the total energy produced, the energy dissipated in distribution lines, the overvoltage measured at grid nodes, and the power factor measured at the PCC are reported. The following aspects are noted:

- The measured distribution losses in case B are reduced by 20% with respect to case A. A further reduction by 25% can be achieved with the integration of energy storage devices at active nodes (case C). The maximum registered overvoltage stays within the programmed 4% limit (see Table 4.1) when the corresponding control functionality is active, notably, in cases B and C. On the other hand, if no provisions are taken, case A reveals a maximum reached overvoltage of 5.5%.
- In the application example considered, power-based control accomplishes the full compensation of the reactive power produced by the loads, achieving a unity power factor measured at the PCC. Fig. 4.18 shows the share of reactive power obtained between EG 1, EG 2, and the PCC, together with the reactive power absorbed by the load.

- Thanks to the effective management of the generated and stored energy performed by the proposed control scheme, a reduction of only 6% in the total energy produced can be noticed in spite of the stringent overvoltage limitation of 4% with respect to the nominal value V_G.

9. SUMMARY

This chapter presented a master-slave architecture with a power-based control algorithm to coordinate the DERs in low-voltage smart microgrids. The main elements of the architecture (namely, a UI—embedding the microgrid master controller—and the EGs) were introduced and the fundamental control principle was presented. A narrowband communication system was employed with the purpose of coordinating the constituting devices. By means of the UI and the EGs, the architecture allows the decoupling of the need to ensure high quality of the voltage provided to microgrids loads, which is demanded of the UI, and the need to attain a synergistic operation of DERs, which is demanded of the EGs. Notably, the architecture described allows DERs to be added to the microgrid in a plug-and-play fashion because any specific tuning of the resources that are already integrated is not required. The *power-based control* algorithm employed in the architecture requires only nontime critical power data to be transferred from the active nodes to a centralized controller through a narrowband communication link; the centralized controller, in turn, broadcasts active and reactive power set points for all the active nodes. The control algorithm does not require specific knowledge of the controlled plant (e.g., grid parameters, grid topology), constituting, therefore, a *model-free* approach for the control of DERs in microgrids.

In summary, the system described shows the following features:
- no need for plant models;
- no need for measures from passive nodes;
- no circulation currents among generators;
- simple calculation and communication requirements;
- prompt regulation of the power flow at the interface between the microgrid and the mains;
- regularization of voltage profiles along distribution lines; and
- possibility of operating in both grid-connected mode and islanded mode.

REFERENCES

[1] Z. Liu, J. Liu, Y. Zhao, A unified control strategy for three-phase inverter in distributed generation, IEEE Trans. Power Electron. 29 (3) (2014) 1176–1191.

[2] P. Tenti, T. Caldognetto, S. Buso, A. Costabeber, Control of utility interfaces in low-voltage microgrids, in: 5th International Symposium on Power Electronics for Distributed Generation (PEDG), June, 2014, pp. 1–8.

[3] J. Rocabert, A. Luna, F. Blaabjerg, P. Rodríguez, Control of power converters in AC microgrids, IEEE Trans. Power Electron. 27 (11) (2012) 4734–4749.

[4] A. Angioni, et al., Coordinated voltage control in distribution grids with LTE based communication infrastructure, in: IEEE 15th International Conference on Environment and Electrical Engineering (EEEIC), June, 2015, pp. 2090–2095.

[5] H. Mahmood, D. Michaelson, J. Jiang, A power management strategy for PV/battery hybrid systems in islanded microgrids, IEEE J. Emerg. Sel. Top. Power Electron. 2 (4) (2014) 870–882.

[6] J.M. Guerrero, J.C. Vasquez, J. Matas, L.G. de Vicuna, M. Castilla, Hierarchical control of droop-controlled AC and DC microgrids—a general approach toward standardization, IEEE Trans. Ind. Electron. 58 (1) (2011) 158–172.

[7] K. Jong-Yul, K. Seul-Ki, J. Jin-Hong, Coordinated state-of-charge control strategy for microgrid during islanded operation, in: Proceedings of the 3rd IEEE International Conference Power Electronics for Distributed Generation Systems (PEDG), 2012, pp. 133–139.

[8] K. Jong-Yul, et al., Cooperative control strategy of energy storage system and microsources for stabilizing the microgrid during islanded operation, IEEE Trans. Power Electron. 25 (12) (2010) 3037–3048.

[9] T. Caldognetto, P. Tenti, P. Mattavelli, S. Buso, D.I. Brandao, Cooperative compensation of unwanted current terms in low-voltage microgrids by distributed power-based control, in: IEEE Power Electronics Society-Southern Power Electronics Conference, November, 2015.

[10] P. Carvalho, P.F. Correia, L. Ferreira, Distributed reactive power generation control for voltage rise mitigation in distribution networks, IEEE Trans. Power Syst. 23 (2) (2008) 766–772.

[11] S. Lissandron, P. Mattavelli, A controller for the smooth transition from grid-connected to autonomous operation mode, in: IEEE Energy Conversion Congress and Exposition (ECCE), September, 2014, pp. 4298–4305.

[12] R. Tonkoski, L. Lopes, T. El-Fouly, Coordinated active power curtailment of grid connected PV inverters for overvoltage prevention, IEEE Trans. Sustain. Energy 2 (2) (2011) 139–147.

[13] M. Kabir, Y. Mishra, G. Ledwich, Z. Dong, K. Wong, Coordinated control of grid-connected photovoltaic reactive power and battery energy storage systems to improve the voltage profile of a residential distribution feeder, IEEE Trans. Ind. Inform. 10 (2) (2014) 967–977.

[14] A. Anurag, Y. Yang, F. Blaabjerg, Thermal performance and reliability analysis of single-phase PV inverters with reactive power injection outside feed-in operating hours, IEEE J. Emerg. Sel. Top. Power Electron., 3 (4) (2015) 870–880.

[15] P. Tenti, A. Costabeber, P. Mattavelli, D. Trombetti, Distribution loss minimization by token ring control of power electronic interfaces in residential microgrids, IEEE Trans. Ind. Electron. 59 (10) (2012) 3817–3826.

[16] A. Micallef, M. Apap, C. Spiteri-Staines, J.M. Guerrero, J.C. Vasquez, Reactive power sharing and voltage harmonic distortion compensation of droop controlled single phase islanded microgrids, IEEE Trans. Smart Grid 5 (3) (2014) 1149–1158.

[17] W. Du, Q. Jiang, M.J. Erickson, R.H. Lasseter, Voltage-source control of PV inverter in a CERTS microgrid, IEEE Trans. Power Deliv. 29 (4) (2014) 1726–1734.
[18] M. Erickson, T. Jahns, R. Lasseter, Comparison of PV inverter controller configurations for CERTS microgrid applications, in: IEEE Energy Conversion Congress and Exposition (ECCE), September, 2011, pp. 659–666.
[19] S. Bolognani, S. Zampieri, A distributed control strategy for reactive power compensation in smart microgrids, IEEE Trans. Autom. Control 58 (11) (2013) 2818–2833.
[20] T. Stetz, F. Marten, M. Braun, Improved low voltage grid-integration of photovoltaic systems in Germany, IEEE Trans. Sustainable Energy 4 (2) (2013) 534–542.
[21] P. Tenti, A. Costabeber, F. Sichirollo, P. Mattavelli, Minimum loss control of low-voltage residential microgrids, in: Proceedings of 38th Annual Conference on IEEE Industrial Electronics Society (IECON'12), 2012, pp. 5650–5656.
[22] A. Ahmed, L. Ran, S. Moon, J.-H. Park, A fast PV power tracking control algorithm with reduced power mode, IEEE Trans. Energy Convers. 28 (3) (2013) 565–575.

CHAPTER 5

Load-Frequency Controllers for Distributed Power System Generation Units

M.I. Abouheaf[*], M.S. Mahmoud[†]
[*]Aswan University, Aswan, Egypt
[†]King Fahd University of Petroleum and Minerals (KFUPM), Dhahran, Saudi Arabia

1. INTRODUCTION

Distributed generation (DG) units such as photovoltaic arrays, wind turbine generators, and microturbines are increasingly used to reduce energy prices and environmental problems. High DG penetration levels has brought about the concept of the "microgrid." A microgrid is an integrated energy system consisting of loads, a distribution grid, and DG units which can operate in [1]:

- grid-connected mode;
- islanded (autonomous) mode; and
- transition between the two modes.

Under normal conditions, where a microgrid operates in grid-connected mode, each DG unit within the microgrid uses the well-known dq-current control strategy [2] to regulate its real/reactive power components. Autonomous operation of a microgrid, however, requires sophisticated control strategies and protection systems. Depending on the electrical proximity of the DG units and their dedicated loads, several topologies for microgrids can be defined (e.g., parallel connection, ring, and radial connection of DGs). Each DG unit within a microgrid is connected to the point of common coupling (PCC), where the dedicated loads are also connected. When all the PCCs are along a transmission line with nonzero impedance between the PCCs, the radial configuration is obtained. The control of the microgrid is a key aspect which aims at the stable operation of a microgrid [3] and has been the focus of research in the past few years. During islanded operation, the main task of the microgrid is to deliver quality power by regulating the output voltage. The microgrid, with its own control structure, should be able to regulate any disturbances in the load toward zero to ensure the stability of the system [4]. The dq-current

137

control strategy for multiple DG units in an islanded microgrid, based on frequency-power and voltage-reactive power droop characteristics of each DG unit, is well known and extensively reported [1, 2, 5]. In this approach each DG unit is equipped with two droop characteristics:

1. frequency as a linear function of real power;
2. voltage magnitude as a linear function of reactive power.

On the basis of these droop characteristics, frequency is dominantly controlled by real power flow, and voltage magnitude is regulated by reactive power flow of the DG unit. This approach does not directly incorporate load dynamics in the control loop. Thus large and/or fast load changes can result in either poor dynamic response or voltage/frequency instability. A control strategy for autonomous operation of a DG unit and its dedicated load was introduced in [6]. This method is intended for a fairly fixed load and cannot accommodate large perturbations in the load parameters.

The optimal control theory uses the Hamilton-Jacobi-Bellman (HJB) equation, whose solution is the optimal cost-to-go function [7–9]. The cooperative control problems involve the consensus and the synchronization control problems [10–12]. In the consensus control problem, the agents synchronize to a leader node dynamics. In the synchronization control problem, the control policies are selected such that the agents reach the same dynamics [13–15]. Game theory provides a solution framework for cooperative control problems [16]. Off-line Nash equilibrium solutions are found in terms of the coupled Hamilton-Jacobi equations (which are difficult to solve) [17]. The noncooperative dynamic game type provides an environment for formulating multiplayer decision control problems [17]. Each agent finds its optimal control policy by optimizing its performance index independently. Off-line solutions for the games are given in terms of the respective coupled Hamilton-Jacobi equations [17]. Continuous-time differential graphical games were developed in [18].

The dynamic graphical game is a special class of the standard dynamic game and explicitly captures the structure of a communication graph, where the information flow between the agents is governed by the communication graph topology [17, 19–21]. Herein, an online adaptive learning solution for the dynamic graphical game is developed in terms of the underlying coupled HJB equations. This adaptive learning solution will be used to design controllers for a network of distributed photovoltaic cells that are interacting in the frame of a communication graph. This adaptive learning solution finds the online Nash equilibrium solution for the dynamic graphical game in real time. The convergence proof for the adaptive learning algorithm is given under mild assumptions about the graph interconnectivity properties. Critic neural network structures are used to implement the online adaptive learning solution. Only partial knowledge of the dynamics (input control matrices) is required, and the tuning is done in a distributed fashion in terms of the local information available to each agent. This chapter brings together cooperative control, optimal control, game theory, and reinforcement learning techniques to design controllers for a network of photovoltaic cells.

Dynamic programming is solved with use of approximate dynamic programming [20, 22]. Reinforcement learning, adaptive critics, and dynamic programming techniques are used to solve the optimal control problems [23–26]. Reinforcement learning is concerned with learning from interaction in a dynamic environment [27, 28]. The associated reinforcement learning algorithms are used to solve multiplayer games for finite-state systems in [29] and to learn online in real time the solutions for the optimal control problems of the differential games in [30, 31]. Heuristic dynamic programming (HDP) is used to solve the graphical game in real time using actor-critic neural network structures [20]. Differential dynamic programming is solved by means of Q-learning [32]. Action-dependent HDP is used to solve the optimal control problem, and the solution is equivalent to iterating the underlying Riccati equations [33]. In [29], the multi-agent systems converge in behavior and the Q-learning update rule converges to the optimal response.

Online policy iteration employed approximate dynamic programming to find online solutions for the Riccati equations in [34]. A policy iteration solution for the adaptive optimal control problem can be obtained by relaxation of the HJB equation to the equivalent optimization problem [35]. Actor-critic networks are temporal difference methods with separate structures that explicitly represent the policies apart from the value structures [36]. These structures involve forward-in-time algorithms for computing optimal decisions that are implemented online in real time. The actor component applies control policies to its environment, while the critic rewards some decisions and punishes other decisions [27].

The chapter is organized as follows. Section 2 provides a brief background to the dynamic model of the islanded microgrid used to conduct this study. The control scheme based on an online value iteration algorithm is formulated in Section 3. Section 4 presents simulation results to verify the performance of the proposed control scheme. Section 5 briefly introduces the dynamic model of a network of photovoltaic cells. Section 6 reviews the cooperative control problem for dynamic games on graphs. In this section the mathematical setup of the dynamic graphical game is introduced in terms of sets of coupled Bellman equations and their respective Hamiltonian functions. This section provides the solution for the graphical game in terms of the solution of a set of coupled HJB equations. An online Nash equilibrium solution for the graphical game is given in terms of the underlying coupled HJB equations. An online adaptive learning iteration algorithm to solve the dynamic graphical game in real time is developed in Section 7. The convergence proof is provided under mild assumptions about the graph interconnectivity properties. The online adaptive learning algorithm using critic neural network structures is implemented in Section 8. Section 9 tests the validity of the proposed algorithm.

2. AUTONOMOUS MICROGRID SYSTEM

Autonomous or islanded operation of a microgrid can be caused by network faults/failures in the utility grid, scheduled maintenance [5, 37], and economical optimization or management constraints [38]. In this work the dynamic model of the autonomous microgrid proposed in [6, 39] is adopted to conduct the research. The model consists of a DG unit electronically coupled to its local load.

A schematic single-line diagram of an electronically coupled microgrid model is shown in Fig. 5.1. A switch at the PCC will isolate the microgrid from the utility grid [1]. The islanded system consists of inverter-based DG units supplying a load via a series filter and step-up transformer. The DC voltage source represents the generating unit, and R_t and L_t represent the series filter. A local load, modeled by a three-phase parallel RLC network, is connected at the PCC. The system parameters are shown in Table 5.1.

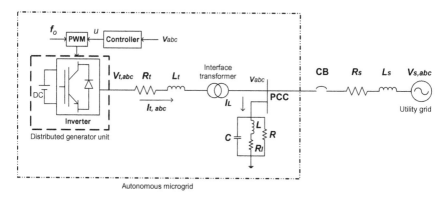

Fig. 5.1 The autonomous microgrid electrical model.

Table 5.1 System parameters

Quantity	Value
R_t	1.5 mΩ
L_t	300 μH
V_{dc}	1500 V
PWM carrier frequency	1980 Hz
Load parameters	
R	76 Ω
L	111.9 mH
C	62.855 μF
R_l	0.3515 Ω
Grid parameters	
R_s	1 Ω
L_s	10 μH
Nominal frequency f_o	60 Hz
Nominal voltage (rms)	13.8 kV
Interface transformer parameters	
Type	Wye-delta
Rating	2.5 MVA
Voltage ratio (n)	0.6/13.8 kV

During islanded operation, the main task of the microgrid is to deliver quality power by regulating any disturbances in the load. The microgrid, with its own control structure, should be able to maintain the load voltage level at a desired prespecified set point.

2.1. State-Space Model of the Autonomous Microgrid

Consider the system described in Fig. 5.1 to be balanced. Then the equations governing the microgrid are [6]

$$V_{t,abc} = L_t \frac{dI_{t,abc}}{dt} + R_t I_{t,abc} + V_{abc}, \tag{5.1}$$

$$I_{t,abc} = \frac{1}{R} V_{abc} + I_{L,abc} + C \frac{dV_{abc}}{dt}, \tag{5.2}$$

$$V_{abc} = L \frac{dI_{L,abc}}{dt} + R_l I_{L,abc}. \tag{5.3}$$

These equations are in the *abc* frame, and $V_{t,abc}$, $I_{t,abc}$, and V_{abc} are 3×1 vectors with individual phase quantities. Under balanced conditions, each three-phase variable x_{abc} in Eqs. (5.1)–(5.3) can be transferred to a stationary $\alpha\beta$ reference frame system by application of the following *abc* to $\alpha\beta$ transformation [6]:

$$x_{\alpha\beta} = x_a e^{j0} + x_b e^{j\frac{2\pi}{3}} + x_c e^{j\frac{4\pi}{3}}, \tag{5.4}$$

where $x_{\alpha\beta} \triangleq x_\alpha + jx_\beta$. In the $\alpha\beta$ frame, the resulting dynamic model is given by

$$\frac{dI_{t,\alpha\beta}}{dt} = -\frac{R_t}{L_t} I_{t,\alpha\beta} - \frac{V_{\alpha\beta}}{L_t} + \frac{V_{t,\alpha\beta}}{L_t}, \tag{5.5}$$

$$\frac{dV_{\alpha\beta}}{dt} = \frac{1}{C} I_{t,\alpha\beta} - \frac{1}{RC} V_{\alpha\beta} - \frac{1}{C} I_{L,\alpha\beta}, \tag{5.6}$$

$$\frac{dI_{L,\alpha\beta}}{dt} = \frac{1}{L} V_{\alpha\beta} - \frac{R_l}{L} I_{L,\alpha\beta}. \tag{5.7}$$

This can be transferred to a rotating reference frame by means of the following transformation:

$$x_{\alpha\beta} = x_{dq} e^{j\theta} = (x_d + jx_q e^{j\theta}), \tag{5.8}$$

where θ is the phase angle of an arbitrary reference vector $x_\alpha^{ref} + jx_\beta^{ref}$ in the $\alpha\beta$ frame and is given by

$$\theta = \arctan\left(\frac{x_\beta^{ref}}{x_\alpha^{ref}}\right), \tag{5.9}$$

where $V_{\alpha\beta}$ is taken as the reference vector such that $V_q = 0$. In autonomous mode, the system frequency is controlled in an open-loop manner as a voltage source converter generates three-phase voltages at frequency ω_0 by employing an internal oscillator of constant frequency of $\omega_0 = 2\pi f_0$. Moreover, the steady-state voltage and current signals are at frequency ω_0 if the local load is passive. Therefore, the dq state variables are given by

$$\frac{dI_{td}}{dt} = -\frac{R_t}{L_t} I_{t,d} + \omega_0 I_{tq} - \frac{1}{L_t} V_d + \frac{1}{L_t} V_{td},$$

$$\frac{dI_{tq}}{dt} = \omega_0 I_{td} - \frac{R_l}{L} I_{tq} - 2\omega_0 I_{Ld} + \left(\frac{R_l C \omega_0}{L} - \frac{\omega_0}{R}\right) V_d,$$

$$\frac{dI_{Ld}}{dt} = \omega_0 I_{tq} - \frac{R_l}{L} I_{Ld} + \left(\frac{1}{L} - \omega_0^2 C\right) V_d, \tag{5.10}$$

$$\frac{dV_d}{dt} = \frac{1}{C} I_{td} - \frac{1}{C} I_{Ld} - \frac{1}{RC} V_d,$$

$$V_{tq} = L_t \left[2\omega_0 I_{td} + \left(\frac{R_t}{L_t} - \frac{R_l}{L}\right) I_{tq} - 2\omega_0 I_{Ld} + \left(\frac{R_l \omega_0 C}{L} - \frac{\omega_0}{R}\right) V_d \right]. \tag{5.11}$$

If we put the foregoing autonomous microgrid system into the standard time state-space representation

$$\dot{x}(t) = A_c x(t) + B_c u(t), \quad y(t) = C_c x(t), \quad u(t) = v_{td},$$

it follows that the system matrices are given by

$$
A_c = \begin{bmatrix}
-\frac{R_t}{L_t} & \omega_0 & 0 & \frac{1}{L_t} \\
\omega_0 & \frac{R_l}{L} & -2\omega_0 & \frac{R_l C \omega_0}{L} - \frac{\omega_0}{R} \\
0 & \omega_0 & -\frac{R_l}{L} & \frac{1}{L} - \omega_0^2 C \\
\frac{1}{C} & 0 & -\frac{1}{C} & -\frac{1}{RC}
\end{bmatrix}, \quad
B_c = \begin{bmatrix} \frac{1}{L_t} \\ 0 \\ 0 \\ 0 \end{bmatrix}, \quad
C_c^t = \begin{bmatrix} 0 \\ 0 \\ 0 \\ 1 \end{bmatrix},
$$

$$(5.12)$$

where the state vector is

$$x^t = \begin{bmatrix} I_{td} & I_{tq} & I_{Ld} & V_d \end{bmatrix}. \tag{5.13}$$

For the purpose of convenience later, we will discretize the model given by Eqs. (5.10)–(5.12).

3. REINFORCEMENT LEARNING TECHNIQUES

Reinforcement learning [40] refers to interactions of an agent with its environment so as to improve its actions/control policies depending on evaluation of the information received from the environment. It is also known as *action-based learning* [23, 41]. In these reinforcement learning techniques, the agent is goal directed and has an understanding of reward versus lack of reward, successful control decisions remembered to be used a second time. The key feature of reinforcement learning is that it provides an adaptive control which converges to the optimal control [42].

Reinforcement learning techniques can be implemented with use of actor-critic architecture. The actor applies a control (action) to the environment, and the value of this control is assessed by the critic. The results of current actions observed from the environment are used by the critic to perform policy evaluation. The critic assesses the value of current policies on the basis of optimality criteria [43, 44]. Depending on this assessment, different schemes can then be used to modify the action to yield an improved value.

In this chapter we will consider HDP. A simple HDP system consists of two subnetworks (i.e., actor and critic networks). These networks have feedforward and feedback components.

3.1. Heuristic Dynamic Programming

HDP is based on adaptive critics [36] and uses value function approximation to solve dynamic programming problems. Fig. 5.2 shows the structure of the HDP design; it consists of a system to be controlled and two subnetworks (i.e., *actor* and *critic* networks) [42].

The actor is mathematically described as having state $x(t)$ as an input and its output is control $u(t)$. The system is considered as part of the environment as everything outside the actor is considered to be the environment. The control structure does not require the desired control signals to be known. Reinforcement learning techniques are successful with complex systems with unknown or partially known dynamics. The actor-critic networks are tuned sequentially with use of the data observed along the system trajectory. The weights of one network are tuned until convergence, while the weights of the other network are held constant. The process is repeated until both networks have converged. Both the cost function and the control policy are approximated at each step by these two networks [45].

The actor network selects the control policy to minimize the value function. For each iteration, in feedforward mode the output of the actor network is a series of control signals and in feedback mode it adjusts the internal critic network weights. The critic network establishes a relationship between the control policies and the value function. The critic function is twofold: in the feedforward mode it predicts the value function for an initial set of control policies, and in the feedback mode it assists the actor network to generate a control policy which minimizes the value function [46].

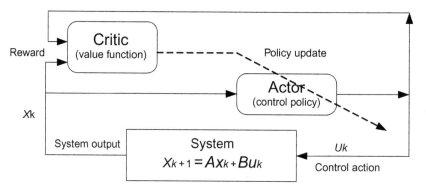

Fig. 5.2 Block diagram of the actor-critic structure for heuristic dynamic programming.

3.2. Discrete-Time Bellman Equation

Consider the following discrete-time system in state–space form:

$$x_{k+1} = Ax_k + Bu_k, \tag{5.14}$$

which is a discrete version of system (5.12), where the states $x_k \in \mathbb{R}^n$ and control input $u_k \in \mathbb{R}^m$ and k is the discrete time index. Assume that system (5.14) is stabilizable on some set $\Omega \in \mathbb{R}^n$.

Definition 5.1 (Stabilizable System). *A system is said to be stabilizable on a set $\Omega \in \mathbb{R}^n$ if there exists a control input $u \in \mathbb{R}^m$ such that the closed-loop system given by x_{k+1} is asymptotically stable on Ω.*

A function $h(\cdot): \mathbb{R}^n \to \mathbb{R}^m$ is a mapping from state space to control space such that $u_k = h(x_k)$. This describes the actor mathematically as it is the one generating control policy in reinforcement learning techniques. It is desired to find the control policy $u(x_k)$ which minimizes the following performance measure/value function:

$$V(x_k) = \sum_{i=k}^{\infty} \frac{1}{2}(x_i^t Q x_i + u_i^t R u_i), \tag{5.15}$$

where the matrices $Q = Q^t > 0 \in \mathbb{R}^{n \times n}$ and $R = R^t > 0 \in \mathbb{R}^{m \times m}$ so that the performance measure is well defined.

Definition 5.2 (Admissible Control [47]). *A control policy $u_k = h(x_k)$ is said to be admissible if it stabilizes system (5.14) and yields a finite performance $V(x_k)$.*

For any admissible control $u_k = h(x_k)$, $V(x_k)$ is known as the *cost function* or *value function* and it can be selected on the basis of the minimum energy, minimum cost, etc. We can write Eq. (5.15) as follows:

$$V(x_k) = \frac{1}{2}(x_k^t Q x_k + u_k^t R u_k) + V(x_{k+1}), \quad V(0) = 0. \tag{5.16}$$

Therefore, by using the current control policy u_k, we can evaluate the cost by solving the difference equation above. The Bellman equation is a functional equation consisting of system dynamics and a value function.

According to Bellman's optimality principle [8], the optimal value and the optimal policy can be obtained by

$$V^*(x_k) = \min_{u_k}\left[\frac{1}{2}\left(x_k^t Q x_k + u_k^{*t} R u_k^*\right) + V^*(x_{k+1})\right],$$

$$u_k^* = -R^{-1}B^t \nabla V^*(x_{k+1}). \tag{5.17}$$

In this chapter we are interested in a value iteration technique, an iterative method for determining the optimal control and optimal value function. This technique does not require an initial stabilizing policy.

3.3. Value Iteration Algorithm

In this section a value iteration algorithm for an autonomous microgrid system (see Eq. 5.12) is developed and used to solve the discrete-time Bellman equation (see Eq. 5.17). The value iteration algorithm is summarized by Algorithm 5.1.

Algorithm 5.1   Value Iteration Algorithm for an Autonomous Microgrid

1. **Initialization:** Select arbitrary initial values for the policy u_k and $V(x_k)$, not necessarily admissible or stabilizing.
2. **Value update:** Solve the Bellman equation to get $V^{\ell+1}(x_k)$ as follows:

$$V^{\ell+1}(x_k) = \frac{1}{2}(x_k' Q x_k + u_k^{\ell t} R u_k^\ell) + V^\ell(x_{k+1}), \qquad (5.18)$$

where ℓ is the iteration index.

3. **Policy improvement:** The control policy u_k is updated as follows:

$$u_k^{\ell+1} = -R^{-1} B' \, \nabla \, V(x_{k+1})^{\ell+1}, \qquad (5.19)$$

where the gradient is defined as $\nabla V(x_{k+1}) = \frac{\partial V(x_{k+1})}{\partial x_{k+1}}$.
4. **Convergence:** The above steps are repeated until

$$\| V(x_k)^{\ell+1} - V(x_k)^\ell \|$$

converges.

Remark 5.1. *It is important to note that the value iteration depends on solution of the simply recursive equation (5.18), which is easy to compute and is called partial backup in reinforcement learning. Value iteration successfully mixes one sweep of policy evaluation and one sweep of policy improvement in each of its sweeps.*

3.4. Actor-Critic Neural Networks

The performance function (5.15) or (5.18) is now approximated by a critic network and the control policy (5.19) is approximated by an actor network. Let $W_c \in \mathbb{R}^{n \times n}$ and $W_a \in \mathbb{R}^{n \times m}$ be the critic and actor weights, respectively. Therefore the performance function and control policy approximations can be written as follows:

$$\hat{V}_k(W_c) = \frac{1}{2} x_k^t W_c^t x_k, \tag{5.20}$$

$$\hat{u}_k(W_a) = W_a^t x_k. \tag{5.21}$$

Hence the actor network approximation error is given as

$$\zeta_{u_k}^{V(x_k)} = \hat{u}_k(W_a) - u_k \tag{5.22}$$

and the control policy (5.19) is given in terms of the critic network such that

$$\tilde{u}_k = -R^{-1} B^t \nabla \hat{V}(x_{k+1}). \tag{5.23}$$

On expressing this target control in terms of critic weights, one obtains

$$\tilde{u}_k = -R^{-1} B^t W_c^t x_k. \tag{5.24}$$

The squared approximation error is given by $\frac{1}{2}(\zeta_{u_k}^{V(x_k)})^t \zeta_{u_k}^{V(x_k)}$, and the change in the actor weights is given by the gradient descent method. Therefore the actor update rule is given as follows:

$$W_a^{(\ell+1)^t} = W_a^{\ell^t} - \lambda_a[(W_a^{\ell^t} x_k - \tilde{u}_k^\ell)(x_k^t)], \tag{5.25}$$

where $0 < \lambda_a < 1$ is the actor learning rate.

Let $\psi_{x_k}^{V(x_k)}$ be the target value of the critic network and let the value update rule be given by Eq. (5.18). Therefore we have

$$\psi_{x_k}^{V(x_k)} = \frac{1}{2}\left[\left(x_k^t Q x_k + u_k^{\ell^t} R u_k^\ell\right)\right] + \hat{V}^\ell(x_{k+1}). \tag{5.26}$$

The network approximation error of the critic is

$$\zeta_{x_k}^{V(x_k)} = \psi_{x_k}^{V(x_k)} - \hat{V}_k(W_c).$$

Similarly, the squared approximation error is given by $\frac{1}{2}(\zeta_{x_k}^{V(x_k)})^t \zeta_{x_k}^{V(x_k)}$, and the change in the critic weights is given by the gradient descent method. Therefore the critic update rule is given as follows:

$$W_c^{(\ell+1)^t} = W_c^{\ell^t} - \lambda_c\left[\psi_{x_k}^{V(x_k)} - \frac{1}{2} x_k^t W_c^{\ell^t} x_k\right] \frac{1}{2} x_k x_k^t, \tag{5.27}$$

where $0 < \lambda_c < 1$ is the critic learning rate.

Remark 5.2. *It is important to note that the value iteration depends on solution of simply recursive equation (5.27), which is easy to compute and is called partial backup in reinforcement learning. Value iteration successfully mixes one sweep of policy evaluation and one sweep of policy improvement in each of its sweep.*

4. ONLINE ACTOR-CRITIC NEURAL NETWORK IMPLEMENTATION

Algorithm 5.2 is used to solving the microgrid control problem by *online* tuning of actor-critic neural networks. The algorithm makes use of real-time data measured along the system trajectories and tunes the actor-critic network weights in real-time.

Algorithm 5.2   Actor-Critic Implementation of Algorithm 5.1

1. Initialize the actor W_a and critic W_c weights.
2. **Loop 1**: Begin with q as an iteration index.
   Start with random initial values for the system state (i.e., initialize $x_0 = rand_{4 \times 1}$).
   Loop 2:
   a. The iteration loop begins with ℓ as iteration index.
   b. The control policy $\hat{u}_k^\ell$ is evaluated with the equation $\hat{u}_k(W_a) = W_a^t x_k$.
   c. The dynamics of the system x_{k+1}^ℓ are evaluated with $x_{k+1}^\ell = Ax_k^\ell + Bu_k$.
   d. The performance measure $\hat{V}_{k+1}^\ell$ is calculated.
   e. The critic network is updated with use of Eq. (5.27).
   f. The actor network is updated with use of Eq. (5.25).
   g. On convergence of the actor-critic weights, end loop 2.
3. Calculate the difference $\hat{V}(x_k)^{\ell+1} - \hat{V}(x_k)^\ell$.
4. On convergence of $\| \hat{V}(x_k)^{\ell+1} - \hat{V}(x_k)^\ell \|$, end loop 2.
   Transfer the actor-critic weights to the next iterations (i.e., $q + 1$ as initialization for the next iteration). End loop 1.

4.1. Performance Evaluation of the Proposed Controller

The parameters for online Algorithm 5.2 were chosen as $\mu_a = 0.2$, $\mu_c = 0.1$, $Q = I_{4 \times 4}$, and $R = I$. Fig. 5.3 describes the Simulink structure for implementation of Algorithm 5.2 for the system. The control generated is fed online to the system, and at time $t = 0.1$ s, a pulse disturbance is introduced in the system states. Fig. 5.4 shows the response of all four states. From Fig. 5.4 it can be concluded that online Algorithm 5.2 yields stability and proves the synchronization of the weights. Figs. 5.5–5.7 show the online tuning of the critic and actor weights, and the agents' dynamics.

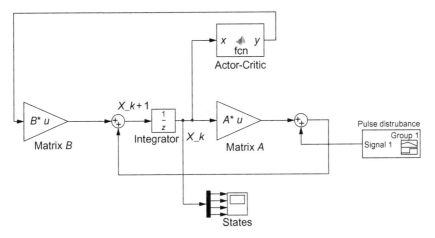

Fig. 5.3 Simulink implementation of Algorithm 5.2.

5. PHOTOVOLTAIC SOLAR CELLS

Solar energy is a clean source of energy that is integrated into the power system network by use of photovoltaic cells and conditioning circuits. The general I-V characteristics and the equivalent models of the photovoltaic cells have been studied extensively [48]. The solar cell model includes a photodiode, a shunt resistor depicting leakage current, and a series resistance that represents the internal resistance to the current flow [49]. In grid-connected mode the conditioning circuits involve a DC-DC converter, a DC-link capacitor, an inverter, and an output filter circuit [50]. A photovoltaic cell model with a DC-DC converter is shown in Fig. 5.8 [50].

The dynamic model of a photovoltaic cell is given as follows:

$$\frac{dI_{pv}}{dt} = \frac{1}{L_{dc}}[V - (1 - d_c)V_{dcp}],$$

$$\frac{dV_{dcp}}{dt} = \frac{1}{C_{dc}}[I_{dc1} - I_{dc2}], \qquad (5.28)$$

where I_{pv} is the net output current from the photovoltaic cell, V_{pv} is the voltage of the photovoltaic cell, L_{dc} is the inductance of the DC-DC converter, d_c is the duty cycle, V_{dcp} is the voltage across the capacitor C_{dc}, $I_{dc1} = (1 - d_c)I_{pv}$, and I_{dc2} is the input current to the inverter stage.

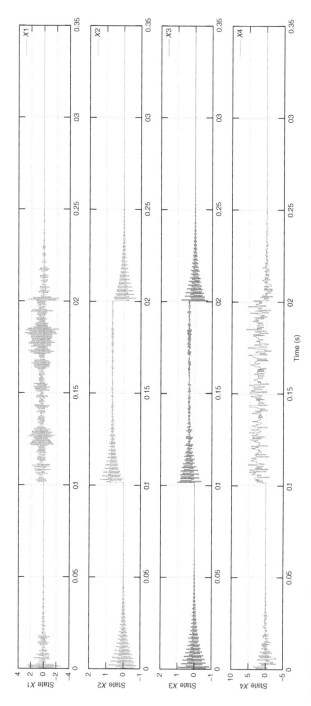

Fig. 5.4 Response of the system states.

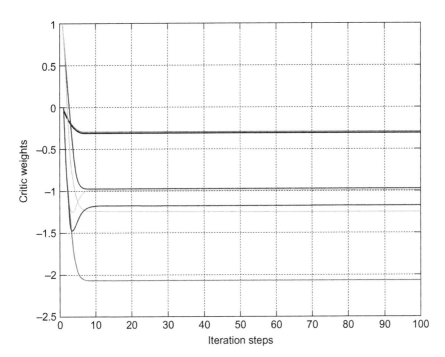

Fig. 5.5 Online critic weights.

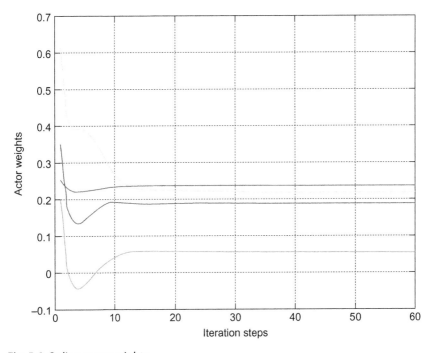

Fig. 5.6 Online actor weights.

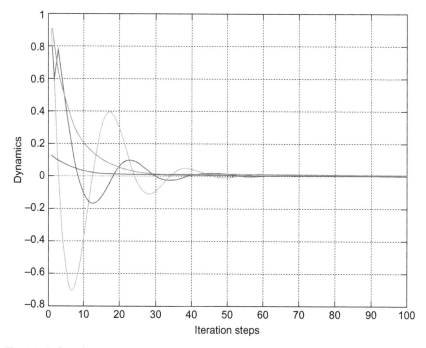

Fig. 5.7 Online dynamics.

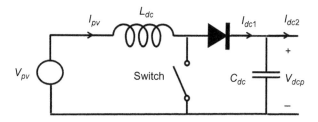

Fig. 5.8 Dynamic model of a photovoltaic cell.

6. COOPERATIVE CONTROL FOR DYNAMIC GAMES OVER GRAPHS

In this section, cooperative control ideas are used to achieve synchronization among agents distributed on a communication graph structure to a leader's dynamics [19]. The graphical game coupled Hamilton functions and Bellman equations are introduced herein.

6.1. Graphs

The graph $\check{G}$ is composed of N vertices or agents $V = \{v_1, \ldots, v_n\}$ with a number of connecting edges $E \subseteq V \times V$ [11]. The connectivity matrix E is defined such that $E = [e_{ij}]$, with $e_{ij} > 0$ if there is a connection between the agents $(v_j, v_i) \in E$ and $e_{ij} = 0$ otherwise. We define the diagonal in-degree matrix as $D = \text{diag}\{d_i\}$, with $d_i = \sum_{j \in N_i} e_{ij}$ the weighted in-degree of agent i. The graph Laplacian matrix L is defined as $L = D - E$ [11]. Directed graphs are considered in the sequel.

6.2. Formulation of the Cooperative Control Problem

The dynamics of each agent i is given by

$$x_{i(k+1)} = Ax_{ik} + B_i u_{ik}, \tag{5.29}$$

where $x_{ik} \in \mathfrak{R}^n$ is the state vector of agent i, $A \in \mathfrak{R}^{n \times n}$ is the agent dynamics, $B_i \in \mathfrak{R}^{n \times m_i}$ is the control input matrix for agent i, and $u_{ik} \in \mathfrak{R}^{m_i}$ is the control input vector for agent i.

The leader's dynamics is given by

$$x_{o(k+1)} = Ax_{ok}. \tag{5.30}$$

The objective of the cooperative control problem is to design and select the control policies u_{ik} for all i with use of local neighbor information so that all agents synchronize to the leader dynamics; that is, $\lim_{k \leftarrow \infty} \|x_{ik} - x_{ok}\| = 0$ for all i. The leader is pinned to a small percentage of the agents [51].

The local neighborhood tracking error $\varepsilon_{ik} \in \mathfrak{R}^n$ for each agent i is given such that [51]

$$\varepsilon_{ik} = \sum_{j \in N_i} e_{ij}[x_{jk} - x_{ik}] + g_i[x_{ok} - x_{ik}], \tag{5.31}$$

where g_i is the pinning gain of agent i, which is nonzero ($g_i > 0$) if agent i is coupled to the leader agent x_o [14].

The overall tracking error vector $\varepsilon_k = [\varepsilon_{1k}^t \varepsilon_{2k}^t \ldots \varepsilon_{Nk}^t]^t$ is given by

$$\varepsilon_k = -[(L + G) \otimes I_n]\eta_k, \tag{5.32}$$

where $G = \text{diag}\{g_i\} \in \mathfrak{R}^{N \times N}$ is a diagonal matrix of the pinning gains. The global synchronization error vector η [51, 52] is given by

$$\eta_k = [\bar{x}_k - \check{x}_{ok}] \in \mathfrak{R}^{nN}, \tag{5.33}$$

where $\bar{x}_k \in \mathfrak{R}^{nN}$ is the global state vector with $\check{x}_{ok} = \check{I}x_o \in \mathfrak{R}^{nN}$, $\check{I} = \alpha \otimes I_n$, and α is the N-vector of 1's.

The graph is assumed to be directed and strongly connected and the pinning gain is for at least one node i. Then the graph matrix is nonsingular [51]. Thus the synchronization error is bounded such that

$$\|\eta_k\| \leq \|\varepsilon_k\|/\sigma_m(L + G), \tag{5.34}$$

where $\sigma_M(\cdot)$ and $\sigma_m(\cdot)$ are the maximum and minimum singular values of a matrix, respectively.

The objective of the control problem is to minimize the local neighborhood tracking errors between the agents or the power systems units. Inequality Eq. (5.34) will guarantee synchronization.

With use of Eq. (5.31), the tracking error dynamics for each agent i is

$$\varepsilon_{i(k+1)} = A\varepsilon_{ik} - (d_i + g_i)B_i u_{ik} + \sum_{j \in N_i} e_{ij} B_j u_{jk}. \tag{5.35}$$

6.3. Formulation of Dynamic Graphical Games

The interacting agents or units with error dynamics (Eq. 5.35) form a dynamic game. The dynamics for each unit are locally coupled in the sense that they are driven by the unit's control action and those of its neighbors. Discrete-time Bellman and Hamiltonian equations represent the mathematical frame to solve the dynamic graphical game [7, 8, 19, 21, 25].

6.3.1. Evaluation of the Dynamic Graphical Game

The graphical game is formed by the interactions among the agents or players on a communication graph. The following definitions and notation are used to facilitate the mathematical setup of the graphical game problem:

- $u_{-i} = \{u_j | j \in N_i\}$: The control actions of the neighbors of each agent i.
- $u_{-i,-\{-i\}} = \{u_j | j \in N_i, N_{\{-i\}}\}$: The group of control actions of the neighbors of each agent i and the control actions of the neighbors of the neighbors of each agent i.
- $u_{\check{i}} = \{u_j | j \in N_i, j \neq i\}$: The actions of all the agents in the graph excluding agent i.

The evaluation of the dynamic graphical game captures the nature of the synchronization problem. The local neighborhood dynamics (Eq. 5.35) shows the interaction between agents or units on a communication graph.

Therefore, to define a dynamic graphical game, the local performance index for each agent i is chosen such that

$$J_i = \sum_{k=0}^{\infty} U_i(\varepsilon_{ik}, u_{ik}, u_{-ik}) \tag{5.36}$$

and the utility function U_i for each agent i is given by

$$U_i(\varepsilon_{ik}, u_{ik}, \varepsilon_{-ik}) = \frac{1}{2}\left[\varepsilon_{ik}^t Q_{ii} \varepsilon_{ik} + u_{ik}^t R_{ii} u_{ik} + \sum_{j \in N_i} u_{jk}^t R_{ij} \varepsilon_{jk} \right], \tag{5.37}$$

where $0 \leq Q_{ii} \in \Re^{n \times n}, 0 < R_{ii} \in \Re^{m_i \times m_i}$, and $0 < R_{ij} \in \Re^{m_i \times m_j}$ are symmetric weighting matrices.

The objective is to find the optimal solutions so that

$$J_i^* = \min_{u_{ik}} \sum_{k=0}^{\infty} U_i(\varepsilon_{ik}, u_{ik}, u_{-ik}^*), \quad \forall i \in N, \tag{5.38}$$

where u_{-ik}^* denotes the optimal policies of the neighbors.

Given arbitrary policies $\mu_{i\ell}, \mu_{-i\ell}$ of each agent i and its neighbors, the value function for each agent i is given in terms of the utility function such that

$$V_i(\bar{\varepsilon}_{ik}) = \sum_{\ell=k}^{\infty} U_i(\varepsilon_{i\ell}, \mu_{i\ell}, \mu_{-i\ell}), \tag{5.39}$$

where $\bar{\varepsilon}_{ik}$ is a vector of the state ε_{ik} of agent i and the states of its neighbors ε_{-ik}.

The value function (5.39) captures local information and reflects the design of the performance index. This means that the solution structure of the value function will depend on the vector $\bar{\varepsilon}_{ik}$ (local error dynamics).

Definition 5.3. *The control policies $\bar{u}_i = \{u_{ik}\}_{k=0}^{\infty}$ for all $i \in N$ are said to be admissible if they stabilize Eq. (5.35) and guarantee that $V_i(\bar{\varepsilon}_{ik})$ for all i are finite.*

6.3.2. Discrete-Time Coupled Hamiltonian Functions

The Hamiltonian function for each agent i, considering the dynamics (Eq. 5.35) and the performance index (Eq. 5.36), is given by

$$H_i(\bar{\varepsilon}_{ik}, \lambda_{i(k+1)}, u_{ik}, u_{-ik}) = \lambda_{i(k+1)}^t \psi_i(\bar{\varepsilon}_{ik}, u_{ik}, u_{-ik,-(-ik)})$$
$$+ U_i(\varepsilon_{ik}, u_{ik}, u_{-ik}), \tag{5.40}$$

where λ_{ik} is the costate variable of each agent i and

$$\bar{\varepsilon}_{ik} = \tilde{Z}_{ik}\varepsilon_k = \begin{bmatrix} 0 & \cdots & [I_n]_{ii} & \cdots & 0 \\ 0 & \cdots & [I_n]_{ij} & \cdots & 0 \end{bmatrix} \varepsilon_k \in \mathfrak{R}^{nN_{i,j}},$$

where $N_{i,j}$ denotes the number of each agent i and its neighbors j. The Hamiltonian constraints' function $\psi_i(\bar{\varepsilon}_{ik}, u_{ik}, u_{-ik,-(-ik)})$ is defined by

$$\psi_i(\bar{\varepsilon}_{ik}, u_{ik}, u_{-ik,-(-ik)}) \equiv \bar{\varepsilon}_{i(k+1)} = [\varepsilon_{i(k+1)}^t \quad \bar{\varepsilon}_{-i(k+1)}^t]^t. \tag{5.41}$$

Denote the stationary admissible policy for each agent i by $u_{ik} = \pi_{ik} = \pi_i(\bar{\varepsilon}_{ik})$. Thus the Hamiltonian given fixed stationary policies for the neighbors is written such that

$$H_i^{\pi}(\bar{\varepsilon}_{ik}, \lambda_{i(k+1)}, u_{ik}, \pi_{-ik}) = \lambda_{i(k+1)}^t(\bar{\varepsilon}_{i(k+1)})$$
$$+ U_i(\varepsilon_{ik}, u_{ik}, \pi_{-ik}). \tag{5.42}$$

The optimal control policy based on the Hamiltonian equation (5.42) is given by application of the necessity optimality condition [8]:

$$\frac{\partial H_i}{\partial u_{ik}} = 0,$$

so

$$u_{ik}^* = \arg\min_{u_{ik}}(H_i^{\pi}(\bar{\varepsilon}_{ik}, \lambda_{i(k+1)}, u_{ik}, \pi_{-ik})). \tag{5.43}$$

Then

$$u_{ik}^* = M_i \lambda_{i(k+1)}, \tag{5.44}$$

where $M_i = R_{ii}^{-1}([\cdots (g_i + d_i) \cdots - e_{ji} \cdots] \otimes B_i^t)$.

6.3.3. Discrete-Time Bellman Equations

The value function (5.39) given stationary admissible policies yields the coupled Bellman equation for each agent such that

$$V_i^{\pi}(\bar{\varepsilon}_{ik}) = \frac{1}{2}\left[\varepsilon_{ik}^t Q_{ii}\varepsilon_{ik} + \Pi_{ik}^t R_{ii}\Pi_{ik} + \sum_{j\in N_i} u_{jk}^t R_{ij}u_{jk}\right]$$
$$+ V_i^{\pi}(\bar{\varepsilon}_{i(k+1)}) \tag{5.45}$$

with initial conditions $V_i^{\pi}(0) = 0$ for all i.

Define $\Delta V_i^{\pi}(\bar{\varepsilon}_{ik})$ and $\nabla V_i^{\pi}(\bar{\varepsilon}_{i(k+1)})$ as

$$\Delta V_i^{\pi}(\bar{\varepsilon}_{ik}) = V_i^{\pi}(\bar{\varepsilon}_{i(k+1)}) - V_i^{\pi}(\bar{\varepsilon}_{ik}),$$

$$\nabla V_i^\pi (\bar{\varepsilon}_{i(k+1)}) = \frac{\partial V_i^\pi (\bar{\varepsilon}_{i(k+1)})}{(\partial \bar{\varepsilon}_{i(k+1)})}.$$

Finding a solution for the graphical games is based on finding the optimal value function $V_i^0(\bar{\varepsilon}_{ik})$ for each agent i so that

$$V_i^0(\bar{\varepsilon}_{ik}) = \min_{u_{ik}} (V_i(\bar{\varepsilon}_{ik})) = \min_{u_i} \left(\sum_{\ell=k}^{\infty} U_i(\varepsilon_{i\ell}, u_{i\ell}, u_{-i\ell}) \right). \tag{5.46}$$

Consider admissible polices for the neighbors of agent i. Then application of the Bellman optimality necessity condition yields

$$V_i^0(\bar{\varepsilon}_{ik}) = \min_{u_{ik}} (U_i(\varepsilon_{ik}, u_{ik}, u_{-ik})) + V_i^0(\bar{\varepsilon}_{i(k+1)}). \tag{5.47}$$

Consequently, the optimal control policy for each agent i is given by

$$u_{ik}^0 = \arg \min_{u_{ik}} (U_i(\varepsilon_{ik}, u_{ik}, \pi_{-ik})) + V_i^0(\bar{\varepsilon}_{i(k+1)}). \tag{5.48}$$

Then

$$\begin{aligned} \pi_{ik} &= u_{ik}^0 \\ &= R_{ii}^{-1}([\cdots(g_i + d_i)\cdots - e_{ji}\cdots] \otimes B_i')\nabla V_i^0(\bar{\varepsilon}_{i(k+1)}) \\ &= M_i \nabla V_i^0(\bar{\varepsilon}_{i(k+1)}). \end{aligned} \tag{5.49}$$

Substituting Eq. (5.49) into Eq. (5.47) yields the coupled graphical game Bellman optimality equations:

$$\begin{aligned} V_i^0(\bar{\varepsilon}_{ik}) = &\frac{1}{2}[\varepsilon_{ik}^t Q_{ii}\varepsilon_{ik} + \nabla V_i^{0^t}(\bar{\varepsilon}_{i(k+1)})M_i^t R_{ii}M_i \nabla V_i^0(\bar{\varepsilon}_{i(k+1)}) \\ &\sum_{j\in N_i} \nabla V_j^{0^t}(\bar{\varepsilon}_{j(k+1)})M_j^t R_{ij}M_i \nabla V_j^0(\bar{\varepsilon}_{j(k+1)})] \\ &+ V_i^0(\bar{\varepsilon}_{i(k+1)}). \end{aligned} \tag{5.50}$$

6.3.4. Coupled HJB Optimality Equations

The discrete Hamiltonian mechanics relates Eqs. (5.42), (5.45). The Hamilton-Jacobi theory relates the Hamiltonian functions (5.42) and the respective Bellman equations (5.45) [9].

Consequently, the discrete-time Hamilton-Jacobi equation is given by

$$\Delta V_i^\pi (\bar{\varepsilon}_{ik}) - \nabla V_i^\pi (\bar{\varepsilon}_{i(k+1)})^t \bar{\varepsilon}_{i(k+1)} + H_i^\pi (\bar{\varepsilon}_{ik}, \nabla V_i^\pi (\bar{\varepsilon}_{i(k+1)}), u_{ik}, \pi_{-ik}) = 0. \tag{5.51}$$

This equation relates the value function for each agent to its Hamiltonian function and costate variable. Thus the costate variable is given in terms of the value function such that

$$\lambda_{i(k+1)} = \nabla V_i^\pi(\bar{\varepsilon}_{i(k+1)}). \tag{5.52}$$

The discrete-time coupled HJB equation describes the Hamiltonian function (5.42) along the optimal trajectories and the Bellman optimality equation (5.50) such that

$$H_i^\pi(\bar{\varepsilon}_{ik}, \nabla V_i^{\pi 0}(\bar{\varepsilon}_{i(k+1)}), u_{ik}^0, u_{-ik}^0) = \nabla V_i^{\pi 0'}(\bar{\varepsilon}_{i(k+1)})\bar{\varepsilon}_{i(k+1)}$$
$$+ U_i(\varepsilon_{ik}, u_{ik}^0, u_{-ik}^0)$$
$$= 0 \tag{5.53}$$

with initial condition $V_i^{\pi 0}(\mathbf{0}) = 0$, where

$$\Pi_{ik} = u_{ik}^0 = M_i \nabla V_i^{\pi 0}(\bar{\varepsilon}_{i(k+1)}). \tag{5.54}$$

6.3.5. Nash Solution for the Dynamic Graphical Game

The dynamic game is solved in the scope of finding the optimal value function (5.46). The Nash equilibrium condition for the graphical game is defined as follows [17].

Definition 5.4. *The N-player game with N-tuple of optimal control policies*

$$\{u_1^*, u_2^*, \ldots, u_N^*\}$$

is said to have a Nash equilibrium solution if for all $i \in N$

$$J_i^0 \equiv J_i(u_i^0, u_{-i}^0) \le J_i(u_i, u_{-i}^0). \tag{5.55}$$

The Nash equilibrium solution for the dynamic graphical game is proved to be equivalent to the solution of the underlying coupled Bellman optimality equations (5.50) [21].

7. ONLINE ADAPTIVE LEARNING SOLUTION

An online adaptive reinforcement learning algorithm (policy iteration) is developed to solve the dynamic graphical games in real time. This algorithm solves the dynamic graphical games online without knowing the full dynamics of the agents, where only the control matrices are required. This algorithm is based on the underlying coupled Bellman equations (5.45) and solves the respective coupled Bellman optimality equations (5.50) for the optimal values and policies.

Theorem 5.1 provides the convergence proof for the online policy iteration (Algorithm 5.3) when all agents update their policies simultaneously.

The policy iteration convergence proof is given under mild assumptions about the graph interconnectivity properties.

Algorithm 5.3   Online Policy Iteration Algorithm

- **Step 1:** Start with arbitrary initial admissible policies u_{ik}^0 for all i and values $\tilde{V}_i^0(\bar{\varepsilon}_{ik})$ for all i.
- **Step 2:** (Value function evaluation). Solve for $\tilde{V}_i^\ell(\bar{\varepsilon}_{ik})$ for all i.

$$\tilde{V}_i^\ell(\bar{\varepsilon}_{ik}) = U_i\left(\varepsilon_{ik}, u_{ik}^\ell, u_{-ik}^\ell\right) + \tilde{V}_i^\ell\left(\bar{\varepsilon}_{i(k+1)}^{(u_{ik}^\ell, u_{-ik}^\ell)}\right) \quad \forall i, \qquad (5.56)$$

where ℓ is the iteration index.

- **Step 3:** (Policy improvement)

$$u_{ik}^{\ell+1} = R_{ii}^{-1}([\cdots(g_i + d_i)\cdots - e_{ji}\cdots] \otimes B_i^t)\nabla\tilde{V}_i^\ell(\bar{\varepsilon}_{i(k+1)}) \quad \forall i. \qquad (5.57)$$

- **Step 4:** On convergence of $\|\tilde{V}_{ik}^{\ell+1}(\cdot) - \tilde{V}_i^\ell(\cdot)\|$ for all i, end.

Theorem 5.1. *Let all agents perform Algorithm 5.3 simultaneously. Assume all initial policies u_{ik}^0 for all i are admissible. Suppose that $\sigma_M(R_{jj}^{-1}R_{ij})$ is small. Then*

1. *u_{ik}^ℓ for all $i, \ell > 0$ are stabilizing and hence admissible policies.*
2. *Policy iteration of Algorithm 5.3 generates a monotonically decreasing sequence of value functions $\tilde{V}_{ik}^\ell(\cdot)$, such that*

$$0 \le \cdots \tilde{V}_{ik}^{\ell+1}(\cdot) \le \tilde{V}_{ik}^\ell(\cdot)\forall i,$$

and this sequence converges to the best response value functions $\tilde{V}_i^*(\cdot)$ for all i that satisfy Eq. (5.50).

Proof.

1. Eq. (5.56) yields

$$\tilde{V}_i^\ell\left(\bar{\varepsilon}_{i(k+1)}^{(u_{ik}^\ell, u_{-ik}^\ell)}\right) - \tilde{V}_i^\ell\left(\bar{\varepsilon}_{ik}^{(u_{ik}^\ell, u_{-ik}^\ell)}\right) < 0, \quad \forall i, \ell > 0. \qquad (5.58)$$

Thus the value functions $\tilde{V}_i^\ell$ for all $i, \ell > 0$ are Lyapunov functions for arbitrary policies u_i^ℓ for all $i, \ell > 0$.

Assuming u_{ik}^0 for all i are admissible, then there exist value functions $\tilde{V}_i^\ell(\cdot)$ for all i that satisfy Eq. (5.39) or (5.56) such that

$$\tilde{V}_i^\ell\left(\bar{\varepsilon}_{ik}^{(u_{ik}^\ell, u_{-ik}^\ell)}\right) = \sum_{m=k}^\infty U_i(\varepsilon_{im}, u_{im}^\ell, u_{-im}^\ell) = \tilde{V}_i^\ell\left(\bar{\varepsilon}_{ik}^{(u_{ik}^{\ell+1}, u_{-ik}^\ell)}\right)$$

$$+ \Delta\tilde{U}_k(u_i^\ell, u_i^{\ell+1}), \qquad (5.59)$$

where

$$\Delta \tilde{U}_k(u_i^\ell, u_i^{\ell+1}) = \sum_{m=k}^{\infty} \left(\frac{1}{2}(u_{im}^\ell - u_{im}^{\ell+1})^t R_{ii}(u_{im}^\ell - u_{im}^{\ell+1}) \right.$$

$$\left. + u_{im}^{\ell+1^t} R_{ii}(u_{im}^\ell - u_{im}^{\ell+1}) \right).$$

The policies $u_{ik}^{\ell+1}$ for all i, $\ell > 0$ are given by Eq. (5.57). Therefore $\Delta \tilde{U}_k(u_i^\ell, u_i^{\ell+1}) > 0$ and Eq. (5.59) yields

$$\tilde{V}_i^\ell \left(\bar{\varepsilon}_{ik}^{(u_{ik}^\ell, u_{-ik}^\ell)} \right) > \tilde{V}_i^\ell \left(\bar{\varepsilon}_{ik}^{(u_{ik}^{\ell+1}, u_{-ik}^\ell)} \right). \tag{5.60}$$

Similarly,

$$\tilde{V}_i^\ell \left(\bar{\varepsilon}_{ik}^{(u_{ik}^{\ell+1}, u_{-ik}^\ell)} \right) = \sum_{m=k}^{\infty} U_i \left(\varepsilon_{im}, u_{im}^{\ell+1}, u_{-im}^\ell \right) = \tilde{V}_i^\ell \left(\bar{\varepsilon}_{ik}^{(u_{ik}^{\ell+1}, u_{-ik}^{\ell+1})} \right)$$

$$+ \Delta \tilde{U}_k(u_{-i}^\ell, u_{-i}^{\ell+1}). \tag{5.61}$$

The assumption that $\Delta \tilde{U}_k(u_i^\ell, u_i^{\ell+1}) > 0$ guarantees that

$$\tilde{V}_i^\ell \left(\bar{\varepsilon}_{ik}^{(u_{ik}^{\ell+1}, u_{-ik}^\ell)} \right) > \tilde{V}_i^\ell \left(\bar{\varepsilon}_{ik}^{(u_{ik}^{\ell+1}, u_{-ik}^{\ell+1})} \right). \tag{5.62}$$

Thus $\Delta \tilde{U}_k(u_i^\ell, u_i^{\ell+1}) \geq 0$ is a sufficient condition for stabilization, which is guaranteed by

$$\sum_{j \in N_i} \frac{1}{2}(u_{jk}^\ell - u_{jk}^{\ell+1})^t R_{ij}(u_{jk}^\ell - u_{jk}^{\ell+1}) - u_{jk}^{\ell+1^t} R_{ij}(u_{jk}^{\ell+1} - u_{jk}^\ell) > 0. \tag{5.63}$$

Use of the norm properties on this inequality yields

$$\sum_{j \in N_i} \frac{1}{2}\sigma_m(R_{ij})\|\Delta u_{jk}^\ell\| > \sum_{j \in N_i}(g_j + d_j)\sigma_M(R_{jj}^{-1} R_{ij})$$

$$\times \left[(g_j + d_j) \left\| \nabla_j \tilde{V}_j^\ell \left(\bar{\varepsilon}_{j(k+1)}^{(u_{jk}^\ell, u_{-jk}^\ell)} \right) \right\| \right.$$

$$\left. + \sum_{\phi \in N_i} \left(e_{j\phi} \left\| \nabla_\phi \tilde{V}_j^\ell \left(\bar{\varepsilon}_{j(k+1)}^{(u_{jk}^\ell, u_{-jk}^\ell)} \right) \right\| \right) \right] \|B_{jk}\|,$$

$$\tag{5.64}$$

where $\Delta u_{jk}^\ell = (u_{jk}^{\ell+1} - u_{jk}^\ell)$.

Under assumption (5.64), inequalities (5.60) and (5.62) yield

$$\tilde{V}_i^\ell \left(\bar{\varepsilon}_{ik}^{(u_{ik}^\ell, u_{-ik}^\ell)} \right) > \tilde{V}_i^\ell \left(\bar{\varepsilon}_{ik}^{(u_{ik}^{\ell+1}, u_{-ik}^\ell)} \right) > \tilde{V}_i^\ell \left(\bar{\varepsilon}_{ik}^{(u_{ik}^{\ell+1}, u_{-ik}^{\ell+1})} \right). \qquad (5.65)$$

The assumption $\Delta \tilde{U}_k(u_i^\ell, u_i^{\ell+1}) \geq 0$ indicates that choosing small values of $\sigma_M(R_{jj}^{-1} R_{ij})$ guarantees that Eq. (5.63) is satisfied. This can be guaranteed for any choice of R_{ij} of agent i by selection of R_{jj} large enough. Therefore $u_{ik}^{\ell+1}$ for all i, $\ell > 0$ are stabilizing policies and hence admissible.

Eq. (5.58) yields

$$\tilde{V}_i^\ell \left(\bar{\varepsilon}_{i(k+1)}^{(u_{ik}^{\ell+1}, u_{-ik}^{\ell+1})} \right) - \tilde{V}_i^\ell \left(\bar{\varepsilon}_{ik}^{(u_{ik}^{\ell+1}, u_{-ik}^{\ell+1})} \right) < 0. \qquad (5.66)$$

2. Eq. (5.56) yields

$$\tilde{V}_i^{\ell+1} \left(\bar{\varepsilon}_{i(k+1)}^{(u_{ik}^{\ell+1}, u_{-ik}^{\ell+1})} \right) - \tilde{V}_i^{\ell+1} \left(\bar{\varepsilon}_{ik}^{(u_{ik}^{\ell+1}, u_{-ik}^{\ell+1})} \right) + U_i(\varepsilon_{ik}, u_{ik}^{\ell+1}, u_{-ik}^{\ell+1}) = 0. \qquad (5.67)$$

Eqs. (5.66), (5.67) and assumption (5.64) yield

$$\tilde{V}_i^\ell \left(\bar{\varepsilon}_{i(k+1)}^{(u_{ik}^{\ell+1}, u_{-ik}^{\ell+1})} \right) - \tilde{V}_i^\ell \left(\bar{\varepsilon}_{ik}^{(u_{ik}^{\ell+1}, u_{-ik}^{\ell+1})} \right) \leq \tilde{V}_i^{\ell+1} \left(\bar{\varepsilon}_{i(k+1)}^{(u_{ik}^{\ell+1}, u_{-ik}^{\ell+1})} \right) \\ - \tilde{V}_i^{\ell+1} \left(\bar{\varepsilon}_{ik}^{(u_{ik}^{\ell+1}, u_{-ik}^{\ell+1})} \right). \qquad (5.68)$$

Application of the summation on Eq. (5.68) yields

$$\sum_{k=\tilde{K}}^{\infty} \left[\tilde{V}_i^\ell \left(\bar{\varepsilon}_{i(k+1)}^{(u_{ik}^{\ell+1}, u_{-ik}^{\ell+1})} \right) - \tilde{V}_i^\ell \left(\bar{\varepsilon}_{ik}^{(u_{ik}^{\ell+1}, u_{-ik}^{\ell+1})} \right) \right] < \sum_{k=\tilde{K}}^{\infty} \left[\tilde{V}_i^{\ell+1} \left(\bar{\varepsilon}_{i(k+1)}^{(u_{ik}^{\ell+1}, u_{-ik}^{\ell+1})} \right) \\ - \tilde{V}_i^{\ell+1} \left(\bar{\varepsilon}_{ik}^{(u_{ik}^{\ell+1}, u_{-ik}^{\ell+1})} \right) \right].$$

This reduces to

$$\tilde{V}_i^\ell \left(\bar{\varepsilon}_{i(\infty)}^{(u_{i\infty}^{\ell+1}, u_{-i\infty}^{\ell+1})} \right) - \tilde{V}_i^\ell \left(\bar{\varepsilon}_{i\tilde{K}}^{(u_{i\tilde{K}}^{\ell+1}, u_{-i\tilde{K}}^{\ell+1})} \right) < \tilde{V}_i^{\ell+1} \left(\bar{\varepsilon}_{i(\infty)}^{(u_{i\infty}^{\ell+1}, u_{-i\infty}^{\ell+1})} \right) \\ - \tilde{V}_i^{\ell+1} \left(\bar{\varepsilon}_{i\tilde{K}}^{(u_{i\tilde{K}}^{\ell+1}, u_{-i\tilde{K}}^{\ell+1})} \right).$$

The first part of the proof implies that

$$\tilde{V}_i^{\ell}\left(\bar{\varepsilon}_{i(\infty)}^{(u_{i\infty}^{\ell+1}, u_{-i\infty}^{\ell+1})}\right) \to 0, \quad \tilde{V}_i^{\ell+1}\left(\bar{\varepsilon}_{i(\infty)}^{(u_{i(\infty)}^{\ell+1}, u_{-i\infty}^{\ell+1})}\right) \to 0$$

so

$$\tilde{V}_i^{\ell+1}\left(\bar{\varepsilon}_{i\tilde{K}}^{(u_{i\tilde{K}}^{\ell+1}, u_{-i\tilde{K}}^{\ell+1})}\right) < \tilde{V}_i^{\ell}\left(\bar{\varepsilon}_{i\tilde{K}}^{(u_{i\tilde{K}}^{\ell+1}, u_{-i\tilde{K}}^{\ell+1})}\right). \tag{5.69}$$

Therefore, by induction, Eq. (5.69) yields

$$0 < \cdots \tilde{V}_i^{\ell+1} < \tilde{V}_i^{\ell} < \tilde{V}_i^{0}, \quad \forall i, \ell > 0. \tag{5.70}$$

The stabilizing policies (Eq. 5.57) form a Nash equilibrium tuple for the dynamic graphical game. The decreasing sequence (Eq. 5.70) is bounded by

$$\left\{0, \tilde{V}_i^{0}\left(\bar{\varepsilon}_{i0}^{(u_{i0}^{0}, u_{-i0}^{0})}\right)\right\}.$$

Then the best response value function solutions $\tilde{V}_i^*$ for all i exist and satisfy Eq. (5.50) or equivalently Eq. (5.53), so

$$0 < \cdots \tilde{V}_i^* < \tilde{V}_i^{\ell+1} < \tilde{V}_i^{\ell} \cdots < \tilde{V}_i^{0}, \quad \forall i, \ell > 0. \tag{5.71}$$

This result shows that Algorithm 5.3 converges when the performance indices are suitably chosen. □

8. CRITIC NEURAL NETWORK IMPLEMENTATION FOR ONLINE ADAPTIVE LEARNING ALGORITHM 5.3

The adaptive learning solution is implemented with use of a critic neural network structure instead of a full actor-critic structure. The given critic neural network implementation approach is useful compared with the difficulties associated with the least squares approaches. A gradient descent technique is used to tune the weights. Each power system unit or agent i has its own critic to perform the evaluation of the value function for this agent. Then the policy of each unit i is improved with Eq. (5.57), which depends on the evaluated critic value. The implementation of this structure is done with use of the local neighborhood information available to each agent or unit.

The value function $V_i(\bar{\varepsilon}_{ik})$ for each agent i is approximated by a critic network structure $\hat{V}_i(\cdot|\tilde{W}_{ic})$ such that

$$\hat{V}_{ik}(\cdot|\tilde{W}_{ic}) = \frac{1}{2}L_{ik}^t \tilde{W}_{ic} L_{ik}, \tag{5.72}$$

where $\tilde{W}_{ic} \in \mathfrak{R}^{nN_{i,j} \times nN_{i,j}}$ for all i are the weights of the approximated value structure $\hat{Q}_{ik}(\cdot|\tilde{W}_{ic})$, and $L_{ik} = [\varepsilon_{ik}^t \cdots \varepsilon_{-ik}^t]^t$ is an evaluated vector of the states ε_{ik} and the states ε_{-ik} for each agent i.

The control policy is given in terms of the critic neural network approximation $\hat{V}_{ik}(\cdot|\tilde{W}_{ic})$. The update of the control policy $\hat{u}_{ik}$ is given by Eq. (5.57) such that

$$\hat{u}_{ik} = R_{ii}^{-1}([\cdots (g_i + d_i) \cdots - e_{ji} \cdots] \otimes B_i^t)\tilde{W}_{ic}^t L_{ik}, \quad \forall i. \tag{5.73}$$

The policy iteration of Algorithm 5.3 requires that the approximation of the value function (5.72) be written as

$$\hat{V}_{ik}(\cdot|\bar{W}_{ic}) = \bar{W}_{ic}^t \bar{L}_{ik}, \tag{5.74}$$

where $\bar{L}_i \in \mathfrak{R}^{(nN_{i,j})(nN_{i,j}+1)/2 \times 1}$ denotes the Kronecker product quadratic polynomial basis vector with elements $\{(\bar{L}_i)_q(\bar{L}_i)_r\}_{(q=1:nN_{i,j})r=1:nN_{i,j}}$ and $\bar{W}_{ic} = \upsilon(\tilde{W}_{ic})$, where υ is a vector-valued matrix function that acts on the symmetric matrices and returns a column vector by stacking the elements of the diagonal and upper triangular part of the symmetric matrix into a vector where the off-diagonal elements are taken as the qrth element of $\tilde{W}_{ic}$. Redundant terms of the weights matrix $\tilde{W}_{ic}$ are removed.

Using Eq. (5.74), we can write the Bellman equation (5.56) in terms of the critic weights as

$$\bar{W}_{ic}^T (\bar{L}_{ik} - \bar{L}_{i(k+1)}) = \frac{1}{2} \left[\varepsilon_{ik}^t Q_{ii} \varepsilon_{ik} + \hat{u}_{ik}^t R_{ii} \hat{u}_{ik} + \sum_{j \in N_i} \hat{u}_{jk}^t R_{ij} \hat{u}_{jk} \right]. \tag{5.75}$$

The vectors $\bar{W}_{ic}$ for all i are updated online in real time as will be shown.

The critic network structure for each agent i performs the value function evaluation (Eq. 5.56). The policy improvement (Eq. 5.57) depends on the evaluated value function $\hat{V}_{ik}(\cdot|\tilde{W}_{ic})$.

Let $\mathfrak{I}_{\bar{\varepsilon}_{ik}}^{\tilde{V}_i(\bar{\varepsilon}_{ik})}$ be the target value of the critic neural network structure

$$\tilde{V}_i(\bar{\varepsilon}_{ik}) = [\hat{V}_{ik}(\cdot|\tilde{W}_{ic}) - \hat{V}_{i(k+1)}(\cdot|\tilde{W}_{ic})]$$

such that

$$\mathfrak{I}_{\bar{\varepsilon}_{ik}}^{\tilde{V}_i(\bar{\varepsilon}_{ik})} = \frac{1}{2} \left[\varepsilon_{ik}^t Q_{ii} \varepsilon_{ik} + \hat{u}_{ik}^t R_{ii} \hat{u}_{ik} + \sum_{j \in N_i} \hat{u}_{jk}^t R_{ij} \hat{u}_{jk} \right]. \tag{5.76}$$

The critic neural network approximation error is given by

$$\zeta_{\bar{\varepsilon}_{ik}}^{\tilde{V}_i(\bar{\varepsilon}_{ik})} = \Im_{\bar{\varepsilon}_{ik}}^{\tilde{V}_i(\bar{\varepsilon}_{ik})} - \tilde{V}_i(\bar{\varepsilon}_{ik}). \tag{5.77}$$

The square sum of the approximation error for each critic network i can be written as

$$Err_i = \frac{1}{2}\left(\zeta_{\bar{\varepsilon}_{ik}}^{\tilde{V}_i(\bar{\varepsilon}_{ik})}\right)^t \zeta_{\bar{\varepsilon}_{ik}}^{\tilde{V}_i(\bar{\varepsilon}_{ik})} = \frac{1}{2}\left\| \Im\Im_{\bar{\varepsilon}_{ik}}^{\tilde{V}_i(\bar{\varepsilon}_{ik})} - \bar{W}_{ic}^t \Im(\bar{L}_{i(k,k+1)}) \right\|_2^2, \tag{5.78}$$

where $\Im\Im_{\bar{\varepsilon}_{ik}}^{\tilde{V}_i(\bar{\varepsilon}_{ik})} \in Re^{1 \times (n \times N_{i,j})(nN_{i,j}+1)/2}$ is a row vector of the target values (Eq. 5.76) for $(n \times N_{i,j})(nN_{i,j} + 1)/2$ samples, and $\Im(\bar{L}_{i(k,k+1)})$ is a square matrix of $(nN_{i,j})(nN_{i,j} + 1)/2$ samples of $(\bar{L}_{ik} - \bar{L}_{i(k+1)})$.

The change in the network weights is given by gradient descent on this function, whose gradient is

$$-\Delta \bar{W}_{ic}^{\ell t} = \left[\frac{\partial Err_i}{\partial \zeta_{\bar{\varepsilon}_{ik}}^{\tilde{V}_i(\bar{\varepsilon}_{ik})}} \right] \left[\frac{\partial \zeta_{\bar{\varepsilon}_{ik}}^{\tilde{V}_i(\bar{\varepsilon}_{ik})}}{\partial \bar{W}_{ic}^t} \right] \Bigg|_{\bar{W}_{ic}^t = W_{ic}^{\ell t}}$$

$$= \left(\Im\Im_{\bar{\varepsilon}_{ik}}^{\tilde{V}_i(\bar{\varepsilon}_{ik})} - \bar{W}_{ic}^{\ell t} \Im(\bar{L}_{i(k,k+1)}) \right) \times \Im(\bar{L}_{i(k,k+1)})^t. \tag{5.79}$$

Therefore the update rules for the critic weights are given by

$$\bar{W}_{ic}^{(\ell+1)T^t} = \bar{W}_{ic}^{\ell T^t} - \tilde{\mu}_{ic}\left[\Im\Im_{\bar{\varepsilon}_{ik}}^{\tilde{V}_i(\bar{\varepsilon}_{ik})} - \bar{W}_{ic}^{\ell T^t} \Im(\bar{L}_{i(k,k+1)}) \right]$$
$$\times \Im(\bar{L}_{i(k,k+1)})^t, \tag{5.80}$$

where $0 < \tilde{\mu}_{ic} < 1$ is the network-learning rate.

It is worthwhile noting that the control policy is updated only after the network weights have converged to the solution of Eq. (5.56).

9. ONLINE CRITIC-NETWORK TUNING IN REAL TIME

Herein, an algorithm is developed to tune the critic network weights in real time with use of data measured along the system trajectories. This is done with use of the local neighborhood information.

Algorithm 5.4   Online Tuning of the Critic Weights

1. Initialize the critic weights $\bar{W}_{ic}^0$ for all i.
2. Do loop (ℓ iterations):
    a. Start with given initial state ε_i^0 for all i on the system trajectory.

b. Do loop (s iterations):
   - The network weights are given by $\bar{W}_{ic}^{s} = \bar{W}_{ic}^{\ell}$.
   - Calculate $\hat{u}_i^s$ for all i with use of Eq. (5.73).
   - Measure the dynamics $\varepsilon_{i(k+1)}^{s}$ for all i.
   - Evaluate the value $\tilde{V}^s(\varepsilon_{ik}^s)$ for all i with use of Eq. (5.74).
   - End loop when $s = (nN_{i,j})(nN_{i,j} + 1)/2$.

c. Network weights update rule:

$$\bar{W}_{ic}^{(\ell+1)'} = \bar{W}_{ic}^{\ell'} - \tilde{\mu}_{ic}\left[\Im\Im_{\bar{\varepsilon}_{ik}}^{\tilde{V}_i(\bar{\varepsilon}_{ik})} - \bar{W}_{ic}^{\ell'}\Im(\bar{L}_{i(k,k+1)})\right] \times \Im(\bar{L}_{i(k,k+1)})',$$

where $\Im_{\bar{\varepsilon}_{ik}}^{\tilde{V}_i(\bar{\varepsilon}_{ik})}$ is given by Eq. (5.76).

d. On convergence of $\| \tilde{V}_i^{\ell+1} - \tilde{V}_i^{\ell} \|$, end loop.

---

10. SIMULATION RESULTS

In this section, simulation is preformed to test the validity of Algorithm 5.4 (online adaptive learning solution for dynamic graphical games). Consider the four-agent directed graph shown in Fig. 5.9.

In the sequel, the online multiagent adaptive learning control scheme will be developed for a network of photovoltaic cells that are distributed on a graph. For the purpose of simulation, the photovoltaic example has the following structure:

- Agents' dynamics:

$$A = \begin{bmatrix} -0.22 & 0 \\ 0.5 & -0.15 \end{bmatrix}, \quad B_1 = \begin{bmatrix} 1 \\ 1 \end{bmatrix}, \quad B_2 = \begin{bmatrix} 0.695 \\ 0.680 \end{bmatrix},$$

$$B_3 = \begin{bmatrix} -0.25 \\ 0.32 \end{bmatrix}, \quad B_4 = \begin{bmatrix} 0.7 \\ -0.1 \end{bmatrix}.$$

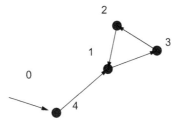

Fig. 5.9 Graphical game example.

- Pinning gains:

$$g_1 = 0, \quad g_2 = 0, \quad g_3 = 0, \quad g_4 = 1.$$

- Graph connectivity matrix:

$$e_{12} = 0.8, \quad e_{14} = 0.7, \quad e_{23} = 0.6, \quad e_{31} = 0.8.$$

- Performance index weighting matrices:

$$Q_{11} = Q_{22} = Q_{33} = Q_{44} = 0.001 * I_{22}, \quad R_{11} = R_{22} = R_{33} = R_{44} = 1,$$
$$R_{13} = R_{21} = R_{32} = R_{41} = 0, \quad\quad\quad R_{12} = R_{14} = R_{23} = R_{31} = 1.$$

- The critic learning rates are ($\tilde{\mu}_{ic} = 0.0001$) for all i.

Figs. 5.10–5.12 show that the dynamics of the photovoltaic cells synchronize to the leader's dynamics, where smooth behavior is observed.

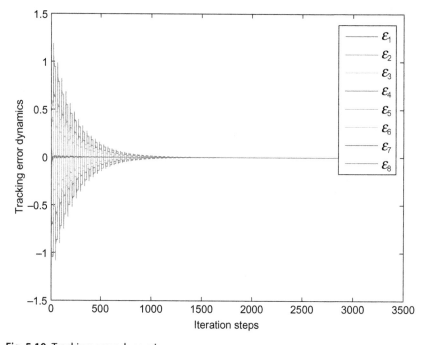

Fig. 5.10 Tracking error dynamics.

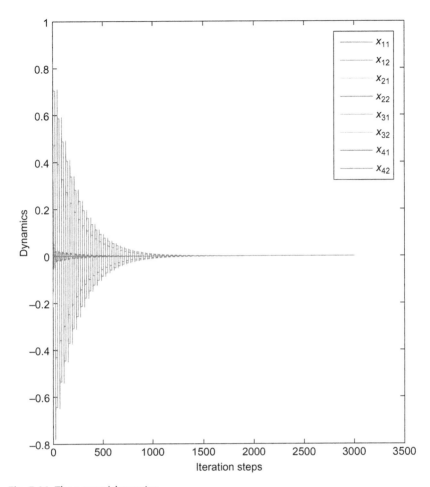

Fig. 5.11 The agents' dynamics.

11. CONCLUSIONS

Reinforcement learning techniques allow the development of algorithms to learn the solutions to the optimal control problems for dynamic systems that are described by difference equations. These involve two-step techniques known as *policy iteration* or *value iteration*. Herein, online value iteration and policy iteration learning techniques were developed for single-agent and multiagent systems, respectively. Value iteration was used to control an autonomous microgrid using online actor-critic structures. The control

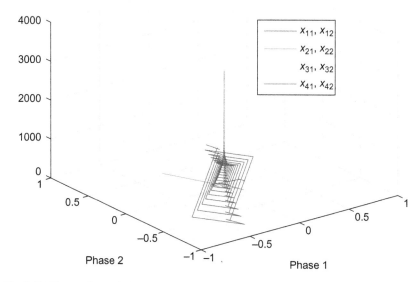

Fig. 5.12 Phase plane plot.

strategy was designed with use of a dynamic model of an islanded microgrid and made use of an internal oscillator for frequency control. The simulation results show that the proposed control technique stabilizes the system and is robust to load disturbances. Furthermore, an online adaptive learning structure (policy iteration) was used to design controllers for a network of islanded photovoltaic cells that are distributed on a communication graph. A new class of control policies was developed to solve the dynamic graphical game with an innovative performance index that is used to measure the system performance. The online policy iteration structure was implemented by means of critic neural network structures. Only partial knowledge of the system's dynamics is required to select the controller's policies, where each agent depends on the local information available to that agent.

ACKNOWLEDGMENTS

This work is supported by the Deanship of Scientific Research (DSR) at King Fahd University of Petroleum and Minerals (KFUPM) through book writing research project no. BW 151004.

REFERENCES

[1] F. Katiraei, M.R. Iravani, P.W. Lehn, Microgrid autonomous operation during and subsequent to islanding process, IEEE Trans. Power Deliv. 20 (1) (2005) 248–257.

[2] P. Piagi, R.N. Lasseter, Autonomous control of microgrids, in: IEEE Power Engineering Society General Meeting, June 18–22, 2006, pp. 139–147.

[3] M.S. Mahmoud, S.A. Hussain, M.A. Abido, Modeling and control of microgrid: an overview, J. Franklin Inst. 351 (5) (2014) 2822–2859.

[4] A.M. Bouzid, et al., A survey on control of electric power distributed generation systems for microgrid applications, Renew. Sustain. Energy Rev. 44 (2015) 751–766.

[5] J.A.P. Lopes, L.C. Moreira, A.G. Madureira, Defining control strategies for microgrids islanded operation, IEEE Trans. Power Syst. 21 (2) (2006) 916–924.

[6] H. Karimi, H. Nikkhajoei, M.R. Iravani, Control of an electronically-coupled distributed resource unit subsequent to an islanding event, IEEE Trans. Power Deliv. 23 (1) (2008) 493–501.

[7] A. Bryson, Optimal control—1950 to 1985, IEEE Trans. Control Syst. 16 (3) (1996) 26–33.

[8] F. Lewis, D. Vrabie, V. Syrmos, Optimal Control, third ed., John Wiley, New York, 2012.

[9] S. Lall, M. West, Discrete variational Hamiltonian mechanics, J. Phys. A Math Gen. 39 (19) (2006) 5509–5519.

[10] R. Beard, V. Stepanyan, Synchronization of information in distributed multiple vehicle coordination control, in: Proceedings of the IEEE Conference on Decision and Control, Maui, HI, 2003, pp. 2029–2034.

[11] W. Ren, R. Beard, E. Atkins, A survey of consensus problems in multi-agent coordination, in: Proceedings of the American Control Conference, 2005, pp. 1859–1864.

[12] J. Tsitsiklis, Problems in decentralized decision making and computation, PhD dissertation, Department of Electrical Engineering and Computer Science, MIT, Cambridge, MA, 1984.

[13] Z. Li, Z. Duan, G. Chen, L. Huang, Consensus of multi-agent systems and synchronization of complex networks: a unified viewpoint, IEEE Trans. Circuits Syst. 57 (1) (2010) 213–224.

[14] X. Li, X. Wang, G. Chen, Pinning a complex dynamical network to its equilibrium, IEEE Trans. Circuits Syst. 51 (10) (2004) 2074–2087.

[15] W. Ren, K. Moore, Y. Chen, High-order and model reference consensus algorithms in cooperative control of multivehicle systems, J. Dyn. Syst. Meas. Control 129 (5) (2007) 678–688.

[16] R. Gopalakrishnan, J. Marden, A. Wierman, An architectural view of game theoretic control, ACM SIGMETRICS Perform. Eval. Rev. 38 (3) (2011) 31–36.

[17] T. Basar, G.J. Olsder, Dynamic Non-cooperative Game Theory, in: Classics in Applied Mathematics, second ed., SIAM, Philadelphia, 1999.

[18] K. Vamvoudakis, F. Lewis, G. Hudas, Multi-agent differential graphical games: online adaptive learning solution for synchronization with optimality, Automatica 48 (8) (2012) 1598–1611.

[19] M. Abouheaf, F. Lewis, K. Vamvoudakis, S. Haesaert, R. Babuska, Multi-agent discrete-time graphical games and reinforcement learning solutions, Automatica 50 (12) (2014) 3038–3053.

[20] M. Abouheaf, F. Lewis, S. Haesaert, R. Babsuka, K. Vamvoudakis, Multi-agent discrete-time graphical games: interactive Nash equilibrium and value iteration solution, in: 2013 American Control Conference, 2013, pp. 4189–4195.

[21] M. Abouheaf, F. Lewis, Dynamic Graphical Games: Online Adaptive Learning Solutions Using Approximate Dynamic Programming, World Scientific Publishing, Singapore, 2014, pp. 1–46.

[22] P. Werbos, Neural networks for control and system identification, in: Proceedings of the 28th IEEE Conference on Decision and Control, 1989, pp. 260–265.

[23] P. Werbos, Approximate dynamic programming for real-time control and neural modeling, in: D.A. White, D.A. Sofge (Eds.), Handbook of Intelligent Control, Van Nostrand Reinhold, New York, 1992.

[24] A. Barto, R. Sutton, C. Anderson, Neuron like elements that can solve difficult learning control problems, IEEE Trans. Syst. Man Cybern. 13 (1983) 835–846.

[25] R. Bellman, Dynamic Programming, Princeton University Press, Princeton, NJ, 1957.

[26] D. Bertsekas, J. Tsitsiklis, Neuro-Dynamic Programming, Athena Scientific, Belmont, MA, 1996.

[27] R. Sutton, A. Barto, Reinforcement Learning: An Introduction, MIT Press, Cambridge, MA, 1998.

[28] S. Sen, G. Weiss, Learning in multiagent systems, in: Multiagent Systems: A Modern Approach to Distributed Artificial Intelligence, MIT Press, Cambridge, 1999, pp. 259–298.

[29] M. Littman, Value-function reinforcement learning in Markov games, J. Cogn. Sys. Res. 2 (1) (2001) 55–66.

[30] K. Vamvoudakis, F. Lewis, Online actor-critic algorithm to solve the continuous-time infinite horizon optimal control problem, Automatica 46 (5) (2010) 878–888.

[31] K. Vamvoudakis, F. Lewis, Multi-player non-zero sum games: online adaptive learning solution of coupled Hamilton-Jacobi equations, Automatica 47 (8) (2011) 1556–1569.

[32] J. Morimoto, G. Zeglin, C. Atkeson, Minmax differential dynamic programming: application to a biped walking robot, in: IEEE International Conference on Intelligent Robots and Systems, Las Vegas, NV, United States, 2003.

[33] T. Landelius, Reinforcement learning and distributed local model synthesis, PhD dissertation, Linkoping University, Sweden, 1997.

[34] Y. Jiang, Z. Jiang, Computational adaptive optimal control for continuous-time linear systems with completely unknown dynamics, Automatica 48 (10) (2012) 2699–2704.

[35] Y. Jiang, Z.P. Jiang, Global adaptive dynamic programming for continuous-time nonlinear systems, 2013, ArXiv:1401.0020 [math.DS], http://arxiv.org/abs/1401.0020.

[36] B. Widrow, N.K. Gupta, S. Maitra, Punish/reward: learning with a critic in adaptive threshold systems, IEEE Trans. Syst. Man Cybern. 5 (1973) 455–465.

[37] H. Jiayi, J. Chuanwen, X. Rong, A review on distributed energy resources and microgrid, Renew. Sustain. Energy Rev. 12 (9) (2008) 2472–2483.

[38] F. Katiraei, R. Iravani, N. Hatziargyriou, A. Dimeas, Microgrids management, IEEE Trans. Power Energy Mag. 6 (3) (2008) 54–65.

[39] H. Karimi, E.J. Davison, R. Iravani, Multivariable servomechanism controller for autonomous operation of a distributed generation unit: design and performance evaluation, IEEE Trans. Power Syst. 25 (2) (2010) 853–865.

[40] R.S. Sutton, A.G. Barto, Introduction to Reinforcement Learning, MIT Press, Cambridge, MA, 1998.

[41] D.P. Bertsekas, J.N. Tsitsiklis, Neuro-dynamic programming: an overview, IEEE Proc. Decis. Control 1 (1995) 560–564.

[42] F.L. Lewis, D. Vrabie, Reinforcement learning and adaptive dynamic programming for feedback control, IEEE Circuits Syst. Mag. 9 (3) (2009) 32–50.

[43] C.J.H.I. Watkins, Learning from delayed rewards, PhD thesis, King's College, Cambridge, United Kingdom, 1989.

[44] C.J.H.I. Watkins, P. Dayan, Q-learning, Mach. Learn. 8 (1992) 279–292.

[45] C. Szepesvari, M.L. Littman, A unified analysis of value-function-based reinforcement-learning algorithms, Neural Comput. 11 (8) (1999) 2017–2060.

[46] A.L. Strehl, M.L. Littman, Online linear regression and its application to model-based reinforcement learning, in: Advances in Neural Information Processing Systems, 2008.

[47] M. Abu-Khalaf, F.L. Lewis, Nearly optimal control laws for nonlinear systems with saturating actuators using a neural network HJB approach, Automatica 41 (5) (2005) 779–791.

[48] J.C.H. Phang, D.S.H. Chan, J.R. Phillips, Accurate analytical method for the extraction of solar cells model parameters, Electron. Lett. 20 (10) (1984) 406–408.

[49] R.J. Wai, W.H. Wang, Grid connected photovoltaic generation system, IEEE Trans. Circuits Syst. I 55 (2008) 953–964.

[50] M.S. Mahmoud, S.M. Ur-Rahman, F.M. Al-Sunni, Review of microgrid architectures: a system of systems perspective, IET Renew. Power Gen. 9 (8) (2015) 1064–1078.

[51] S. Khoo, L. Xie, Z. Man, Robust finite-time consensus tracking algorithm for multi-robot systems, IEEE Trans. Mechatron. 14 (2009) 219–228.

[52] R. Olfati-Saber, R. Murray, Consensus problems in networks of agents with switching topology and time-delays, IEEE Trans. Autom. Control 49 (9) (2004) 1520–1533.

CHAPTER 6

An Optimization Approach to Design Robust Controller for Voltage Source Inverters

A.H. Syed, M.A. Abido
King Fahd University of Petroleum and Minerals, Dhahran, Saudi Arabia

1. INTRODUCTION

The demand for electric power is predicted to rise significantly soon. This challenge having been foreseen, advancement of the technology for efficient and environmentally friendly power generation has attracted a great deal of interest in recent years. Hence renewable energy sources and microgrids are rapidly populating the modern power generation network [1]. This work focuses on a specific topic in the domain of renewable energy systems, which is control system design for the voltage source inverter (VSI), the power electronic interface between a renewable energy source or a distributed generation unit and the main utility grid [2].

VSIs have attracted a lot of interest for its control design with an output filter. Its applications range from distributed generation units [3], active power filters [4], and power electronic variable speed motor drives [5], to uninterruptable electric power supply systems [6, 7]. VSIs with an output filter [8] convert the DC power output of the renewable energy source to AC. Because of the inherent fluctuation in the power generation of renewable energy sources, their output electric power is converted to DC, if it is not already DC. In this way, the electric power can be stored in batteries or other energy storage units. To supply power to the main utility grids, it is converted back to AC by use of VSIs and thus it becomes possible to supply uninterrupted power from these sources to a load according to their capacity [9].

The VSI needs to be controlled for either its output voltage or its output power, which will be transferred to the main grid, depending on which mode of operation it is working in. There are two possible modes: stand-alone mode and grid-connected mode. In stand-alone mode a reference tracking controller needs to be designed for the output voltage and frequency of the inverter. In grid-connected mode the output voltage and frequency are dictated by the main grid and the power output of the inverter is controlled on the basis of demand and capacity [10]. There is much literature on the control of VSIs [11]. Conventional methods, such as the proportional resonant controller [12], the deadbeat predictive controller [13], the active damping algorithm with an LCL filter [14], and feedback linearization [6, 15], have been extensively applied, but are reported to be vulnerable in systems with parameter variations.

Generally there is a lot of research work in the area of control of VSIs. Controllers such as the proportional plus integral controller, which are relatively simple in design, have been applied to VSIs but they have the disadvantage of having fixed steady-state error when tracking a sinusoidal reference. The proportional resonant [16] method has been developed to overcome this problem; however, it requires careful design as it can adversely affect the bandwidth and phase margin. Also there are other problems associated with this method such as slow response and sensitivity to parameter variations. Another simple control design method, based on hysteresis-band control, was presented in [17, 18], but has problems such as variable switching frequency and high current ripple. Three direct power control algorithms are compared in [19]: state vector modulation direct power control is shown to give better performance in terms of lower tracking error and lower total harmonic distortion. A disadvantage, however, is the rise in computational difficulty and cost. A VSI with an LCL filter is also considered in [8]. A new control method using a control Lyapunov function is presented and its performance is compared with that of the commonly used proportional resonant plus state feedback controller. The control Lyapunov function controller alleviates the major problems associated with the proportional resonant controller, with good disturbance rejection, and also deals well with parameter uncertainties. Deadbeat predictive controllers [20] have also been used with VSIs, offering fast reference tracking, low computational cost, and simplicity of design. However, reference tracking performance and stability are often compromised in the presence of uncertainties and delays. Another deadbeat predictive controller was designed in [21] that improves the robustness of the system is the presence of delays by use of a Luenberger observer despite the presence of fractional delay possibly degrading the deadbeat performance.

VSIs can perform poorly for systems having parameter variations if uncertainties are not accounted for in the control design. During grid operation, a change in line impedance and grid faults often occur, leading to parameter variations which could direct the system toward instability [22, 23]. Some state-of-the-art robust control methods include the adaptive voltage control method [10], which features good robustness to parameter uncertainties and load disturbances and simplicity of implementation. Another recently proposed method is the robust state-feedback method [24], which was designed for a single phase inverter, offering suitable transient response as well. A full state observer predictive current control method is proposed in [9] and provides improvement in system performance in the presence of uncertainties but at a price of reduction in control bandwidth. Another new sliding mode robust control method is presented in [25] and offers overshoot-free and fast tracking control performance. A synthesis-based new robust control method [2] that provides robust stabilization and reference tracking with fast response and high control bandwidth was also developed recently. Similarly there are several robust control design methods that exist in theory but have not found any practical application. A class of such methods is robust control design for systems with norm-bounded uncertainties. Some of these methods can be found in [26–29]. Application of these methods requires customization for the plant under consideration.

Recent research in the direction of robust control design for VSIs has been focused on improving the trade-off between robust stability and performance. Some of this research can be found in [3, 30, 31]. The grid-connected VSI is considered with uncertainties in the line impedance of the grid in [3]. The proposed controller is based on H_∞ theory. The controller is designed by use of a standard procedure in which weighting functions are selected for robust stability and tracking performance of the uncertain system. The designed controller is of high order; thus a reduced-order approximation is obtained for implementation. The results of the proposed method are very good as the controller demonstrates good robustness against uncertain parametric variations. The performance of the controller is compared with that of a proportional plus integral controller, and improvement is shown in terms of total harmonic distortion values for different reference levels of the output current.

In this chapter a novel controller design method is proposed for VSIs in the grid-connected mode of operation with uncertainties in the grid line impedance. The proposed design method is applicable with any theory from the class of robust control design that considers structured norm-bounded uncertainties in the system. One of the major contributions of this work is the derivation of control design criteria with use of constraints on parameter variations. These criteria mainly provide a way to choose the arbitrary uncertainty matrices, thus facilitating the application of robust control design methods. Moreover, these criteria introduce a set of parameters that can be selected within a certain range. By tuning these parameters intelligently, one designs a controller with enhanced performance. The overall procedure for control design is orderly and systematic, which makes it user-friendly. For the purpose of reference tracking, the integral controller is also augmented in the stabilizing control law. The rest of the chapter is organized as follows. The preliminaries are formalized along with the development of the mathematical model of the system in Section 2. In Section 3, details regarding the development of the proposed controller are presented. Section 4 covers the description of implementation of the control methods on the VSI system and their results along with discussion and analysis. Conclusions are presented in Section 5.

2. SYSTEM MODELING

The dynamic model of the VSI with an LC filter connected to the grid is developed in this section. The model is developed in the dq reference frame. First some relevant preliminaries are discussed in the following section.

2.1. Preliminaries

A brief description of space phasors is now presented. Consider the following [32]:

$$x_a(t) = x_p \cos \theta(t), \tag{6.1}$$

$$x_b(t) = x_p \cos \left(\theta(t) - \frac{2\pi}{3} \right), \tag{6.2}$$

$$x_c(t) = x_p \cos \left(\theta(t) + \frac{2\pi}{3} \right), \tag{6.3}$$

with

$$\theta(t) = \theta_0 + \int_0^t \omega'(\tau) \, d\tau, \tag{6.4}$$

where $x_{abc} = [x_a(t) \, x_b(t) \, x_c(t)]$ is a vector of a balanced three-phase function, with x_p, $\omega'(t)$, and θ_0 being its amplitude, frequency in radians per second, and phase angle in radians, respectively. For x_{abc} the space phasor is defined as

$$\vec{x}(t) = \frac{2}{3}\left[x_a(t) + e^{\frac{j2\pi}{3}}x_b(t) + e^{\frac{j4\pi}{3}}x_c(t)\right] \tag{6.5}$$

or

$$\vec{x}(t) = \overline{x}e^{j\theta(t)}. \tag{6.6}$$

In the case of constant angular frequency ω,

$$\vec{x}(t) = x_p e^{j\theta_0} e^{j\omega t} = \overline{x}e^{j\omega t}. \tag{6.7}$$

Now consider a function $\rho(t)$ as

$$\rho(t) = \rho_0 + \int_0^t \omega'(\tau)\, d\tau. \tag{6.8}$$

The space phasor $\vec{x}(t)$ is now shifted by the angle $-\rho(t)$ to give

$$\vec{x}(t)e^{-j\rho(t)} = x_p e^{j(\theta_0 - \rho_0)} = x_d + jx_q, \tag{6.9}$$

$$\vec{x}(t) = \left(x_d + jx_q\right)e^{j\rho(t)} = x_{dq}e^{j\rho(t)}, \tag{6.10}$$

where x_d and x_q represent the components of x_{abc} in the rotating dq reference frame. The angles $\theta(t)$ and $\rho(t)$ may not be same but their time rate of change must remain equal.

2.2. Derivation of the Mathematical Model of the System

The VSI and LC filter system considered here is shown in Fig. 6.1. The following space phasor equations represent the dynamics of the system:

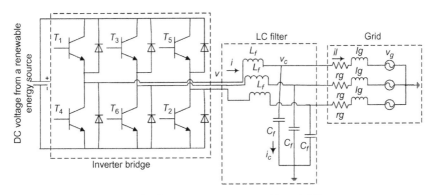

Fig. 6.1 The grid-connected voltage source inverter with an LC filter system.

$$\vec{v}(t) - L_f \frac{d\vec{i}(t)}{dt} - \vec{v}_c(t) = 0, \tag{6.11}$$

$$\vec{i}(t) = \vec{i}_c(t) + \vec{i}_l(t), \tag{6.12}$$

$$\vec{v}_c(t) + r_g \vec{i}_l(t) + l_g \frac{d\vec{i}_l(t)}{dt} + \vec{v}_g(t) = 0. \tag{6.13}$$

If we substitute $\vec{v}(t) = v_{dq}(t)e^{j\rho(t)}$, $\vec{i}(t) = i_{dq}(t)e^{j\rho(t)}$, $\vec{v}_c(t) = v_{cdq}(t)e^{j\rho(t)}$, $\vec{i}_l(t) = i_{ldq}(t)e^{j\rho(t)}$, and $\vec{v}_g(t) = v_{gdq}(t)e^{j(\omega t+\theta_0)}$, Eq. (6.11) becomes

$$\vec{v}_{dq}(t)e^{j\rho(t)} - L_f \frac{di_{dq}(t)e^{j\rho(t)}}{dt} - v_{cdq}(t)e^{j\rho(t)} = 0,$$

$$(v_d(t) + jv_q(t))e^{j\rho(t)} - L_f \left(\left(\frac{di_d(t)}{dt} + j\frac{di_q(t)}{dt} \right) e^{j\rho(t)} + j\omega (i_d(t) + ji_q(t))e^{j\rho(t)} \right)$$
$$- v_{cdq}(t)e^{j\rho(t)} = 0.$$

Similarly Eqs. (6.12), (6.13) can be worked on and solved for the state derivatives, yielding the dynamic model of the system in Eqs. (6.14)–(6.19):

$$\frac{di_d(t)}{dt} = \frac{1}{L_f}v_d(t) + \omega i_q(t) - \frac{1}{L_f}v_{cd}(t), \tag{6.14}$$

$$\frac{di_q(t)}{dt} = \frac{1}{L_f}v_q(t) - \omega i_d(t) - \frac{1}{L_f}v_{cq}(t), \tag{6.15}$$

$$\frac{dv_{cd}(t)}{dt} = \frac{1}{C_f}i_d(t) + \omega v_{cq}(t) - \frac{1}{C_f}i_{ld}(t), \tag{6.16}$$

$$\frac{dv_{cq}(t)}{dt} = \frac{1}{C_f}i_q(t) - \omega v_{cd}(t) - \frac{1}{C_f}i_{lq}(t), \tag{6.17}$$

$$\frac{di_{ld}(t)}{dt} = \frac{1}{l_g}v_{cd}(t) - \frac{r_g}{l_g}i_{ld}(t) + \omega i_{lq}(t) - \frac{1}{l_g}v_{gd}(t), \tag{6.18}$$

$$\frac{di_{lq}(t)}{dt} = \frac{1}{l_g}v_{cq}(t) - \frac{r_g}{l_g}i_{lq}(t) - \omega i_{ld}(t) - \frac{1}{l_g}v_{gq}(t), \tag{6.19}$$

where $i_d(t)$, $i_q(t)$, $v_d(t)$, $v_q(t)$, $v_cd(t)$, $v_cq(t)$, $i_ld(t)$, $i_lq(t)$, $v_gd(t)$, and $v_gq(t)$ are dq components of the output current and output voltage of the VSI, the capacitor voltage in the LC filter, the current injected into the grid, and the grid voltage, respectively. The filter inductance and capacitance are L_f and C_f, respectively, and the grid input resistance and inductance are r_g and l_g, respectively. The grid impedance is therefore $Z_g = r_g + j\omega l_g$, in which all the resistances and inductances on the grid side are lumped together. These include the impedances of switches, circuit breakers, relays, transformers, transmission lines, etc., which are changing with time depending on the operating conditions and the environment. All of these devices thus contribute to the uncertainty of the grid impedance. Also, in reality the grid impedance is unknown and is estimated to be in a certain range [33, 34]. Thus the need arises for a robust controller.

3. THE PROPOSED CONTROL DESIGN METHOD

Consider the state-space uncertain plant model as follows:

$$\dot{x} = Ax + Bu + B_d d = (A + \Delta A)x + Bu + B_d d \tag{6.20}$$

$$y = Cx, \tag{6.21}$$

where $x \in \mathbb{R}^n$, $y \in \mathbb{R}^p$, $u \in \mathbb{R}^m$, and $d \in \mathbb{R}^q$ are the state, output, control input, and disturbance input vectors of the system, respectively. The system constant matrices are $A \in \mathbb{R}^{n \times n}$, $B \in \mathbb{R}^{n \times m}$, and $A \in \mathbb{R}^{p \times n}$, such that the pair (A, B) is stabilizable and the pair (A, C) is detectable. The uncertainty $\Delta A = M \Lambda N$, where $\Lambda^T \Lambda \leq I$. The mathematical model of the VSI system developed in Eqs. (6.14)–(6.19) can be transformed as Eqs. (6.20), (6.21). The state variables are $i_d(t)$, $i_q(t)$, $v_{cd}(t)$, $v_{cq}(t)$, $i_{ld}(t)$, and $i_{lq}(t)$, while $v_d(t)$ and $v_q(t)$ are the control inputs, $v_{gd}(t)$ and $v_{gq}(t)$ are disturbance inputs, and $i_{ld}(t)$ and $i_{lq}(t)$ are the outputs with the following system matrices:

$$\bar{A} = \begin{bmatrix} 0 & \omega & -\frac{1}{L_f} & 0 & 0 & 0 \\ -\omega & 0 & 0 & -\frac{1}{L_f} & 0 & 0 \\ \frac{1}{C_f} & 0 & 0 & \omega & -\frac{1}{C_f} & 0 \\ 0 & \frac{1}{C_f} & -\omega & 0 & 0 & -\frac{1}{C_f} \\ 0 & 0 & \frac{1}{l_g} & 0 & -\frac{r_g}{l_g} & \omega \\ 0 & 0 & 0 & \frac{1}{l_g} & -\omega & -\frac{r_g}{l_g} \end{bmatrix}, \quad B = \begin{bmatrix} \frac{1}{L_f} & 0 \\ 0 & \frac{1}{L_f} \\ 0 & 0 \\ 0 & 0 \\ 0 & 0 \\ 0 & 0 \end{bmatrix},$$

$$B_d = \begin{bmatrix} 0 & 0 \\ 0 & 0 \\ 0 & 0 \\ 0 & 0 \\ \frac{1}{l_g} & 0 \\ 0 & \frac{1}{l_g} \end{bmatrix}, \quad C = \begin{bmatrix} 0 & 0 & 0 & 0 & 1 & 0 \\ 0 & 0 & 0 & 0 & 0 & 1 \end{bmatrix}. \tag{6.22}$$

A controller is to be designed to robustly stabilize the system and enable the output y to track the reference input $\mathbf{r} \in \mathbb{R}^p$. Uncertainties are considered in the grid input impedance, characterized as follows:

$$r_g = r_{g0} + \Delta r_g, \quad r_g = l_{g0} + \Delta l_g, \tag{6.23}$$

with

$$-\lambda_R \leq \Delta r_g \leq \lambda_R, \quad -\lambda_L \leq \Delta l_g \leq \lambda_L \tag{6.24}$$

If no uncertainty is considered, $\overline{\mathbf{A}} = \mathbf{A}$. However, in the case of uncertainties as given in Eqs. (6.23), (6.24), $\mathbf{\Delta A}$ can be determined with use of $\overline{\mathbf{A}}$ from Eqs. (6.22), (6.23). This structure of the state-space model in Eqs. (6.20), (6.21) with uncertainty only in the $\mathbf{A}$ matrix is considered because the uncertain parameters of the system are the grid resistance r_g and grid inductance l_g, which do not affect the $\mathbf{B}$ matrix as evident in Eq. (6.22). Further, l_g being part of the $\mathbf{B_d}$ matrix does not affect the control design. In fact, the whole term $\mathbf{B_d d} = \overline{\mathbf{d}}$ can be considered as a disturbance signal. The signal $\mathbf{d}$ is used to represent the grid voltage as a known fixed disturbance input signal. With the uncertain $\mathbf{B_d}$, $\overline{\mathbf{d}}$ could be considered as a step disturbance input signal and will be dealt with in later when a tracking controller is designed. To solve for the robust controller gains, the uncertainty coefficient matrices $\mathbf{M}$ and $\mathbf{N}$ are required. These matrices are usually chosen arbitrarily. However, it can be easily observed that the choice of these matrices affects the performance of the designed controller. Thus a careful and appropriate choice of these matrices is required to give optimum performance of the controller. Finding $\mathbf{M}$ and $\mathbf{N}$ by trial and error or even by an intelligent algorithms method for optimum control performance could be computationally highly exhaustive because of the dimensions of these matrices and the unknown search space of the elements. The process could take a long time and the solution is likely to be just a local optimum. Therefore an analytical approach is proposed here to determine these matrices. This approach is derived on the basis of the pattern of uncertainty in the mathematical model of the system. It enables the user to determine the uncertainty coefficient matrices by the selection of just two scalars within a known bounded search space, thus making the control design process efficient and convenient. The controller is designed systematically and with optimum performance.

3.1. Controller for Robust Stabilization

3.1.1. Robust Stabilization

Now a theory of robust control and stabilization is presented [29]. An observer-based robust controller can be designed as in Eqs. (6.25), (6.26) for a given plant in the state-space form of Eqs. (6.20), (6.21):

$$\dot{\hat{\mathbf{x}}} = \mathbf{A}\hat{\mathbf{x}} + \mathbf{B}\mathbf{u} + \mathbf{L}(\mathbf{y} - \mathbf{C}\hat{\mathbf{x}}), \tag{6.25}$$

$$\mathbf{u} = -\mathbf{K}\hat{\mathbf{x}}, \tag{6.26}$$

where $\hat{\mathbf{x}} \in \mathbb{R}^n$ are the states of the observer, and the controller and observer gains are $\mathbf{K}$ and $\mathbf{L}$, respectively. The gains are calculated by the solving of the linear matrix inequality (LMI) in Eq. (6.27) with the equality constraint in Eq. (6.28):

$$\begin{bmatrix} \mathbf{\Phi} & \mathbf{B}\hat{\mathbf{K}} & \mathbf{P}\mathbf{M}_1 \\ * & \mathbf{\Psi} & \mathbf{R}\mathbf{M}_1 \\ * & * & \sigma_1\mathbf{I} \end{bmatrix}, \tag{6.27}$$

$$\mathbf{P}\mathbf{B} = \mathbf{B}\hat{\mathbf{P}}, \tag{6.28}$$

where $\mathbf{\Phi} = \mathbf{A}^T\mathbf{P} + \mathbf{P}\mathbf{A} - \hat{\mathbf{K}}^T\mathbf{B}^T - \mathbf{B}\hat{\mathbf{K}} + \sigma_1\mathbf{N}_1^T\mathbf{N}_1 + \sigma_2\mathbf{I}$, $\mathbf{\Psi} = \mathbf{A}^T\mathbf{R} + \mathbf{R}\mathbf{A} - \hat{\mathbf{L}}\mathbf{C} - \mathbf{C}^T\hat{\mathbf{L}}^T + \sigma_2\mathbf{I}$. The positive definite matrices $\mathbf{P} \in \mathbb{R}^{n \times n}$ and $\mathbf{R} \in \mathbb{R}^{n \times n}$, the positive scalars σ_1 and σ_2 and the matrices $\hat{\mathbf{P}} \in \mathbb{R}^{m \times m}$, $\hat{\mathbf{K}} \in \mathbb{R}^{n \times m}$, and $\hat{\mathbf{L}} \in \mathbb{R}^{p \times n}$ are the LMI variables with $\mathbf{K} = \hat{\mathbf{P}}^{-1}\hat{\mathbf{K}}$ and $\mathbf{L} = \mathbf{R}^{-1}\hat{\mathbf{L}}$. The convergence rate h of the system is given in Eq. (6.29), where the function $\lambda_{\max}(\mathbf{M})$ gives the maximum eigenvalue of any matrix $\mathbf{M}$:

$$h = \sigma_2 / \left(2\max\left(\lambda_{\max}(\mathbf{P}), \lambda_{\max}(\mathbf{R})\right)\right). \tag{6.29}$$

Now the following theorem is presented for the application of this method to design an enhanced performance robust controller for a grid-connected VSI.

Theorem 6.1. *The grid-connected VSI with an LC filter, given in Eqs. (6.20)–(6.22), having uncertainty in the grid line impedance given by Eqs. (6.23), (6.24) is stabilized asymptotically with the convergence rate h by* *the controller given in Eqs. (6.25), (6.26). One designs the controller by solving the LMI in Eq. (6.27) with an equality constraint in Eq. (6.28) by choosing the uncertainty matrices in Eq. (6.29):*

$$M = \begin{bmatrix} I_{4 \times 4} & 0_{4 \times 2} \\ 0_{2 \times 4} & \begin{bmatrix} \frac{1}{\mu} & 0 \\ 0 & \frac{1}{\mu} \end{bmatrix} \end{bmatrix}, \quad N = \begin{bmatrix} I_{4 \times 4} & 0_{4 \times 2} \\ 0_{2 \times 4} & \begin{bmatrix} \frac{\mu}{v} & 0 \\ 0 & \frac{\mu}{v} \end{bmatrix} \end{bmatrix}, \tag{6.30}$$

where μ and v are two constants given as

$$\mu = \alpha \overline{\mu}, \quad \overline{\mu} = \frac{l_{g0}^2}{\lambda_L}, \quad \|\alpha\| \le 1, \tag{6.31}$$

and

$$v = \beta \overline{v}, \quad \overline{v} = \sqrt{1 - \alpha^2} \frac{l_{g0}^2}{l_{g0} \lambda_R + r_{g0} \lambda_L}, \quad \|\beta\| \le 1 \tag{6.32}$$

Proof. The uncertainties in the system given by Eqs. (6.23), (6.24) substituted in Eqs. (6.20)–(6.22), ΔA, can be written as

$$\Delta A = \begin{bmatrix} 0 & 0 & 0 & 0 & 0 & 0 \\ 0 & 0 & 0 & 0 & 0 & 0 \\ 0 & 0 & 0 & 0 & 0 & 0 \\ 0 & 0 & 0 & 0 & 0 & 0 \\ 0 & 0 & \overline{\delta}_1 & 0 & -\overline{\delta}_2 & 0 \\ 0 & 0 & 0 & \overline{\delta}_1 & 0 & -\overline{\delta}_2 \end{bmatrix}, \tag{6.33}$$

where

$$\overline{\delta}_1 = -\frac{\Delta l_g}{l_{g0}(l_{g0} + \Delta l_g)} = -\frac{\Delta l_g}{l_{g0} l_g} \cong -\frac{\Delta l_g}{l_{g0}^2}, \tag{6.34}$$

$$\overline{\delta}_2 = -\frac{l_{g0} \Delta r_g - r_{g0} \Delta l_g}{l_{g0} l_g} \cong \frac{l_{g0} \Delta r_g - r_{g0} \Delta l_g}{l_{g0}^2}. \tag{6.35}$$

Now reordering ΔA in the form of MAN, we obtain

$$
\mathbf{\Delta A} =
\begin{bmatrix}
1 & 0 & 0 & 0 & 0 & 0 \\
0 & 1 & 0 & 0 & 0 & 0 \\
0 & 0 & 1 & 0 & 0 & 0 \\
0 & 0 & 0 & 1 & 0 & 0 \\
0 & 0 & 0 & 0 & \frac{1}{\mu} & 0 \\
0 & 0 & 0 & 0 & 0 & \frac{1}{\mu}
\end{bmatrix}
\begin{bmatrix}
0 & 0 & 0 & 0 & 0 & 0 \\
0 & 0 & 0 & 0 & 0 & 0 \\
0 & 0 & 0 & 0 & 0 & 0 \\
0 & 0 & 0 & 0 & 0 & 0 \\
0 & 0 & \mu\bar{\delta}_1 & 0 & -\nu\bar{\delta}_2 & 0 \\
0 & 0 & 0 & \mu\bar{\delta}_1 & 0 & -\nu\bar{\delta}_2
\end{bmatrix}
$$

$$
\begin{bmatrix}
1 & 0 & 0 & 0 & 0 & 0 \\
0 & 1 & 0 & 0 & 0 & 0 \\
0 & 0 & 1 & 0 & 0 & 0 \\
0 & 0 & 0 & 1 & 0 & 0 \\
0 & 0 & 0 & 0 & \frac{\mu}{\nu} & 0 \\
0 & 0 & 0 & 0 & 0 & \frac{\mu}{\nu}
\end{bmatrix}.
\tag{6.36}
$$

For the LMI to be applicable, $\mathbf{\Lambda}$ should be such that $\mathbf{\Lambda}^{\mathbf{T}}\mathbf{\Lambda} \leq \mathbf{I}$. With $\delta_1 = \mu\bar{\delta}_1$ and $\delta_2 = \nu\bar{\delta}_2$, consider the following:

$$
\mathbf{\Lambda}^{\mathbf{T}}\mathbf{\Lambda} \leq \mathbf{I} =
\begin{bmatrix}
0 & 0 & 0 & 0 & 0 & 0 \\
0 & 0 & 0 & 0 & 0 & 0 \\
0 & 0 & 0 & 0 & \delta_1 & 0 \\
0 & 0 & 0 & 0 & 0 & \delta_1 \\
0 & 0 & 0 & 0 & -\delta_2 & 0 \\
0 & 0 & 0 & 0 & 0 & -\delta_2
\end{bmatrix}
\begin{bmatrix}
0 & 0 & 0 & 0 & 0 & 0 \\
0 & 0 & 0 & 0 & 0 & 0 \\
0 & 0 & 0 & 0 & 0 & 0 \\
0 & 0 & 0 & 0 & 0 & 0 \\
0 & 0 & \delta_1 & 0 & -\delta_2 & 0 \\
0 & 0 & 0 & \delta_1 & 0 & -\delta_2
\end{bmatrix},
\tag{6.37}
$$

$$
\mathbf{\Lambda}^{\mathbf{T}}\mathbf{\Lambda} \leq \mathbf{I} =
\begin{bmatrix}
0 & 0 & 0 & 0 & 0 & 0 \\
0 & 0 & 0 & 0 & 0 & 0 \\
0 & 0 & \delta_1^2 & 0 & -\delta_1\delta_2 & 0 \\
0 & 0 & 0 & \delta_1^2 & 0 & -\delta_1\delta_2 \\
0 & 0 & -\delta_1\delta_2 & 0 & \delta_2^2 & 0 \\
0 & 0 & 0 & -\delta_1\delta_2 & 0 & \delta_2^2
\end{bmatrix} \leq \mathbf{I},
\tag{6.38}
$$

or

$$
\mathbf{\Gamma} = \mathbf{I} - \mathbf{\Lambda}^{\mathbf{T}}\mathbf{\Lambda} \leq \mathbf{I} =
\begin{bmatrix}
1 & 0 & 0 & 0 & 0 & 0 \\
0 & 1 & 0 & 0 & 0 & 0 \\
0 & 0 & 1-\delta_1^2 & 0 & \delta_1\delta_2 & 0 \\
0 & 0 & 0 & 1-\delta_1^2 & 0 & \delta_1\delta_2 \\
0 & 0 & \delta_1\delta_2 & 0 & 1-\delta_2^2 & 0 \\
0 & 0 & 0 & \delta_1\delta_2 & 0 & 1-\delta_2^2
\end{bmatrix} \geq \mathbf{0}.
\tag{6.39}
$$

With use of Sylvester's criterion for positive definiteness, $\mathbf{\Gamma} \geq \mathbf{0}$ if and only if all of its principal minors are positive. Say the principal minors of $\mathbf{\Gamma}$ are P_i for $i = 1, 2, 3, \ldots, 6$. Clearly P_1, P_2, and P_4 are positive as follows:

$$P_1 = 1 > 0, \quad P_2 = 1 > 0, \quad P_4 = (1 - \delta_1^2)^2 > 0. \tag{6.40}$$

For the rest, consider the following:

$$P_3 = 1 - \delta_1^2 = 1 - \left(-\mu \frac{\Delta l_g}{l_{g0}^2}\right)^2. \tag{6.41}$$

Say

$$\mu = \alpha \overline{\mu} = \alpha \frac{l_{g0}^2}{\lambda_L} \Rightarrow P_3 = 1 - \left(\alpha \frac{\Delta l_g}{\lambda_L}\right)^2 = 1 - \alpha^2 \frac{\Delta l_g^2}{\lambda_L^2}. \tag{6.42}$$

If

$$|\alpha| \leq 1 \Rightarrow P_3 \geq 0. \tag{6.43}$$

Similarly, consider the fifth principal minor P_5:

$$P_5 = \begin{vmatrix} 1 & 0 & 0 & 0 & 0 \\ 0 & 1 & 0 & 0 & 0 \\ 0 & 0 & 1-\delta_1^2 & 0 & \delta_1\delta_2 \\ 0 & 0 & 0 & 1-\delta_1^2 & 0 \\ 0 & 0 & \delta_1\delta_2 & 0 & 1-\delta_2^2 \end{vmatrix} = \begin{vmatrix} 1-\delta_1^2 & 0 & \delta_1\delta_2 \\ 0 & 1-\delta_1^2 & 0 \\ \delta_1\delta_2 & 0 & 1-\delta_2^2 \end{vmatrix}, \tag{6.44}$$

$$P_5 = (1 - \delta_1^2)[(1 - \delta_1^2)(1 - \delta_2^2) - \delta_1^2\delta_2^2], \tag{6.45}$$

$$P_5 = (1 - \delta_1^2)(1 - \delta_1^2 - \delta_2^2) = (1 - \delta_1^2)^2 - \delta_2^2(1 - \delta_1^2) = P_3^2 - \delta_2^2 P_3. \tag{6.46}$$

If $P_3 - \delta_2^2 \geq 0$ or $P_3 \geq \delta_2^2 \Rightarrow P_5 \geq 0$.
Now with $P_3 \geq \delta_2^2$ consider

$$\nu^2 \left(\frac{l_{g0}\Delta r_g - r_{g0}\Delta l_g}{l_{g0}^2}\right)^2 \leq 1 - \alpha^2 \frac{\Delta l_g^2}{\lambda_L^2}, \tag{6.47}$$

$$\nu^2 \leq \left(1 - \alpha^2 \frac{\Delta l_g^2}{\lambda_L^2}\right) \left(\frac{l_{g0}^2}{l_{g0}\Delta r_g - r_{g0}\Delta l_g}\right)^2$$

$$= \left(\frac{\lambda_L^2 - \alpha^2 \Delta l_g^2}{\lambda_L^2}\right) \left(\frac{l_{g0}^2}{l_{g0}\Delta r_g - r_{g0}\Delta l_g}\right)^2 = \Pi \tag{6.48}$$

Now by minimizing Π, we can obtain the upper bound on v^2. For this $\Delta l_g = -\lambda_L$ and $\Delta r_g = \lambda_R$:

$$v^2 \leq \left(1 - \alpha^2\right) \left(\frac{l_{g0}^2}{l_{g0}\lambda_R + r_{g0}\lambda_L} \right)^2, \tag{6.49}$$

or with $|\beta| \leq 1$

$$v^2 = \beta^2 \left(1 - \alpha^2\right) \left(\frac{l_{g0}^2}{l_{g0}\lambda_R + r_{g0}\lambda_L} \right)^2, \tag{6.50}$$

$$v = \beta \sqrt{\left(1 - \alpha^2\right)} \left(\frac{l_{g0}^2}{l_{g0}\lambda_R + r_{g0}\lambda_L} \right) = \beta \bar{v}. \tag{6.51}$$

In a similar manner it can be proved that the sixth principal minor $P_6 \geq 0$. This completes the proof.   □

Remark 6.1. *Theorem 6.1 introduces two tuning parameters α and β in the control design method. These parameters define the uncertainty coefficient matrices M and N and hence the performance of the controller. One can optimize the performance of the controller using any intelligent method by tuning only these two parameters within the search range given in Eqs. (6.31), (6.32). The criterion for optimum performance could be maximization of the system convergence rate, closeness to either the target eigenvalues or damping ratios, or a combination of both based on the specifications of the required transient response.*

Theorem 6.1 provides a method for robust controller design with a certain convergence rate. However, the uncertainty bounds for which this convergence rate is guaranteed remains unknown. Theorem 6.2 is presented now to answer this question.

Theorem 6.2. *The grid-connected VSI system with an LC filter given by Eqs. (6.20)–(6.22), is guaranteed to be robustly asymptotically stable with convergence rate h in Eq. (6.29) by the control law in Eqs. (6.25), (6.26) for all the line impedance uncertainties given by Eqs. (6.52), (6.53). One can obtain the controller gains by solving the LMI in Eq. (6.27) with an equality constraint in Eq. (6.28) with the uncertainty matrices given in Eq. (6.30), where v is a constant given in Eq. (6.32) and μ is another constant given in Eqs. (6.54), (6.55);*

$$-\bar{\lambda}_R \leq \Delta r_g \leq \bar{\lambda}_R, \quad -\bar{\lambda}_L \leq \Delta l_g \leq \bar{\lambda}_L, \tag{6.52}$$

where

$$\overline{\lambda}_L = |\lambda_L/\alpha| \quad and \quad \overline{\lambda}_R = \frac{\lambda_R}{|\beta|\sqrt{1-\alpha^2}} + \frac{r_{g0}\lambda_L}{l_{g0}} \left(\frac{1}{|\beta|\sqrt{1-\alpha^2}} - \frac{1}{|\alpha|} \right),$$

$$(6.53)$$

$$\mu = \alpha\overline{\mu}, \quad \overline{\mu} = \frac{l_{g0}^2}{\lambda_L}, \quad \underline{\alpha} \le |\alpha| \le 1, \tag{6.54}$$

where

$$\underline{\alpha} = \frac{\beta^2}{\left(1 - \frac{\lambda_R l_{g0}}{\lambda_L r_{g0}}\right)^2 + \beta^2}. \tag{6.55}$$

Proof. The conditions required for the validity of Theorem 6.1 can be used to provide the proof for this theorem. From Eqs. (6.41) to (6.43),

$$P_3 = 1 - \alpha^2 \frac{\Delta l_g^2}{\lambda_L^2} \ge 0, \tag{6.56}$$

which gives

$$|\Delta l_g| \le |\lambda_L/\alpha| = \overline{\lambda}_L. \tag{6.57}$$

This means as long as the magnitude of variation of load inductance is less than or equal to $\overline{\lambda}_L$, the conditions in Eq. (6.37) and Theorem 6.1 remains valid. In other words, if Eq. (6.57) holds, the condition on the uncertainty radius of the system is satisfied (i.e., $\mathbf{\Lambda}^T\mathbf{\Lambda} \le \mathbf{I}$). Thus $\overline{\lambda}_L$ is the upper bound for the uncertainty in line inductance, which will be greater than or equal to λ_L since $\underline{\alpha} \le |\alpha| \le 1$. In a similar manner, a condition for variation in line resistance can be derived. From Eq. (6.39), for the matrix to be positive semidefinite, the diagonal elements must be positive semidefinite. Therefore

$$1 - \delta_2^2 \ge 0, \tag{6.58}$$

$$1 - \nu^2 \left(\frac{l_{g0}\Delta r_g - r_{g0}\Delta l_g}{l_{g0}^2} \right)^2 \ge 0. \tag{6.59}$$

The condition in Eq. (6.59) should hold for all possible Δr_g and Δl_g. The worst case is when the negative term is maximum. This occurs when the difference $l_{g0}\Delta r_g - r_{g0}\Delta l_g$ is maximum. For this, Eq. (6.57) implies in Eq. (6.59) that

$$|v| \left| l_{g0} \Delta r_g + r_{g0} \frac{\lambda_L}{|\alpha|} \right| \le l_{g0}^2, \tag{6.60}$$

$$\left| l_{g0} \Delta r_g + r_{g0} \frac{\lambda_L}{|\alpha|} \right| \le \left| l_{g0} \Delta r_g \right| + \left| r_{g0} \frac{\lambda_L}{|\alpha|} \right| \le \frac{l_{g0}^2}{|v|}, \tag{6.61}$$

$$|\Delta r_g| \le \frac{l_{g0}}{|v|} - \frac{r_{g0} \lambda_L}{l_{g0} |\alpha|}. \tag{6.62}$$

Substituting v from Eq. (6.32), we obtain

$$|\Delta r_g| \le \frac{l_{g0} \lambda_R + r_{g0} \lambda_L}{l_{g0} |\beta| \sqrt{1-\alpha^2}} - \frac{r_{g0} \lambda_L}{l_{g0} |\alpha|} \tag{6.63}$$

or

$$|\Delta r_g| \le \frac{\lambda_R}{|\beta| \sqrt{1-\alpha^2}} + \frac{r_{g0} \lambda_L}{l_{g0}} \left(\frac{1}{|\beta| \sqrt{1-\alpha^2}} - \frac{1}{|\alpha|} \right) = \bar{\lambda}_R, \tag{6.64}$$

where $\bar{\lambda}_R$ is the upper bound on the line resistance uncertainty. Since $|\alpha| \le 1$, $\bar{\lambda}_R$ tends to be negative as α becomes small. Thus a lower bound on α must be enforced for $\bar{\lambda}_R$ to remain nonnegative, which is given in Eq. (6.55). This condition allows the variation in the magnitude of the uncertainty of line resistance up to $\bar{\lambda}_R$ for robust stability with a guaranteed convergence rate. An interesting thing to observe from Eqs. (6.57), (6.64) is that as α decreases, $\bar{\lambda}_L$ increases, while $\bar{\lambda}_R$ decreases, and vice versa. Thus α acts as a tuning parameter for the controller to provide a trade-off between allowable variations in line resistance or inductance. For β, $\bar{\lambda}_R$ is inversely related to it, and thus smaller values will yield bigger allowable variations in the line resistance. This completes the proof. $\square$

Theorem 6.2 can be used to design a robust controller with a guaranteed convergence rate for a certain range of uncertainty, although this uncertainty range is conservative because of the initial assumption that the deviations in the uncertain parameters from their nominal values are much smaller than the nominal values themselves. The system will be robustly stabilized for a larger uncertainty range. This is acceptable because the objective of the method is to design an enhanced performance robust controller, which is achieved by the derived theorems.

3.2. Integral Control for Set Point Tracking

To control the output power of the inverter, a tracking controller for the output current is designed. The output real and reactive powers in terms of the dq components are given as follows:

$$P_O(t) = \frac{3}{2} \left(v_{gd}(t) i_{ld}(t) + v_{gq}(t) i_{lq}(t) \right), \tag{6.65}$$

$$Q_O(t) = \frac{3}{2} \left(-v_{gd}(t) i_{lq}(t) + v_{gq}(t) i_{ld}(t) \right). \tag{6.66}$$

The grid voltage is known and is assumed to be $[v_{gd}(t)\, v_{gq}(t)] = [V_{GD}\, 0]$, where V_{GD} is the peak level of the three-phase grid voltage. The reference currents can then be calculated as

$$i_{ldref}(t) = \frac{2}{3 V_{GD}} P_{ref}(t), \tag{6.67}$$

$$i_{lqref}(t) = -\frac{2}{3 V_{GD}} Q_{ref}(t), \tag{6.68}$$

where $P_{ref}(t)$ and $Q_{ref}(t)$ are the reference real and reactive powers. One can design the reference tracking controller for the output current of the inverter by augmenting the stabilizing control law with integral control. With $\mathbf{y}(t) = [i_{ld}(t)\, i_{lq}(t)]$ and $\mathbf{r}(t) = [i_{ldref}(t)\, i_{lqref}(t)]$, the integrator states are defined as follows:

$$\dot{\mathbf{v}} = \mathbf{e} = \mathbf{r} - \mathbf{y} = \mathbf{r} - \mathbf{C}\mathbf{x}, \tag{6.69}$$

with

$$\mathbf{v} = \int_0^t \mathbf{e}\, d\tau. \tag{6.70}$$

If Eq. (6.71) holds, the controller for reference tracking is obtained by our modifying Eq. (6.26) as follows:

$$rank \begin{bmatrix} \mathbf{A} & \mathbf{C} \\ \mathbf{B} & \mathbf{0} \end{bmatrix} = n + p, \tag{6.71}$$

$$\mathbf{u} = -[\mathbf{K}\ \ \mathbf{K_I}] \begin{bmatrix} \hat{\mathbf{x}} \\ \mathbf{v} \end{bmatrix} = -\mathbf{K}\hat{\mathbf{x}} - \mathbf{K_I} \int_0^t \mathbf{e}\, d\tau. \tag{6.72}$$

Remark 6.2. *The integral controller gain K_I can be designed with any intelligent technique. The objective function to find the optimum controller gains will be based on the performance of reference tracking, which is the main function of the integral controller.*

Remark 6.3. *With the augmented control law in Eq. (6.72), the robustness of the closed-loop system needs revisiting. A numerical approach is undertaken for this analysis. From Eqs. (6.52), (6.53), which give the range of variation of the grid input impedance, the following are the extrema of the uncertain parameters:*

Table 6.1 A matrices for all combinations of uncertainty extrema

A	A_0	A_1	A_2	A_3	A_4	A_5	A_6	A_7	A_8
r_g	r_{g0}	r_{gm}	r_{gM}	r_{g0}	r_{gm}	r_{gM}	r_{g0}	r_{gm}	r_{gM}
l_g	l_{g0}	l_{g0}	l_{g0}	l_{gm}	l_{gm}	l_{gm}	l_{gM}	l_{gM}	l_{gM}

$$r_{gm} = r_{g0} - \overline{\lambda}_R \leq r_g \leq r_{g0} + \overline{\lambda}_R = r_{gM}, \quad l_{gm} = l_{g0} - \overline{\lambda}_L \leq l_g \leq l_{g0} + \overline{\lambda}_L = l_{gM}. \tag{6.73}$$

The plant matrix A changes with the parameter variations. Table 6.1 lists all the possible combinations of extrema of r_g and l_g and the corresponding A_i matrices.

Use of the controller of Eq. (6.72) in the system in Eq. (6.20) results in the following closed-loop $\mathbf{A_{cli}}$ matrix:

$$\mathbf{A_{cli}} = \begin{bmatrix} \mathbf{A_i} - \mathbf{BK} & -\mathbf{BK_i} \\ -\mathbf{C} & \mathbf{0} \end{bmatrix}. \tag{6.74}$$

The states of the augmented system are $x_{cl}^T = [x^T \ v^T]$. The robustness of the system with the augmented controller of Eq. (6.72) is guaranteed if

$$\mathbf{A_{cli}^T P} + \mathbf{P^T A_{cli}} \leq \mathbf{0} \tag{6.75}$$

for all $i \in [0\,8]$. This test can be performed to ensure the robustness of the closed-loop system after the integral controller has been designed. If the test is not successful, the integral controller gain $\mathbf{K_I}$ is redesigned.

In addition to achieving reference tracking, another important feature of the integral controller is step disturbance rejection. As discussed at the beginning of Section 3, the system faces a step disturbance signal $\overline{\mathbf{d}}(t)$. The designed integral controller will be sufficient to reject this disturbance and provide step reference tracking.

3.3. Controller Optimization

The controller for the grid-connected VSI can be designed with use of the control design scheme presented in the previous section. However, the arbitrary design parameters α and β require optimization to achieve optimum performance of the stabilizing controller in terms of higher convergence rate and uncertainty bounds. Similarly, the integral controller gain also needs to be designed for optimum tracking performance. Thus the controller design requires optimization for both the stabilizing and the tracking controller design. The intelligent technique used for optimization

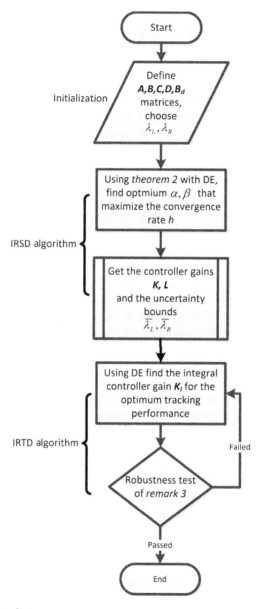

Fig. 6.2 Controller design process.

is differential evolution. Overall, the design process can be divided into three phases as shown in Fig. 6.2. In the first phase the system constants and initialization parameters of differential evolution are defined. The second phase is referred to as the *intelligent robust stabilizer design* (IRSD), where α and β are optimized and robust stabilizing controller gains are obtained for maximized convergence rate. The third phase is for the design of the integral controller gains for optimum tracking performance. This phase is called the *intelligent robust tracker design* (IRTD).

3.4. Controller Implementation

The implementation scheme of the proposed controller is presented in Fig. 6.3. The power system components are connected as described in Fig. 6.1. Measurements are required for only two quantities, one of which is the load current for the controller and other one is the grid voltage which is used by the phase-locked loop to provide the angle $\rho(t)$ for dq transformation. With all the measurements taken in per unit and $V_{base} = V_{DC}/2$, the control signals obtained from Eq. (6.72) act as the dq components of the three-phase modulation signals [32], where V_{DC} is the voltage of the DC source:

$$m_d(t) = \frac{V_{DC}}{2}v_d(t), \tag{6.76}$$

$$m_q(t) = \frac{V_{DC}}{2}v_q(t). \tag{6.77}$$

These signals are transformed to abc signals and given to the pulse width modulation generator, where they are compared with the triangular waves to generate pulses for the control of the inverter bridge.

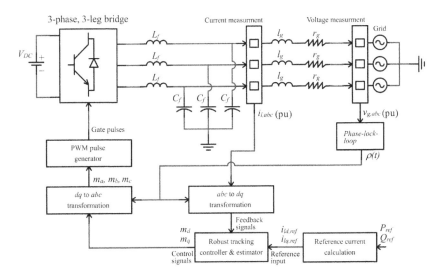

Fig. 6.3 Controller implementation for the grid-connected voltage source inverter.

4. RESULTS AND ANALYSIS

This section comprises the application of the proposed method for the control design of the grid-connected VSI system with certain specifications. First the details of the control design process will be illustrated. Then the performance of the controller will be evaluated under different scenarios and tests.

4.1. Controller Design

The controller is designed according to the process described before. The robust stabilizing controller is designed by the IRSD algorithm for the grid-connected case using Theorem 6.2. The freedom in the choice of α and β is used to maximize the convergence rate of the system by means the intelligent technique of differential evolution. The cost function is defined as the inverse of the convergence rate. To solve the LMIs, the MATLAB LMI Toolbox and YALMIP [35] are used. Starting with the IRSD results, we defined the system constants according to Table 6.2. The initial uncertainty bounds chosen are also shown in Table 6.2 The algorithm was run several times with various random seeds. The cost function minimization plot for the best run is given in Fig. 6.4. The controller gains and other results are provided in Table 6.3.

The next phase is the IRTD, where the tracking controller is designed with use of differential evolution again for optimum tracking performance. The cost function is defined as the integral of time multiplied by the square

Table 6.2 System constants for the grid-connected voltage source inverter

Frequency	f	60 Hz
DC voltage	V_{DC}	600 V
Filter capacitance	C_f	75 μF
Filter inductance	L_f	0.8 mH
Nominal line resistance	R_0	0.4 Ω
Nominal line inductance	L_0	1 μH
Chosen line resistance uncertainty	λ_R	80 mΩ
Chosen line inductance uncertainty	λ_L	0.02 μH
Grid voltage	$[V_{GD} \quad V_{GQ}]$	[220 0] V
Base voltage (phase neutral)	V_B	220 V
Base three-phase power	S_B	100 kVA
Carrier frequency	f_C	12 kHz

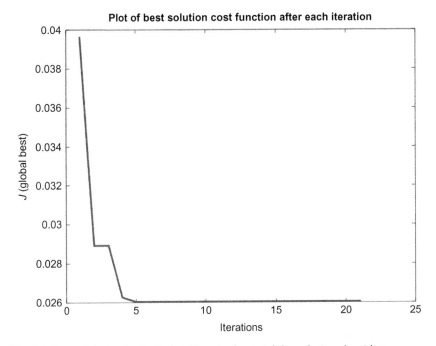

Fig. 6.4 Cost minimization in the intelligent robust stabilizer design algorithm.

Table 6.3 Results of intelligent robust stabilizer design algorithm

Parameters	Symbol	Value
Tuning parameters	$[\alpha\ \beta]$	$[0.236\ 0.99]$
Lower bound of α	$\underline{\alpha}$	0.0904
Stabilizing controller gain	$\mathbf{K}$	$\begin{bmatrix} 213 & 0 & 2.1 & 0 & -8 & 0 \\ 0 & 213 & 0 & 2.1 & 0 & -8 \end{bmatrix} \times 10^{-3}$
Observer gain	$\mathbf{L}$	$\begin{bmatrix} 3.1 & 0 & -26.4 & 0 & -5705 & 0 \\ 0 & 3.1 & 0 & -26.4 & 0 & -5705 \end{bmatrix}$
Convergence rate	h	38.413
Line impedance uncertainty bounds	$[\bar{\lambda}_R\quad \bar{\lambda}_L]$	$[57.6\,\mathrm{m}\Omega\quad 0.085\,\mu\mathrm{H}]$

of the error. The plot of the cost function minimized during the iterations of the algorithm is shown in Fig. 6.5. The designed tracking controller also successfully passes the robustness test, which is performed as mentioned in Remark 6.3. This establishes guaranteed robustness of controller for

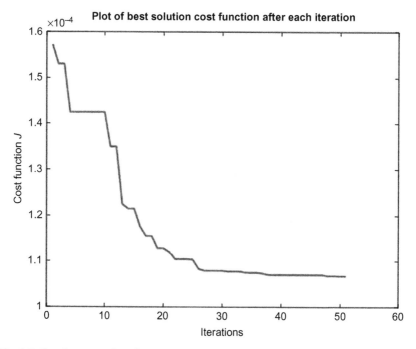

Fig. 6.5 Cost function plot of intelligent robust tracker design algorithm.

the given uncertainty. The resulting controller gains are provided in Eq. (6.78):

$$\mathbf{K_I} = \begin{bmatrix} -1055.6 & 6081.8 \\ -42 & -8473.6 \end{bmatrix}. \tag{6.78}$$

The performance of the designed controller is tested in three cases and a different property is analyzed in each case. The tests and their specifications are given in Table 6.4. The first case is the robustness test, where the performance is observed in the presence of small variations in line resistance and inductance. In the second case the performance of the system is observed for larger disturbances, when a relatively large change occurs in the line parameters. The third case is dedicated to the tracking performance of the system. Different reference inputs for real and reactive power are provided and the results are analyzed. For all cases the system starts at an operating condition of real power output of 20 kW and unity power factor.

Table 6.4 Simulation tests variations

Robustness test

Time (ms)	0	5	15	25	35
r_g (Ω)	0.4	0.45	0.45	0.4	0.4
l_g (μH)	1	1	1.08	1.08	1

Disturbance test

Time (ms)	0	5	25	45	65
r_g (Ω)	0.4	0.6	0.6	0.4	0.4
l_g (μH)	1	1	500	500	1

Tracking performance test

Time (ms)	0	5	25	45	65
P_{ref} (kW)	20	20	15	15	20
Q_{ref} (kVAR)	0	1	1	0	0

4.2. Robustness Test

The purpose of this test is to see the effect of small line impedance variations on the performance of the system. The variations in the grid line impedance are as given in Table 6.4. The changes in the resistance and inductance values are according to the results of the IRSD algorithm, for which robustness is guaranteed with the convergence rate obtained. Thus excellent performance of the system is expected in this test. As seen in Fig. 6.6, the power output of the system is regulated at the desired value during the test. The magnitude of the deviation is very small and the settling time is less than 5 ms. A plot of the current injected into the grid is shown in Fig. 6.7. The robustness of the system is clear from the three-phase sinusoidal waves of the current signal as a very small spike is observed at the time the uncertainty acts. A very small deviation can be observed in i_{ld} as well and it returns to the desired value in less than 5 ms.

4.3. Disturbance Test

In this test large disturbances are applied to the system by means of big changes in the line impedance parameters as given in Table 6.4. The purpose is to see whether the system can maintain stability in the event of such huge variations. As the results in Fig. 6.8 show, the system remains stable in the face of the disturbances and the power output is regulated at the reference value, although the deviations are large, which is expected. The settling time for each deviation is still impressive, around 5 ms. The same can be inferred from the plot of the current signal in Fig. 6.9.

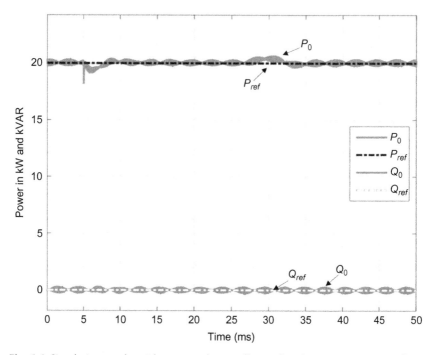

Fig. 6.6 Simulation results with proposed controller: real and reactive power in robustness test.

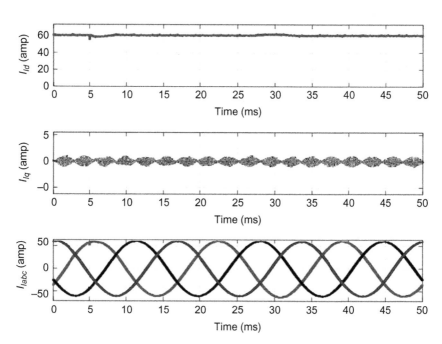

Fig. 6.7 Simulation results with the proposed controller: load current in the robustness test.

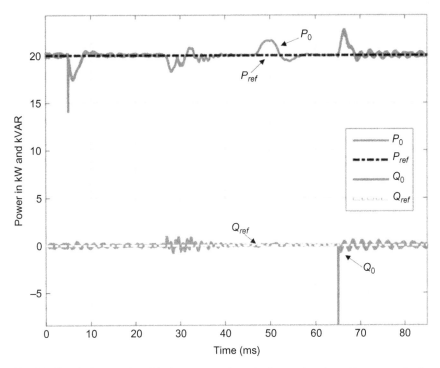

Fig. 6.8 Simulation results with the proposed controller: real and reactive power in the disturbance test.

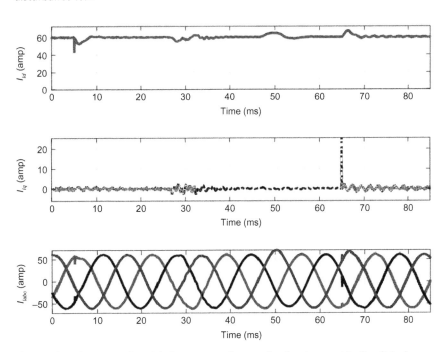

Fig. 6.9 Simulation results with the proposed controller: load current in the disturbance test.

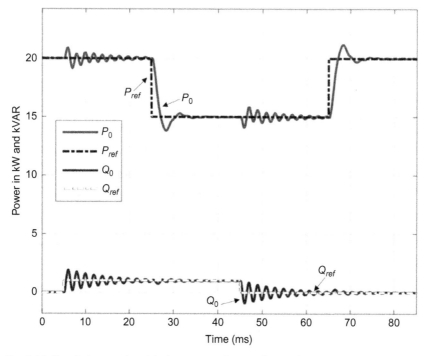

Fig. 6.10 Simulation results with the proposed controller: real and reactive power in the tracking performance test.

4.4. Tracking Performance Test

In this case the output real power and reactive power are given several reference inputs to follow to observe the tracking performance of the system. Fig. 6.10 shows the result. The real power and reactive power of system are successfully following the reference inputs, which are according to Table 6.4. A few oscillations are observed in the response but they settle quickly. The output current behavior is shown in Fig. 6.11, which shows transitions between the reference levels with some oscillations, but settling quickly in a period of about 25 ms.

5. CONCLUSION

A new approach for reference tracking robust control design for a grid-connected VSI was proposed in this chapter. The method was implemented and tested on the VSI system with simulations. The results obtained are

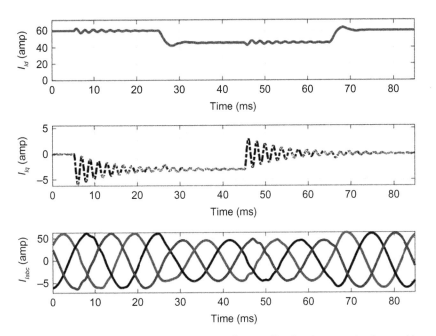

Fig. 6.11 Simulation results with the proposed controller: load current in the tracking performance test.

satisfactory. As demonstrated by the robustness test, the controller was able to robustly stabilize the system with excellent performance in terms of negligible deviation from the steady-state values, if the uncertain parameters stayed in the range of the allowable limit. For greater deviations the performance would deteriorate, which is expected. However, the controller was able to keep the system stable with fairly good performance during large disturbances too as the system converges back to the steady state in a few milliseconds. The tracking controller was also effective as it was able to make the system's output power follow the reference value efficiently. There are several directions to extend this work in the future such as controller design with other robust control theories for norm-bounded uncertainties, application for control of a microgrid, and unified stabilization and tracking controller design.

ACKNOWLEDGMENTS

The authors thank the Deanship of Scientific Research at King Fahd University of Petroleum and Minerals for support through Power Research Group project no. RG 1420. The work reported herein was published in *Control Eng. Pract.* 53 (2016) 92–108, but has been expanded. The permission granted by the journal, as the original source of the publication, to publish this work is hereby acknowledged.

REFERENCES

[1] M. Liserre, T. Sauter, J. Hung, Future energy systems: integrating renewable energy sources into the smart power grid through industrial electronics, IEEE Ind. Electron. Mag. 1 (4) (2010) 18–37.

[2] M. Davari, M.I.Y. Abdel-Rady, Robust multi-objective control of VSC-based DC-voltage power port in hybrid AC/DC multi-terminal microgrids, IEEE Trans. Smart Grid 3 (4) (2013) 1597–1612.

[3] S. Yang, L. Qin, F.Z. Peng, Z. Qian, A robust control scheme for grid-connected voltage-source inverters, IEEE Trans. Ind. Electron. 58 (1) (2011) 202–212.

[4] J.W. He, Y.W. Li, M.S. Munir, A flexible harmonic control approach through voltage-controlled DG-grid interfacing converters, IEEE Trans. Ind. Electron. 59 (1) (2012) 444–455.

[5] R.R. Errabelli, P. Mutschler, Fault-tolerant voltage source inverter for permanent magnet drives, IEEE Trans. Power Electron. 27 (2) (2012) 500–508.

[6] D.-E. Kim, Feedback linearization control of three-phase UPS inverter systems, IEEE Trans. Ind. Electron. 57 (3) (2010) 963–968.

[7] B. Tamyurek, A high-performance SPWM controller for three-phase UPS systems operating under highly nonlinear loads, IEEE Trans. Power Electron. 28 (8) (2013) 3689–3701.

[8] S. Eren, A. Bakhshai, P. Jain, A CLF-based nonlinear control technique for a grid-connected voltage source inverter with LCL filter used in renewable energy power conditioning systems, in: Proceedings of the 4th International Conference on Power Engineering, Energy and Electrical Drives, May, 2013, pp. 840–845.

[9] J.R. Fischer, S.A. Gonzalez, S. Alejandro, M.A. Herran, M.G. Judewicz, D.O. Carrica, Calculation-delay tolerant predictive current controller for three-phase inverters, IEEE Trans. Ind. Inform. 10 (1) (2014) 233–242.

[10] J.W. Jung, N.T. Vu, D.Q. Dang, T.D. Do, Y.S. Choi, H.H. Choi, A three-phase inverter for a standalone distributed generation system: adaptive voltage control design and stability analysis, IEEE Trans. Energy Convers. 29 (1) (2014) 46–56.

[11] D.E. Kim, D.C. Lee, Inverter output voltage control of three-phase UPS systems using feedback linearization, in: Proceedings of the 33rd Annual Conference of the IEEE Industrial Electronics Society (IECON 2007), 2007, pp. 1737–1742.

[12] G.S. Shen, D. Xu, L. Cao, X. Zhu, An improved control strategy for grid-connected voltage source inverters with an LCL filter, IEEE Trans. Power Electron. 23 (4) (2008) 1899–1906.

[13] M.A. Herran, J.R. Fischer, S.A. Gonzalez, M.G. Judewicz, D.O. Carrica, Adaptive dead-time compensation for grid-connected PWM inverters of single-stage PV systems, IEEE Trans. Power Electron. 28 (6) (2013) 2816–2825.

[14] J. Dannehl, M. Liserre, F.W. Fuchs, Filter-based active damping of voltage source converters with LCL filter, IEEE Trans. Ind. Electron. 58 (8) (2011) 3623–3633.

[15] X. Bao, F. Zhuo, Y. Tian, P. Tan, Simplified feedback linearization control of three-phase photovoltaic inverter with an LCL filter, IEEE Trans. Power Electron. 28 (6) (2013) 2739–2752.

[16] S. Jiang, D. Cao, Y. Li, J. Liu, F.Z. Peng, Low THD, fast transient, and cost-effective synchronous-frame repetitive controller for three-phase UPS inverters, in: Proceedings of the IEEE Energy Conversion Congress and Exposition, September, 2011, pp. 2819–2826.

[17] C.N.M. Ho, V.S.P. Cheung, H.S.H. Chung, Constant-frequency hysteresis current control of grid-connected VSI without bandwidth control, in: Proceedings of the IEEE Energy Conversion Congress and Exposition, September, 2009, pp. 2949–2956.

[18] Z. Yao, L. Xiao, Two-switch dual-buck grid-connected inverter with hysteresis current control, IEEE Trans. Power Electron. 27 (7) (2012) 3310–3318.

[19] N. Mendoza, J. Pardo, M. Mantilla, J. Petit, A comparative analysis of direct power control algorithms for three-phase power inverters, in: Proceedings of the 2013 IEEE Power & Energy Society General Meeting, July, 2013, pp. 1–5.

[20] K.J. Lee, B.G. Park, R.Y. Kim, D.S. Hyun, Robust predictive current controller based on a disturbance estimator in a three-phase grid-connected inverter, IEEE Trans. Power Electron. 27 (1) (2012) 276–283.

[21] J.M. Espi, J. Castello, R. Garca-Gil, G. Garcera, E. Figueres, An adaptive robust predictive current control for three-phase grid-connected inverters, IEEE Trans. Ind. Electron. 58 (8) (2011) 3537–3546.

[22] G. Wenming, H. Shuju, X. Honghua, Robust current control design of voltage source converter under unbalanced voltage conditions, in: Proceedings of the IEEE PES Asia-Pacific Power and Energy Engineering Conference (APPEEC), December, 2013, pp. 1–5.

[23] M.A. Hassan, M.A. Abido, Optimal design of microgrids in autonomous and grid-connected modes using particle swarm optimization, IEEE Trans. Power Electron. 26 (3) (2011) 755–769.

[24] J.R. Massing, L.A. Maccari, V.F. Montanger, H. Pinheiro, C. Rech, R.C.L.F. Oliveira, Robust state feedback current controller applied to converters connected to the grid through LCL filters, in: Anais do XIX Congresso Brasileiro de Automatica, CBA, 2012, pp. 1039–1046.

[25] A. Hajizadeh, Robust power control of microgrid based on hybrid renewable power generation systems, Iran. J. Electr. Electron. Eng. 9 (1) (2013) 44–57.

[26] M.S. Mahmoud, S.A. Hussain, Improved resilient feedback stabilization method for uncertain systems, IET Control Theory Appl. 6 (11) (2012) 1654–1660.

[27] H. Kheloufi, A. Zemouche, F. Bedouhene, M. Boutayeb, On LMI conditions to design observer-based controllers for linear systems with parameter uncertainties, Automatica 49 (12) (2013) 3700–3704.

[28] J. Wang, L. Huang, H. Ouyang, Robust output feedback stabilization for uncertain systems, IEE Proc. Control Theory Appl. 150 (5) (2003) 477–482.

[29] C.H. Lien, An efficient method to design robust observer-based control of uncertain linear systems, Appl. Math. Comput. 158 (1) (2004) 29–44.

[30] G. Willmann, D.F. Coutinho, L.F.A. Pereira, F.B. Libano, Multiple-loop H_∞ control design for uninterruptible power supplies, IEEE Trans. Ind. Electron. 54 (3) (2007) 1591–1602.

[31] J.S. Lim, C. Park, J. Han, Y.I. Lee, Robust tracking control of a three-phase DC-AC inverter for UPS applications, IEEE Trans. Ind. Electron. 61 (81) (2014) 4142–4151.

[32] A. Yazdani, R. Iravani, Voltage-Sourced Converters in Power Systems, John Wiley & Sons, Inc., 2010.

[33] L.A. Maccari, C.L. do Amaral Santini, H. Pinheiro, R.C.L.F. de Oliveira, V.F. Montagner, Robust optimal current control for grid-connected three-phase pulse-width modulated converters, IET Power Electron. 8 (8) (2015) 1490–1499.

[34] M. Liserre, R. Teodorescu, F. Blaabjerg, Stability of photovoltaic and wind turbine grid-connected inverters for a large set of grid impedance values, IEEE Trans. Power Electron. 11 (1) (2006) 263–272.

[35] J. Lofberg, YALMIP: a toolbox for modeling and optimization in MATLAB, in: Proceedings of the IEEE International Conference on Robotics and Automation, 2004.

CHAPTER 7

Demand Side Management in Microgrid Control Systems

D. Li*, W.-Y. Chiu†, H. Sun*
*Durham University, Durham, United Kingdom
†Yuan Ze University, Taoyuan, Taiwan

1. INTRODUCTION

Demand-side management (DSM) is a powerful tool that facilitates the process of transforming conventional microgrids into green systems. In this chapter, DSM in microgrid control systems is investigated from various perspectives. First, the history of DSM is briefly presented and basic concepts are introduced. Second, a case study of the California electricity crisis is presented. Next, a detailed classification of demand response is examined. Finally, state-of-the-art technologies for DSM are discussed. Several examples of DSM are then examined. Numerical simulations are provided to illustrate the effectiveness of DSM in microgrid control systems.

2. DEMAND-SIDE MANAGEMENT

Electricity demand always fluctuates dramatically during some short time frames. Generally, to meet the demand, a power system adjusts the supply by increasing/decreasing the generation or adding/curtailing additional resources (e.g., renewable resources and energy storage). The standby generators can incur additional costs and yield system instability, and there may still exist a power shortage during the peak period [1]. Because of severe climate change, carbon dioxide emission reduction is urgent [2]. For these reasons, the idea of DSM has emerged.

Fig. 7.1 gives an example of the annual electricity demand in the United Kingdom. The demand varies by day and season. Winter days require more electricity than summer days. In 2015 the highest daily demand in the United Kingdom was 12.7 TW, while the lowest demand was 7 TW. Fig. 7.2 gives an example of the daily electricity demand in the United Kingdom. Apparently, there are peak hours (i.e., 17:00 to 22:00) and off-peak hours (i.e., 0:00 to 6:00). The highest demand on that day is 50 GW,

Microgrid
http://dx.doi.org/10.1016/B978-0-08-101753-1.00007-3

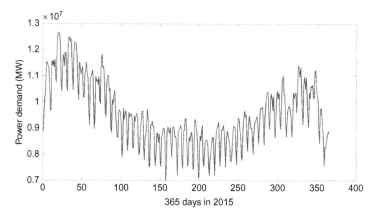

Fig. 7.1 Annual electricity demand in the United Kingdom, 2015. *Data from [3].*

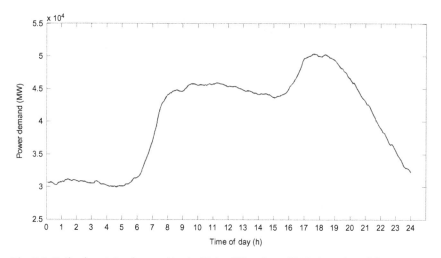

Fig. 7.2 Daily electricity demand in the United Kingdom, 2015. *Data from [3].*

while the lowest demand is 30 GW. Fig. 7.3 gives an example of the share of power generation in the United Kingdom in 2015. It is clear that conventional energy resources still account for a large portion.

2.1. Definition

The term "demand-side management" (DSM), also known as "energy demand management," stands for a variety of activities that are related to energy consumption. It includes not only the control and modification

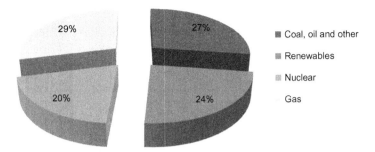

29%   27%   20%   24%

■ Coal, oil and other
■ Renewables
▨ Nuclear
  Gas

Fig. 7.3 The share of power generation in the United Kingdom, 2015. *Data from [4].*

of energy use (e.g., energy conservation, energy efficiency and energy storage) but also the behaviors that are involved in these processes (e.g., device installations, policies and regulation formulation, promotion, and education) [5].

2.2. History

DSM originated from the energy crises [6]. The first energy crisis (also called the "first oil shock") happened in October 1973. During the 1973 Arab-Israeli War, to beat Israel and its allies, the Organization of Arab Petroleum Exporting Countries (OAPEC) announced an oil embargo and export suspension, causing a rise in oil prices. Crude oil prices increased by almost four times from $3 per barrel to nearly $12 per barrel, which caused recessions in Western developed countries [7]. This situation brought energy management into the public consciousness. In response to that, the US Congress legislated the National Energy Act of 1978. As part of it, the National Energy Conservation Policy Act and the Power Plant and the Industrial Fuel Use Act ware enacted, which took DSM into consideration [8].

The second energy crisis in 1979 and the third energy crisis in 1990 sped up the development of DSM. The outbreak of the Iranian revolution and the Iran-Iraq War caused a sharp drop in crude oil production. The crude oil price increased dramatically from about $15 per barrel in 1979 to $39 per barrel in 1981. Then the Gulf War in 1990 stimulated the international market [9]. To deal with this, the Energy Policy Act of 1992 was passed. It addressed the importance of energy efficiency, energy conservation, and energy management, and also prompted the use of renewable energy.

DSM became well known to the public in the 1980s, popularized by the Electric Power Research Institute [10]. The California electricity crisis in

2001 rang alarm bells worldwide, and proved the importance of and need for DSM, especially in the electricity market [11]. Since then, DSM has become a hot issue, attracting more and more attention.

2.3. Advantages

DSM has an important role in power industry development, energy planning, and environmental protection. The introduction of DSM can bring the following advantages to the electricity market:

- It can promote efficient operation of the market and effectively restrain market power.
- It can realize instant information exchange about supply and demand, produce more reasonable and transparent transactions, and speed up and improve the formation of an electricity price mechanism.
- It can effectively relieve demand congestion during peak hours and improve the reliability of the power system.
- It can effectively alleviate investment pressure on power generation, transmission, and distribution.
- It can facilitate the creation of new prospects for the realization of energy conservation and reduction of emissions.

3. RELATED HISTORICAL EVENT: THE CALIFORNIA ELECTRICITY CRISIS

From August 2010 to September 2011, California suffered a serious power crisis. California's electricity market mainly relied on three investor-owned utilities and two municipality-owned utilities. Pacific Gas and Electric (PG&E), Southern California Edison, and San Diego Gas and Electric are private utilities, accounting for 80% of the total supply. The Los Angeles Department of Water and Power and Sacramento Municipal Utility District accounted for the remaining power demand.

3.1. Description of Events

The specific consequences of the crisis are listed below:

- Independent system operators (ISO) frequently issued energy alerts. (The definition of energy alerts is given in Table 7.1.) There were 92 power emergency alarms in 2000, including one stage 3 alarm and 91 stage 1 or stage 2 alarms. In early 2001 the situation became worse. In the first 3 months there were 161 power emergency alarms. The number of stage 3 alarms dramatically increased to 34, and the number of stage 1 or stage 2 alarms increased to 127 [12].

Table 7.1 Stages of emergency alerts

Stage	Description
1	Operating reserves are currently or forecast to be less than the minimum
2	Operating reserves are currently or forecast to be below 5%
3	Operating reserves are currently or forecast to be below 1.5%

- The power system broke down. Rolling blackouts occurred in a large area, which had a strong impact on citizens' daily lives and normal business of industries. The worst case happened on March 19 and 20, 2001. The blackouts affected 1.5 million customers. The crisis caused a total financial loss of about $40 billion to $45 billion [13].

- The wholesale prices and the retail prices dramatically increased. The wholesale price on the California Power Exchange (CalPX) was $29.71 per megawatt hour in December 1999. It reached $376.99 per megawatt hour in December 2000, more than 11 times higher than the level in 1999. The trend was beyond expectation. In sequence, there was an increase in retail prices. In May 2000 the average residential electricity bill in San Diego increased from less than $50 per month to about $100 per month. In August 2000, the average monthly bill increased to $150, causing strong customer dissatisfaction [14].

- The electricity market collapsed. On September 6, 2000, the governor of California signed a decree to limit the retail price to 6.5 cents per kilowatt hour, and decided to freeze the price for three years. (The retail price at that time was 21.4 cents per kilowatt hour.) On January 30, 2001, Power Exchange Corporation announced the suspension of trading, terminating the operation of the current market and day-ahead market. Because of heavy losses, in April 2001, the state's largest utility, PG&E, declared bankruptcy [13].

The California Independent System Operator defines the emergency alerts shown in Table 7.1 [15].

3.2. Analysis of the Reasons

The occurrence of this unprecedented crisis was inevitable but was also inevitable. This crisis had been brewing for years. The reasons for it were varied and complicated. The major causes can be grouped into four factors [16–18]:

1. Rapid economic growth leads to rapid electricity demand growth. In recent years, the Californian economy has become the "engine" of the

US economy. The expansion of high technologies, the increase of job opportunities, the booming of the population, and the development of business in California resulted in an alarming rate of electricity demand growth.

2. Green environmental protection hindered the construction of power plants. To prevent air pollution, old thermal power plants were closed, but new power plants could not be built because of environmental protection restrictions. In recent years, California had not built any new plants. The development of power generation could not keep pace with the demand.

3. The new regulations and market reform were shortsighted. The aim of these processes was to promote competition and reduce the electricity price. The deregulation of the energy control policy removed the cap on the wholesale price. Utilities could trade only on the day-ahead power market (CalPX), and they were not allowed to have a long-term contract with power suppliers. These factors impeded access of new providers to the market and stunted business growth of the three major utilities.

4. Primary energy prices rose. California's power generation mainly relied on traditional resources. In 2000 the prices of fossil oil and natural gas rose suddenly and sharply. This increased the wholesale electricity price but at the same time the retail price remained the same. The three major utilities had no choice but to spend billions of dollars to fill the gap that was caused by a high purchasing price and a low selling price. This led to heavy deficits.

5. Imports from other sources were locked. Hydroelectric power plants in the northwestern United States gave significant support to northern California. But the unusually low water levels caused a reduction in power generation, reducing the surplus to California. And the congestion of high-voltage transmission line path 15 (from southern California to northern California) also contributed to the import shortages.

3.3. Remarks

This serious electricity crisis is unprecedented in history. It caused billions of dollars of financial losses and had a significant effect on California's development. This crisis attracted attention for various reasons. It created the idea of DSM. During this crisis, many approaches to DSM were used, especially the demand response. The next section explains the demand response technologies and divides them into several categories [19].

4. DEMAND RESPONSE

4.1. Definition

"Demand response" mainly refers to the actions taken on the customer side that use the market price to influence the level and time of electricity demand.

According to the Federal Energy Regulatory Commission [20], demand response is:

Changes in electric usage by end-use customers from their normal consumption patterns in response to changes in the price of electricity over time, or to incentive payments designed to induce lower electricity use at times of high wholesale market prices or when system reliability is jeopardized.

In general, the introduction of demand response into the power market requires a precondition: the electricity market must achieve tentative liberalization or full liberalization, which means some kind of real-time market prices and effective market mechanism mist exist in the electricity market. Meanwhile, demand response will accelerate the formation of the real-time market pricing mechanism. And with the high penetration of demand response into the market, it can provide economic incentives to promote other projects such as energy efficiency and energy storage in DSM. But DSM does not need this mechanism. Even without it, DSM can realize some of its projects. At the same time, DSM can fully boost and amplify the economic effectiveness of demand response.

The term "negawatt power" is a derivative term from DSM and demand response. It is coined for the way of supplying additional electrical energy to consumers without adding generation capacity. Even though it is a theoretical unit for electrical energy, it can still be traded as a commodity in the electricity market [21].

4.2. Services Category

Typically, demand response can provide five services to the system: (1) peak clipping, (2) valley filling, (3) load shifting, (4) strategic conservation, and (5) strategic load growth [22, 23]. The first three can be grouped as load management, and the last two can be grouped as load-shape change. Load management is normally related to deliberate behaviors enforced by utilities. In contrast, the load-shape change can be both natural behaviors of customers and deliberate behaviors enforced by utilities [24].

4.2.1. Peak Clipping

When the demand approaches the threshold of the supply capacity or the transmission system approaches the threshold of the thermal requirements, this peak load demand must be reduced. This can be realized by the direct load control in the residential sector (e.g., turning down the thermostat of heaters and increasing the temperature of refrigerators). This can also be achieved by interruption in the industrial and commercial sectors. Fig. 7.4 shows a peak reduction from 12 MW to 10 MW during the period from 18:00 to 20:00. This service can help to release the stress on the system during the peak period. However, because it curtails the consumption of certain loads, it can cause customer dissatisfaction.

4.2.2. Valley Filling

When the demand is manifestly low at off-peak time, which is also not favorable for system stability, the demand should be increased. The commonest method is to add storage devices (e.g., thermal storage for heaters and plug-in electrical vehicles). Fig. 7.5 shows a valley filling from 4 MW to 6 MW during the period from 2:00 to 6:00. This service increases the total power consumption of customers but may not necessarily increase the bill.

4.2.3. Load Shifting

When the load is apparently higher than the average level in a certain period, a certain amount of load must be moved from that period to other periods. This primarily relies on the deferrable appliances, which can justify the time of use (e.g., washing machines). In the short term, load shifting can be achieved on a daily basis from peak time to off-peak time. In the long term, load shifting can be achieved on a seasonal basis. Fig. 7.6 shows a daily load shifting in which part of the peak demand is shifted from 18:00–20:00

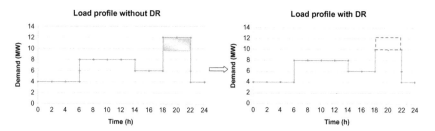

Fig. 7.4 Peak clipping.

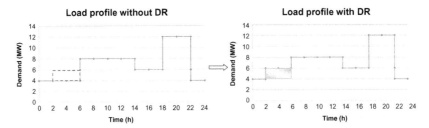

Fig. 7.5 Valley filling.

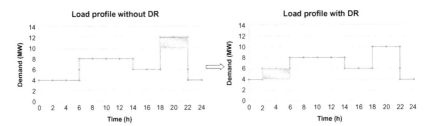

Fig. 7.6 Load shifting.

to 2:00–6:00. It does not reduce the total consumption but only changes the time of use. Therefore this service does not cause customers substantial inconvenience.

4.2.4. Strategic Conservation

When the overall load exceeds the supply level, customers are encouraged to reduce their overall consumption. One basic method is to improve energy efficiency. This can be done on a small scale by the replacement of traditional devices with energy-efficient devices (e.g., changing filament lamps to fluorescent lamps). It also can be done on a large scale (e.g., weatherization program). Besides the technical improvements, the information support is also important. In general, providing consumption and cost details to customers can facilitate power reduction. Fig. 7.7 shows strategic conservation from a high power level to a low level.

4.2.5. Strategic Load Growth

When the demand falls below the normal level of supply, customers are encouraged to increase their overall consumption. The electrification technology has the potential to achieve this (e.g., the popularization of

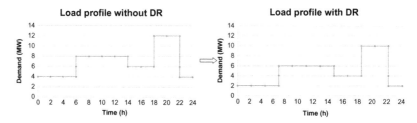

Fig. 7.7 Strategic conservation.

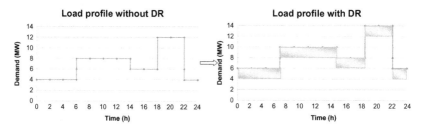

Fig. 7.8 Strategic load growth.

electrical vehicles). Fig. 7.8 shows strategic load growth from a low power level to a high level.

4.3. Customers Category

Demand response is primarily focused on the consumer side. Detailed analysis of customers can facilitate the understanding and design of demand response. Generally, customers can be classified into four sectors [25]:

1. industrial sector
2. residential sector
3. commercial sector
4. transportation sector

Fig. 7.9 shows the portion of electricity consumption for each sector in the United Kingdom in 2014. As for demand response, the industrial, residential, and commercial sectors are mainly concerned.

4.3.1. Residential Sector

The usage patterns in the residential sector are more complicated than in the other sectors. Firstly, the number of customers is much higher. The distribution of customers is wide and scattered. Secondly, the types of appliances used by customers are diverse. Even for the same type of

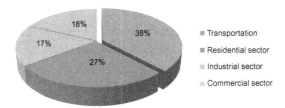

Fig. 7.9 The market share by sector in the United Kingdom in 2014. *Data from [25].*

appliances, power consumption of different brands can differ. Thirdly, every customer has his or her own personal usage preference. That means each customer needs to be treated specifically rather than identically [26].

Customers can be divided into five types [27]:

1. Long-range customers: their elasticity of electricity use is relatively high. They are able to modify their use over a wide time range.
2. Real-world postponing customers: they consider the current and future electricity prices, and give certain responses to utilities,
3. Real-world advancing customers: they focus on the past and future electricity prices, and also give certain responses to utilities,
4. Real-world mixed customers: they are a combination of both postponing customers and advancing customers.
5. Short-range customers: they pay attention only to the current electricity price. They are not willing to change their consumption pattern.

4.3.2. Industrial Sector

This has high electricity consumption, especially at a high voltage level. In addition, the peak load is significant. However, the adaption of demand response in this sector is challenging [28]. Firstly, information on the usage pattern and the operation of appliances is confidential. To some extent, it can reflect the manufacturing process, which is classified in a few industries. Therefore access to this information is limited. Secondly, even if there is sufficient information, the modification of electricity use is still difficult because many procedures are time sensitive. They require a precise order and duration, which means they are less likely to be shifted. In this situation, a proper choice for industries is to improve energy efficiency.

4.3.3. Commercial Sector

The usage pattern in the commercial sector is quite typical and identical. The common and main loads for commercial customers come from the use of heating, ventilation, and air conditioning systems and lighting systems. The modification of these systems is relatively easy. Firstly, in general, these systems are autonomously controlled according to the preset requirements. This makes the systems able to respond quickly to the demand response signals. Secondly, the effects of external factors (e.g., temperature, humidity, and illumination) on these systems are predictable. For example, a lighting system consumes more electricity in winter than in summer [29].

4.3.4. Summary

Among these three sectors it is easier for the commercial and industrial sectors to realize demand response programs. Commercial and industrial customers are distributed regionally and intensively, and the power consumption of these customers is relatively high. What is more, the appliances and control systems for these customers are more advanced. In addition, in an emergency, most of these customers are equipped with backup on-site generators. These appliances can also be used as auxiliary facilities of the demand response programs [30]. Fig. 7.10 shows the potential peak load reduction in three sectors. The commercial and industrial sectors had noticeable growth from 14,362 MW in 2006 to 28,088 MW in 2012. In contrast, the residential sector had slight growth between these years.

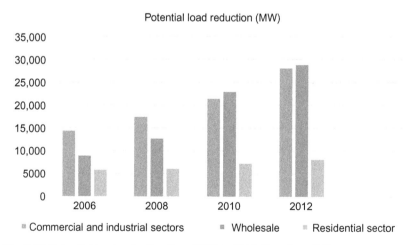

Fig. 7.10 Potential load reduction from 2006 to 2012. *Data from [31].*

4.4. Loads Category

On the basis of the operational characteristics of appliances, the loads can be classified by two standards:

- whether the occupied time duration of appliances can be modified or not;
- whether the total electricity consumption of appliances can be modified or not.

For the first standard, loads can be divided into deferrable loads and nondeferrable loads. For the second standard, loads can be divided into adjustable loads and nonadjustable loads [32].

4.4.1. Deferrable Loads and Nondeferrable Loads

The activation of deferrable loads can be stopped, restarted, or shifted to other time slots (e.g., washing machine and electrical vehicles). Generally, most of the wet loads belong to the deferrable loads. These loads can be scheduled by a demand response program. On the basis of the electricity price or the monetary incentive, they can be shifted from peak hour to off-peak hours, therefore reducing the peak load demand [27]. The modification of these loads needs to abide by the predefined requirements (e.g., deadlines and operation times). In contrast, the nondeferrable loads need to finish the schedule in a specified time, (e.g., lighting systems and kitchen systems). These loads do not allow a time shift and interruption. As such, these loads are not suitable for the demand response program.

4.4.2. Adjustable Loads and Nonadjustable Loads

For the adjustable loads, the consumption can be adjusted to a lower level (e.g., in winter, heaters can be set at 23°C rather than 25°C). Normally, most of the thermal loads are part of the adjustable loads. These loads can be involved in the demand response program. The total consumption can be reduced on the basis of the electricity price or the monetary incentive. However, reducing the consumption can affect customers' comfort as described by the quality of experience (QoE) [33]. QoE refers to the valuation of customers' experiences or degree of satisfaction during a service. When a program is designed, this QoE must be taken into consideration to ensure that the demand response program is executable theoretically and practically [34]. In contrast, for the nonadjustable loads, the total consumption is fixed (e.g., TVs and computers). Same like nondeferrable loads, nonadjustable loads cannot be scheduled by a demand response program.

4.5. Approaches Category

There are a number of motivation methods that encourage customers to participate in a demand response program. These methods can be divided into two groups: time-based demand response and incentive-based demand response. In general, time-based demand response is suitable for the residential sector, and incentive-based demand response is suitable for the industrial and commercial sectors [35, 36].

4.5.1. Incentive-Based Demand Response

In these methods, incentives are offered to customers depending on their behavior in the demand response programs. Normally, customers change their consumption voluntarily. However, in some cases, the failure to meet the requirements will result in a penalty for customers. Generally, there are five types of incentive-based demand response [37]:

1. **Direct load control**. According to the advanced agreement between customers and utilities, utilities can remotely control some customers' appliances (e.g., air conditioners and water heaters). The notices for the operation are normally announced a short time ahead. To participate in this method, customers need to be equipped with a remote control switch system so that utilities can reschedule, turn on, or turn off the appliances [38]. Direct load control is primarily applied to the residential sector or small-scale commercial sector. It is not suitable for the industrial sector because the industrial sector needs a precise process.

2. **Interruptible/curtailable service**. Compared with direct load control, this method is normally applied to the industrial sector and large-scale commercial sector. When the system is congested, customers are asked to reduce some loads to a certain level. By participating in this, customers can receive a rate discount or bill discount. However, if customers fail to respond in the predefined time period, they could receive a penalty. In this method, the operation frequency and the duration are limited [39].

3. **Demand bidding**. Instead of being asked by the utilities to take part in demand response programs, customers can make decisions by themselves in this method. On the basis of the generation and demand situation,

utilities announce the total amount of electricity that must be curtailed. Customers can bid for the amount on the basis of their own situation and the wholesale market. Once the bid has been accepted, customers must provide the specified curtailment, otherwise they will receive a penalty [40]. This method is also suitable for large-scale customers. Small-scale customers can be integrated by aggregators and involved as a unit.

4. **Capacity market program**. When the system is short of reserve, customers are required to reduce their predefined consumption. The announcement is normally released one day ahead. These curtailments are treated as system capacity to replace the conventional generation and delivery resources. By proving their ability for curtailment, customers can receive a reservation payment. And by providing the reduction, customers can receive an incentive. In contrast, if they fail to provide it, they could receive a penalty [37].

5. **Ancillary service market**. Similarly to demand bidding, customers also bid for electricity curtailments. These bids are offered to an independent system operator/regional transmission organization [37]. These curtailments are used as operational reservation. If the bid was accepted, customers need to abide by a standby standard. In this situation, they are paid according to the market price. Once the curtailments have been called, customers are paid according to the spot price.

4.5.2. Time-Based Demand Response

In these methods, electricity prices vary according to the cost of generation and the demand for electricity. On the basis of these prices and other information, customers can decide on their consumption. Generally, there are four types of pricing schemes: (1) flat pricing, (2) time-of-use (ToU) pricing, (3) critical peak pricing, and (4) real-time pricing [30]:

- **Flat pricing**: This is the most traditional and widely used pricing scheme. The electricity price is constant all the time. In this situation the only way to reduce the bill is to reduce the total consumption. The prices can be set seasonally. Within a season, the price is fixed. For another season, a different price is used.

- **ToU pricing**: This is an improvement on flat pricing. The prices are different in different time slots. Within each slot, a flat price is applied. Usually the prices are predefined for 1 day [30]. In this scheme, customers tend to shift their demand to a lower-price period. In this way the ability to reduce the total electricity demand is narrowed. For example, in the United Kingdom, an Economy 7 tariff is applied in some areas. It offers a cheap electricity price for the off-peak time, typically the night. This off-peak time lasts 7 hours in total, as the "7" implies, normally from 0:00 to 07:00. The daytime price is higher, around 13–16 pence per kilowatt hour, while the nighttime price is lower, around 5–7 pence per kilowatt hour. This tariff was first introduced in 1978. To apply for the Economy 7 tariff, customers need to be equipped with a particular meter that can show two different readings: one for daytime electricity consumption and the other for nighttime electricity consumption (Fig. 7.11).

- **Critical peak pricing**: This scheme is derived from the ToU pricing scheme. The extreme peak demand period is picked out. During this period a much higher electricity price is announced. This scheme can effectively bring down the peak demand [26]. The critical peak price can be set on the basis of the demand level or the time of the day. Three types of pricing have been considered: fixed-period critical peak pricing, variable-period critical peak pricing, and variable critical peak pricing. For fixed-period critical peak pricing, a specific period during 1 day is selected and a fixed high electricity price is applied on the basis of the experience accumulated. For variable-period critical peak pricing, the application period is not fixed. The utilities can choose to trigger the critical peak pricing on the basis of the predefined criteria. In this situation the operation frequency and duration are limited. For variable critical peak pricing, the period is fixed but the electricity price can vary on the basis of the current demand situation. Three major California

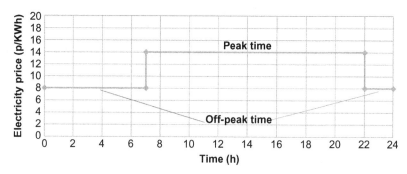

Fig. 7.11 ToU pricing.

utilities, San Diego Gas & Electric, Southern California Edison, and Pacific Gas & Electric in California, have already applied critical peak pricing (Fig. 7.12).

- **Real-time pricing**: The electricity price fluctuates frequently, normally by hours. The change of price can indicate the relationship between supply and demand in the wholesale market [41]. It requires effective two-way communication between utilities and customers. Sometimes, market aggregators also participate in this scheme to deal with data collection and increase the efficiency. Customers are involved mostly in this scheme and are notified of these prices in a day-ahead manner, hour-ahead manner or 15-min-ahead manner. On the basis of the price and their own situation, customers can decide on their consumption pattern. On the basis of the total generation situation, total demand situation, and customers' reactions to the former price, utilities can decide on the prices for the next period. This scheme is more acceptable to the industrial and commercial sectors than to the residential sector. There are two main difficulties for the application of this scheme. Firstly, it relies on continuous real-time data exchange, which is not favorable for customers. Secondly, the large-scale data processing increases the complexity of the whole system. There are some experimental examples conducted in Canada and the United States showing that real-time pricing schemes generally reduce peak demand and provide a flat consumption pattern [30, 35] (Fig. 7.13).

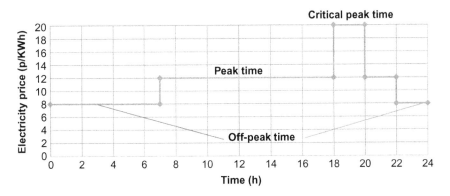

Fig. 7.12 Critical peak pricing.

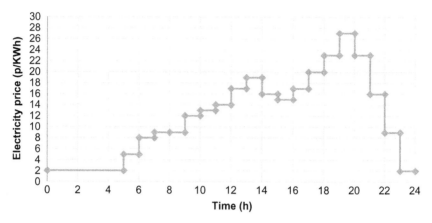

Fig. 7.13 Real-time pricing.

5. DEMAND-SIDE MANAGEMENT METHODS

To effectively design DSM programs, many technologies have been used; for example, a multiobjective optimization method and the linear matrix inequality (LMI) approach. This section gives some examples of DSM programs. The simulation results are provided to verify the effectiveness of the DSM program.

5.1. Multiobjective Optimization Method

This method can take several objectives into consideration and obtain a fair decision that does not favor any specific objective. It has been used in multiple microgrid system design for a market operator and distribution network operator [42]. Three participants are considered: microgrids, a power grid, and an ISO. The relationship among these three participations is shown in Fig. 7.14, in which three microgrids ($N = 3$) and two energy storage systems ($N_s = 2$) are involved. Table 7.2 presents our notation.

For the power grid the objective is to maximize the overall utility: the net gain for providing power to microgrids. The utility function can be defined as

$$\max_{p_{gn}(t),\lambda(t)} U_g(p_g(t), \lambda(t)). \tag{7.1}$$

For microgrids the objective is to maximize the overall utility: the net revenue for consuming power, which is provided by the power grid. The utility function can be defined as:

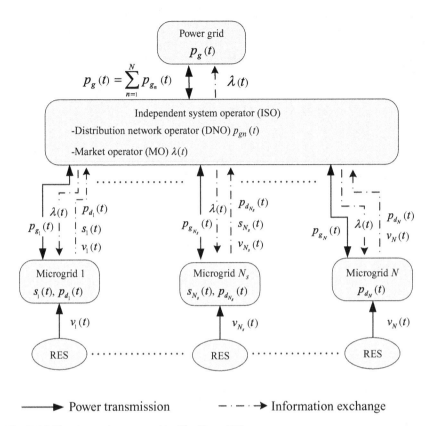

Fig. 7.14 The dynamic system. *Modified from [42].*

Table 7.2 Symbols

Symbol	Description
N	The number of microgrids
t	Time index
$s_n(t)$	Storage level of microgrid n at time t
$\bar{s}_n$	Standard storage level
$p_{g_n}(t)$	Power distribution between the power grid and the nth microgrid at time t
$p_g(t)$	Total power distribution; i.e., $p_g(t) = \sum_{n=1}^{N} p_{g_n}(t)$
$p_{d_n}(t)$	Power demand of the nth microgrid at time t
$\lambda(t)$	Market price at time t
$v_n(t)$	Power supply from renewable energy sources to microgrid n at time t

$$\max_{\lambda(t)} U_d(p_{d_1}(t), \ldots, p_{d_N}(t), \lambda(t)). \tag{7.2}$$

For the ISO, in an emergency situation, the storage must be maintained around a standard level, the closer the better. Therefore the objective is to minimize the sum of differences between the current level and the standard level. The utility function can be defined as

$$\min_{p_{g_n}(t), \lambda(t)} \sum_{n=1}^{N_s} (s_n(t+1) - \bar{s}_n)^2. \tag{7.3}$$

A multiobjective artificial immune system algorithm can be used to find Pareto-optimal solutions to the above problem [43, 44]. The multiobjective artificial immune system algorithm uses the gene operation to maintain diversity. A solution is Pareto dominated if some other solutions can provide better performance for at least one objective without hurting other objectives. First, a group of solutions are generated on the basis of the predefined requirement. Then during the iteration, the dominated solutions are gradually removed, while the nondominated solutions remain. A specific solution that can maximize the minimum improvement in all objectives is selected.

Fig. 7.15 shows the price signal generated from the market operator, which has a significant fluctuation. Fig. 7.16 shows the storage level, which fluctuates in an acceptable range. Fig. 7.17 shows the relationship between the power generation, power demand, price signal, and renewable resources. Power demand is generally larger than the power generation. Renewable resources are used to supply the imbalance between the demand and generation. An increase in use of renewable resources can lead to a decrease of power generation.

This multiobjective optimization method can be applied to other scenarios in which different objectives and constraints are involved. Because this method is not based on a specific model, it represents a framework for searching for Pareto-optimal solutions to multiobjective problems.

5.2. Linear Matrix Inequality Approach

The LMI approach has been used for many situations. Because of the convex property, the associated problem can be solved efficiently [46, 47]. It has been used to design a storage system in smart grid networks [45, 48]. The basic idea is to charge the batteries when the price is lower than the threshold and discharge the batteries when the price is higher than the

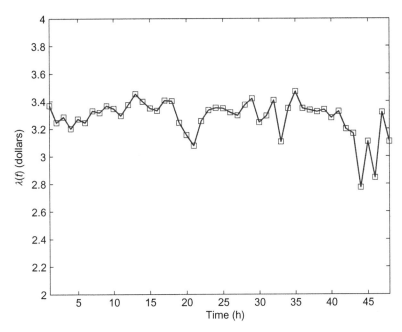

Fig. 7.15 Price signal $\lambda(t)$ generated from the market operator through the use of the multiobjective approach.

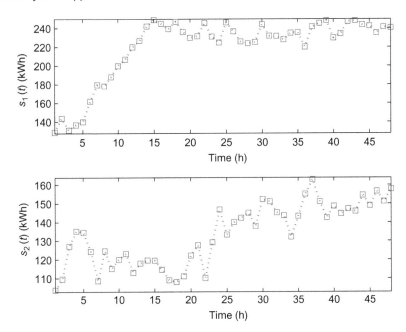

Fig. 7.16 The stored energy levels $s_n(t)$ at the microgrids.

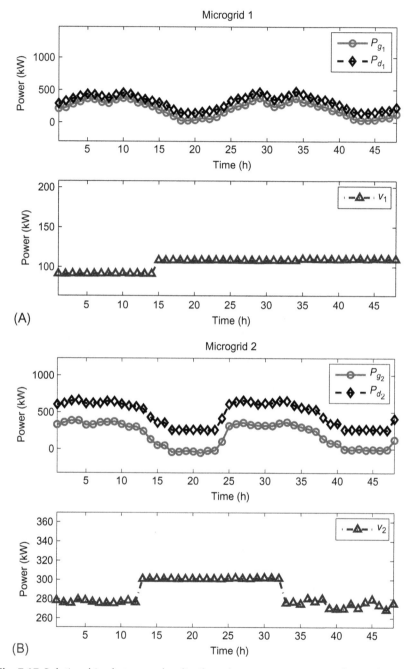

Fig. 7.17 Relationships between the distributed power $p_{g_n}(t)$, power demand $p_{d_n}(t)$, and power input $v_n(t)$ provided by renewable energy sources in (A) microgrid 1, (B) microgrid 2,

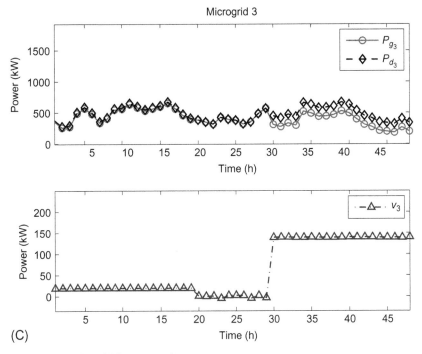

Fig. 7.17, Cont'd and (C) microgrid 3.

threshold. Price signals, system uncertainties, and physical constraints are three important factors that must be considered in this design.

Fig. 7.18 shows an example of the system model. It consists of one conventional power grid and five users. These users are assumed to be smart houses and have storage devices. All users are connected to the power grid so that the power can flow between the grid and all users. Some users are connected to others, and hence the power can also flow among users. This assumption is based on two-way power flow and two-way communication.

Table 7.3 summarizes our notation. The desired storage level $q_r(t)$ is defined as

$$q_r(t + 1) = A_r q_r(t) + r(t). \tag{7.4}$$

The actual storage level $q(t)$ is defined as

$$q(t + 1) = Aq(t) + [Bu(t) - x(t) + v(t)]\Delta t. \tag{7.5}$$

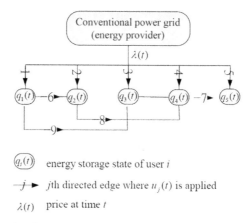

$q_i(t)$ energy storage state of user i

$\xrightarrow{j}$ jth directed edge where $u_j(t)$ is applied

$\lambda(t)$ price at time t

Fig. 7.18 An example of smart grid topology with five users and nine edges. *Modified from [45].*

Table 7.3 Symbols

Symbol	Description
$q_r(t)$	Desired storage level
t	Time index
A	Storage efficiency matrix
$r(t)$	Reference input
$q(t)$	Actual storage level
B	Topology matrix
$u(t)$	Power flow control
$x(t)$	Power demand
$v(t)$	Extra power from renewable sources
$e(t)$	Dynamic error
$\lambda(t)$	Price signal
$u(t)^*$	LMI-based design
$u(t)^c$	Constant-state design
$\Delta\$(t)$	Money saved

LMI, linear matrix inequality.

The system uncertainty $e(t)$ is defined as

$$e(t) = q_r(t) - q(t). \tag{7.6}$$

The money saved is defined as

$$\Delta\$(t) = \$(u(t)^c) - \$(u(t)^*). \tag{7.7}$$

The LMI approach is used to design the power flow control $u(t)$ for the above system. Fig. 7.19 shows that the resulting price signal fluctuates

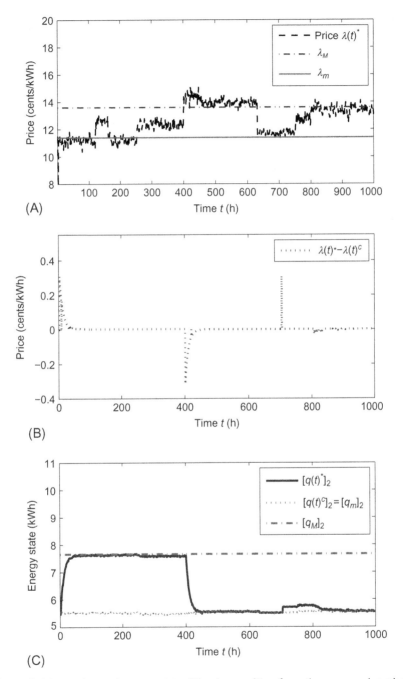

Fig. 7.19 Price and stored energy state: (A) price resulting from the proposed method $u(t)^*$; (B) resultant price differences between the use of $u(t)^*$ and $u(t)^C$; and (C) stored energy states of different energy management systems.

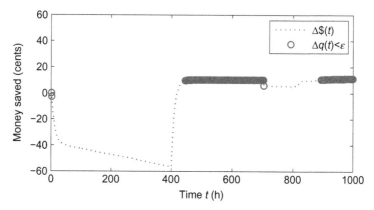

Fig. 7.20 Money saved by using the proposed $u(t)^*$.

around the upper price threshold λ_M and the lower price threshold λ_m. The system charges/discharges batteries when the price is low/high as expected.

Fig. 7.20 shows the money saved with the proposed design. For a fair comparison, the periods where $q(t)^* \approx q(t)^c = q_m$ are mainly concerned. During these periods, the money saved is always positive. As the time increases, the money saved also increases. This means more money can be saved for a longer observation period.

6. CONCLUSIONS

This chapter focused on DSM in the microgrid. The basic concepts and advantages of DSM were examined. The California electricity crisis was then analyzed. The associated events and reasons for this crisis were discussed. Next, the demand response was explained in detail on the basis of the services category, customers category, loads category, and control category. Finally, existing microgrid control technologies were briefly introduced; namely, the multiobjective optimization method and LMI approach.

ACKNOWLEDGMENTS

This work was supported in part by the Ministry of Science and Technology of Taiwan under grant 102-2218-E-155-004-MY3 and in part by the Innovation Center for Big Data and Digital Convergence, Yuan Ze University, Taiwan.

The research leading to these results has received funding from the European Commission's Horizon 2020 Framework Programme (H2020/2014-2020) under grant agreement No. 646470, SmarterEMC2 Project.

REFERENCES

[1] V. Giordano, F. Ganga, G. Fulli, M.S. Jiménez, I. Onyeji, A. Colta, I. Papaioannou, A. Mengolini, C. Alecu, T. Ojala, et al., Smart Grid Projects in Europe: Lessons Learned and Current Developments, Citeseer, 2011.

[2] DECC, The Carbon Plan: Delivering Our Low Carbon Future, Department of Energy & Climate Change, United Kingdom, 2011.

[3] U.K. Gridwatch, National Grid Status, 2015–2016.

[4] DECC, Energy Statistics 2016, Department of Energy & Climate Change, United Kingdom, 2016.

[5] M. Alizadeh, X. Li, Z. Wang, A. Scaglione, R. Melton, Demand-side management in the smart grid: information processing for the power switch, IEEE Signal Process. Mag. 29 (5) (2012) 55–67.

[6] J. Torriti, Peak Energy Demand and Demand Side Response, Routledge, London, 2015.

[7] U.S. Department of State and Office of the Historian, OPEC Oil Embargo 1973–1974, 2012.

[8] J. Richardson, R. Nordhaus, The national energy act of 1978, Nat. Resour. Environ. 10 (1) (1995) 62–68.

[9] U.S. Energy Information Administration, Monthly Energy Review, 2015.

[10] V.S.K.M. Balijepalli, V. Pradhan, S.A. Khaparde, R.M. Shereef, Review of demand response under smart grid paradigm, in: Proceedings of IEEE PES Innovative Smart Grid Technologies-India (ISGT India), 2011, pp. 236–243.

[11] H. Fraser, The importance of an active demand side in the electricity industry, Electr. J. 14 (9) (2001) 52–73.

[12] A.K.N. Reddy, California energy crisis and its lessons for power sector reform in India, Econ. Polit. Wkly. (2001) 1533–1540.

[13] L. Cabral, The California energy crisis, Jpn. World Econ. 14 (3) (2002) 335–339.

[14] P.C. Reiss, M.W. White, Demand and pricing in electricity markets: evidence from San Diego during California's energy crisis, Technical report, National Bureau of Economic Research, 2003.

[15] California Independent System Operator (ISO), System emergency (formerly e-508, e-509), 2015.

[16] J.L. Sweeney, The California electricity crisis: lessons for the future, The Bridge, 2012.

[17] W.W. Clark, T.K. Bradshaw, Agile Energy Systems: Global Lessons from the California Energy Crisis, Elsevier Science, Amsterdam, 2004.

[18] P.T. Duane, Regulation's rationale: learning from the California energy crisis, Yale J. Reg. 19 (2002) 471.

[19] V.L. Smith, S.J. Rassenti, B.J. Wilson, California: energy crisis or market design crisis? 2015.

[20] Federal Energy Regulatory Commission, 2008 assessment of demand response and advanced metering staff report, Federal Energy Regulatory Commission, 2008.

[21] A.B. Lovins, The negawatt revolution, Across Board 27 (9) (1990) 18–23.

[22] C.W. Gellings, The concept of demand-side management for electric utilities, Proc. IEEE 73 (10) (1985) 1468–1470.

[23] C.W. Gellings, J.H. Chamberlin, Demand-Side Management: Concepts and Methods, 1987.

[24] J. Ekanayake, N. Jenkins, K. Liyanage, J. Wu, A. Yokoyama, Smart Grid: Technology and Applications, John Wiley & Sons, New York, 2012.

[25] DECC, Energy Consumption in the UK (2015): Overall Energy Consumption in the UK since 1970, 2015.

[26] K. Herter, Residential implementation of critical-peak pricing of electricity, Energy Policy 35 (4) (2007) 2121–2130.

[27] N. Venkatesan, J. Solanki, S.K. Solanki, Residential demand response model and impact on voltage profile and losses of an electric distribution network, Appl. Energy 96 (2012) 84–91.

[28] M. Paulus, F. Borggrefe, The potential of demand-side management in energy-intensive industries for electricity markets in Germany, Appl. Energy 88 (2) (2011) 432–441.

[29] N. Motegi, M.A. Piette, D.S. Watson, S. Kiliccote, P. Xu, Introduction to commercial building control strategies and techniques for demand response, Lawrence Berkeley National Laboratory LBNL-59975, 2007.

[30] J. Aghaei, M.-I. Alizadeh, Demand response in smart electricity grids equipped with renewable energy sources: a review, Renew. Sustain. Energy Rev. 18 (2013) 64–72.

[31] Federal Energy Regulatory Commission, 2012 assessment of demand response and advanced metering staff report, Federal Energy Regulatory Commission, 2012.

[32] P. Yi, X. Dong, A. Iwayemi, C. Zhou, S. Li, Real-time opportunistic scheduling for residential demand response, IEEE Trans. Smart Grid 4 (1) (2013) 227–234.

[33] L.G.M. Ballesteros, O. Alvarez, J. Markendahl, Quality of experience in the smart cities context: an initial analysis, in: Proceedings of the IEEE First International Smart Cities Conference (ISC2), October, 2015, pp. 1–7.

[34] L. Zhou, J.J.P.C. Rodrigues, L. Oliveira, Quality of experience driven power scheduling in smart grid: architecture, strategy, and methodology, IEEE Commun. Mag. 50 (5) (2012) 136–141.

[35] J.S. Vardakas, N. Zorba, C.V. Verikoukis, A survey on demand response programs in smart grids: pricing methods and optimization algorithms, IEEE Commun. Surv. Tutor. 17 (1) (2015) 152–178.

[36] G. Strbac, Demand side management: benefits and challenges, Energy Policy 36 (12) (2008) 4419–4426.

[37] H.A. Aalami, M.P. Moghaddam, G.R. Yousefi, Modeling and prioritizing demand response programs in power markets, Electr. Power Syst. Res. 80 (4) (2010) 426–435.

[38] B. Ramanathan, V. Vittal, A framework for evaluation of advanced direct load control with minimum disruption, IEEE Tran. Power Syst. 23 (4) (2008) 1681–1688.

[39] H.A. Aalami, M.P. Moghaddam, G.R. Yousefi, Demand response modeling considering interruptible/curtailable loads and capacity market programs, Appl. Energy 87 (1) (2010) 243–250.

[40] H.-S. Oh, R.J. Thomas, Demand-side bidding agents: modeling and simulation, IEEE Trans. Power Syst. 23 (3) (2008) 1050–1056.

[41] P. Samadi, A.-H. Mohsenian-Rad, R. Schober, V.W.S. Wong, J. Jatskevich, Optimal real-time pricing algorithm based on utility maximization for smart grid, in: Proceedings of IEEE International Conference on Smart Grid Communications (Smart Grid Communication), 2010, pp. 415–420.

[42] W.-Y. Chiu, H. Sun, H.V. Poor, A multiobjective approach to multi-microgrid system design, IEEE Trans. Smart Grid 6 (5) (2015) 2263–2272.

[43] R. Shang, L. Jiao, F. Liu, W. Ma, A novel immune clonal algorithm for multi-objective problems, IEEE Trans. Evol. Comput. 16 (1) (2012) 35–50.

[44] G.-C. Luh, C.-H. Chueh, W.-W. Liu, Multi-objective immune algorithm, Eng. Opt. 35 (2) (2003) 143–164.

[45] W.-Y. Chiu, H. Sun, H.V. Poor, Demand-side energy storage system management in smart grid, in: Proceedings of the IEEE International Conference on Smart Grid Communications (SmartGridComm), 2012, pp. 73–78.

[46] S.P. Boyd, L. El Ghaoui, E. Feron, V. Balakrishnan, Linear Matrix Inequalities in System and Control Theory, vol. 15, SIAM, Philadelphia, 1994.

[47] S. Boyd, L. Vandenberghe, Convex Optimization, Cambridge University Press, Cambridge, UK, 2004.

[48] W.-Y. Chiu, H. Sun, H.V. Poor, Energy imbalance management using a robust pricing scheme, IEEE Trans. Smart Grid 4 (2) (2013) 896–904.

CHAPTER 8

Towards a Concept of Cooperating Power Network for Energy Management and Control of Microgrids

H. Dagdougui*, A. Ouammi*,†, R. Sacile‡
*ETS, Montreal, Canada
†The National Center for Scientific and Technical Research, Rabat, Morocco
‡University of Genova, Genova, Italy

1. INTRODUCTION

Nowadays, great attention is being devoted to and there is strong interest in smart grids (SGs) around the world. They can be considered as a new type of power distribution system that combines information communication technologies, control methods, and electric power grids. SGs are intelligent electricity grid platforms that can be considered as one of the current results that enhance the sustainability of energy supplies. SGs are sophisticated power systems where the use of new communication and control technologies allows greater robustness and reliability than the current conventional power systems [1]. The integration of various components such as distributed generation (DG) units, renewable energy sources (RESs), and storage technologies is enabled in an efficient manner. In addition, SGs allow the bidirectional communication of electric power from/to the utility and the consumers [2, 3]. From the demand side, they are more interactive with consumers than in the case of traditional power grids; for instance, SGs allow consumers to reduce the cost of their energy consumption, based on the power real-time cost. One of the main challenges in the development of SGs is how to control and manage effectively the distributed renewable energy generation, while taking into account economic, technical and environmental aspects.

In remote areas and in regions wishing to decrease their dependence on unsustainable energy production, the integration of different RESs has favored the penetration of DG, at reasonable cost and close to the energy consumers, leading to the concept of smart microgrids (SMGs) . In general, the connection of RESs to distribution systems leads to the concept of an active distribution system [4–7], where, over a wide area, adaptive and integrated control strategies coordinate the trade-off between power production and user demand.

SMGs are defined as smart power systems including loads, DG units, and energy storage systems (ESSs) (batteries, electric vehicles, hydraulic storage, etc.) grouped together within a limited geographic area [8, 9]. The main advantage that the SMG offers is to enable customers to have both a bidirectional communication platform and control devices to control their energy needs and excesses. In addition, with an adequate communication structure, it is possible to shape the users' load demand curves by means of demand response (DR) strategies.

SMGs can operate in either grid-connected mode or islanded mode. In grid-connected mode the microgrid (MG) is connected to a highly available power grid that may act as an additional power source for the MG. In this scenario the MG and the distribution network operator have mutual benefits selling/purchasing power. On the other hand, from a sustainable development viewpoint, purchasing energy from the distribution network operator, which mainly produces power from nonrenewable sources, should be regulated within certain limits [10], which, conversely, may affect the quality of service. In islanded mode the MG must keep a sufficient level of distributed power generation and energy storage to enhance system stability and to guarantee the quality of the local load. In an islanded renewable-based MG, the power flow exchanges will bring new challenges regarding the mitigation of renewable power intermittencies, load mitigation, load mismatches, and other key problems [11].

In the literature, several authors have proposed control and management methods for MGs. Some of them are quoted hereinafter. From the MG control perspective, Zhu et al. [12], proposed an optimal charging control problem for electric vehicles in SMGs with RESs. The problem is formulated as a stochastic semi-Markov decision process. An evolutionary strategy is proposed for the optimal selection and sizing of distributed energy resources for integrated MGs in [13]. Logenthiran et al. [13] formulated the problem as a nonlinear mixed-integer optimization problem aiming to minimize the annual capital and operational cost of DG. Bolognani and Zampieri [14] considered a problem of optimal reactive power compensation for the minimization of power distribution losses in a SMG. They proposed an approximate model for the power distribution network, which allows the problem to be cast into the class of convex quadratic, linearly constrained, optimization problems. Zhu et al. [15] proposed a novel multistep coordinated control approach for MG management. It is based on a look-ahead multistep optimization, consisting of a two-layer control algorithm, the component layer, and the top layer controls. Khodaei and Shahidehpour [16] employed a co-optimization approach to the generation and transmission expansion planning in a MG. The approach also considers the most suitable locations for MG installations in a power system. Valverde et al. [17] designed a supervisory model predictive control (MPC) method for optimal power management and control in a hydrogen-based MG. Olivares et al. [18] elaborate on the conceptual design of a centralized energy management system (EMS) and its desirable attributes for a MG in stand-alone operation mode. A game-theoretic communication structure that is a network constructed among distributed controllers in the MG is proposed in [19]. This structure helps in the sharing of local controller information, such as control input and individual objectives among controllers, and finds a better optimized cost for the individual objectives. A three-step control method was described to manage the cooperation among DG, distributed storage, and demand-side load management in [20].

From a SMG energy management perspective, Naraharisetti et al. [21] developed a mathematical model based on mixed integer linear programming for scheduling operations in MGs connected to the national grid. Sanseverino et al. [22] presented a new approach for optimal energy management of electrical distribution in smart grids. Dagdougui et al. [23] studied a dynamic decision model for the real-time control of hybrid renewable energy production systems, which can be particularly suitable for autonomous systems, such as islands or isolated villages. From a system modeling perspective, the problem of how to split an operating grid into islands to best serve current loads with operating sources was addressed in [24], while an algorithm that applies an efficient multilevel and multiobjective graph partitioning technique to a large network was presented in [25].

The issue of reliable integration of DG units has recently gained considerable attention, in particular with the presence of variable and fluctuating RESs. The high intermittency of some RESs as well as the fluctuations in the power output may lead to a mismatch between power production and the load. Under these circumstances, two promising, not mutually exclusive, solutions are proposed to improve the performance of the SMG:

1. the adoption of mixed renewable energy systems with a storage system;
2. cooperation among various MGs to enhance the quality of the power production.

The first option enables the use of diverse RESs—whose different production characteristics are dependent, for example, on weather conditions—to compensate for the difference between the generated power and the required load. This solution minimizes power dissipation due to grid constraints and also avoids the need to build large generators. The storage capacity could also significantly mitigate the drawbacks due to the fluctuating RESs behavior and can provide a cost-effective means to satisfy peak energy demand.

The second option to be considered is the introduction of a cooperative frame among SMGs. The benefits of such cooperation are obvious in the case of clusters of electric loads, where each MG may be located in different areas, having specific RESs and supplying power in real time. The cooperation will enable the exchange of power in real time to satisfy the electricity demand for each cluster of load; leading to global cooperation of the profitability and stability of the grid. In this situation, the MGs may offer great opportunities to optimize the DG use, especially when MGs may share information related to the power based on RES production. However, the control of a cooperative MG network is a more complex task than the control of a single MG, since it has to deal with changes in the state of the charge of the storage system, for each MG in real time.

This new concept is highly interesting in the case of renewable-based MGs. Connecting such MGs together will provide more flexibility for the operation of each MG through the exchanges of power with neighboring MGs, maximizing the use of renewable energy produced at the network level, exploiting the fluctuations of stochastic renewable sources and demands, and ensuring the local loads internally with minimum interaction with the main grid. The challenge is the definition of the global problem at the network level where the components of different MGs' should be modeled and simulated under several conditions and configurations. The challenge that needs to be addressed is how to effectively control and optimize the power flow exchange among MGs, while considering various MG security constraints, storage dynamics, RESs, load uncertainties, etc.

The aim of this chapter is to provide an outline of the operation and control of SMGs as well demand-side management (DSM) and DR programs, in addition to the energy market trading. Furthermore, the chapter introduces the concept of coordination and interconnection of a set of SMGs in a network, showing its benefits, advantages, and outcomes, and presents an overview of the mathematical modeling of the network and its features as well as a description of robust control methods adapted to optimally control energy exchange at the network level considering uncertainties of loads and energy production.

2. TOWARD THE CONCEPT OF A NETWORK OF SMART MICROGRIDS

A network of cooperative power MGs is a recent concept which has been introduced to cope with the uncertainties introduced by DG units. The resulting network is composed of several components depending on the infrastructure and the location. As a result, the operation of such a network becomes more complex with a large number of MGs and a significant number of connections. Consequently, smart control and coordination among various MGs are needed for the effective and efficient operation of such a network. Fig. 8.1 shows a conceptual architecture of a network of MGs.

The control of a network of power MGs is a new research field. This control can be established in a centralized, decentralized, or distributed

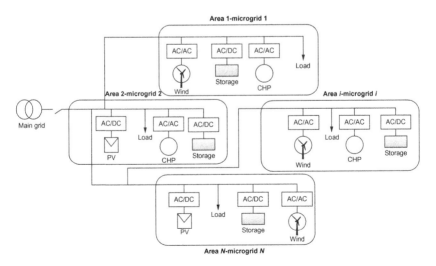

Fig. 8.1 Network of microgrids.

manner. The centralized control approach essentially relies on a central controller available at the network level allowing the implementation of online optimization for a fixed network infrastructure [26]. In this case, the MGs can be controlled by a master controller that is in charge of managing energy in each MG and coordinating the power exchange among MGs. This system is the upper control that governs the hierarchical control system and is primarily responsible for schedule dispatchable DG units, controllable loads (heating, ventilation, and air conditioning systems, electric vehicles, etc.), and ESSs in each MG. It must also avoid undesirable grid power injection and peaks, make full use of local RESs, and manage the power exchanges. Despite the importance of a centralized mechanism, it can sometimes cause a violation of the privacy of MGs and it is hard to implement for a large network because of the high degree of communication infrastructure required.

In the decentralized approach, the objective is to maximize the roles of local controllers. Several intelligent methods based on a multiagent system (MAS) [27, 28], gossip-based algorithms [29], and game theory [30] may be used for decentralized controls. The main advantage of using the decentralized control based on the MAS approach is to avoid manipulating large quantities of data. Typically, the information on each MG is processed by the local agent. Generally, at the unitary MG level, the decentralized control is best suitable for MGs with the following characteristics [31]:

1. Microsources can have different owners, in which case several decisions should be taken locally.
2. MGs operating in a market environment require the action of the controllers of each unit participating in the market to have a certain degree of intelligence.
3. Local microsources may have other tasks besides supplying power to the local distribution networks.

In the distributed configuration, a MAS is used to ensure the proper operation of various components in MG. In this case, an EMS, or "agent," may manage the operation of the MG. The MAS approach tends to be effective, and in this case, the agents of all MGs are independent decision makers. Each MG has an agent that can communicate information as regards its own power demand/offer and/or the allocated local DR and can agree on the proper power exchange policy with the other MGs within a given time horizon. The main challenge in the distributed configuration is to reach a consensus among various MGs' agents regarding power exchange so as to minimize the peak power.

In the specialized literature, a few studies have addressed control strategies related to a network of MGs. In [32], the control approach aims to keep the storage level in each MG around a reference value by exchanging power in a network of MGs. The proposed model is defined according to a linear quadratic formulation, which can allow effective computation of the optimal solution, even for networks with a great number of MGs. Ouammi et al. [33] studied the optimal energy management in a network of MGs on the basis of the use of the original linear quadratic Gaussian (LQG) problem. The proposed problem is a formulation that may be taken into account as a preliminary attempt to model and control the exchange of information and power in a network of MGs. Dagdougui et al. [34] solved a problem of power management at the network level using an original modeling approach based on the mathematical formalization of the Pontryagin minimum principle (PMP). Fathi and Bevrani [35] focused on load demand management of an electric network of interconnected MGs. Carpinelli et al. [36] analyzed potential advantages of distributed energy storage in smart grids with reference to both different possible conceivable regulatory schemes and services. Ouammi et al. [37] proposed an MPC method for optimal power exchanges in a smart network of power MGs. The main purpose was to present a control strategy for a cluster of interconnected MGs to maximize the global benefits. Camponogara and Scherer [38] presented a distributed optimization for MPC of linear dynamic networks with control input and output constraints.

The literature review shows that previous work has attempted to develop both control and optimization methods for MG operation. However, these studies do not consider the study of a network of interconnected MGs. The significance of the current study with respect to actual operation of MGs can be summarized in it providing more flexibility for the operation of each MG through the exchanges of power with neighboring MGs, maximizing the use of renewable energy production, exploiting the fluctuations of stochastic renewable sources and demands, and ensuring the local loads internally while minimizing interaction with the main grid.

3. THE NETWORK MODEL AND ARCHITECTURE

3.1. Microgrid Architecture

The MG can be considered as a small internal grid that can be connected to other MGs and/or connected/disconnected from the main electric grid. Each MG may integrate several units as shown in Fig. 8.2:

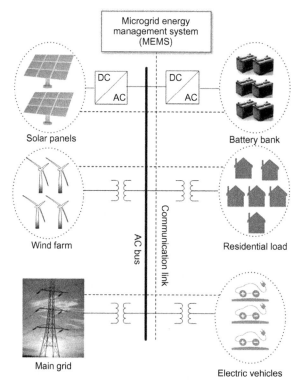

Fig. 8.2 Microgrid architecture.

- DG units that may lead to exploitation of the local RESs.
- An ESS improving the stability, the power quality, and the reliability of the supply.
- Loads representing the energy consumption.
- A point of common coupling, which can be a single, double, or multi-dimensional point depending on the number of incident connections.
- An energy management unit (EMU).

In practical cases, residential, industrial, or commercial buildings integrating distributed power sources, storage systems, controllable and non-controllable loads, a building EMS, and an advanced metering infrastructure may be considered as a MG.

3.2. Wind Power Generation Modeling

It is assumed that wind turbines are available in each MG. The Weibull probability distribution function is generally used to represent the frequencies of the wind speed. It also represents the most frequent starting point of stochastic analysis, simulation, and forecasting of wind speed. Its general formulation is given as follows:

$$f(v) = \left(\frac{\bar{k}}{\bar{c}}\right)\left(\frac{v}{\bar{c}}\right)^{\bar{k}-1} \exp\left(-\left(\frac{v}{\bar{c}}\right)^{\bar{k}}\right), \tag{8.1}$$

where $f(v)$ is the probability of occurrence of wind speed v (m/s), $\bar{k}$ is the dimensionless Weibull shape parameter, high values of which imply that the distribution is concentrated around a given value, whereas low values mean that there is a wide distribution of values, and $\bar{c}$ (m/s) is the Weibull scale parameter that fixes the position of the curve, with higher values for the sites with strong wind and lower values for still sites. These two parameters must be identified for each given location. Their identification can be performed with use of some classic techniques such as the minimum square error, the least squares method, and the standard deviation method.

In the literature, numerous models for the output power of wind turbines are available. The probabilistic power output of the wind turbine can be modeled as follows:

$$u_w(t) = \begin{cases} u_r\left(\frac{v^2(t)-v_c^2}{v_r^2-v_c^2}\right) & v_c \leq v(t) \leq v_r, \\ u_r & v_r \leq v(t) \leq v_f, \\ 0 & v(t) < v_c, \quad v(t) > v_f, \end{cases} \tag{8.2}$$

where $v(t)$ (m/s) is the probabilistic wind speed at the hub height of the wind turbine, v_c (m/s) is the cut-in wind speed, v_r (m/s) is the rated wind speed, v_f (m/s) is the cut-out wind speed, and u_r (kW) is the rated power.

3.3. Solar Photovoltaic Generator Modeling

The optimum operating point current I_{pv} and voltage V_{pv} can be determined as follows assuming that the maximum power point tracker is used and the photovoltaic (PV) module is always working at the maximum power point [39]:

$$I_{pv}(t) = I_{sc,ref}\left\{1 - A\left[\exp\left(\frac{V_{mp,ref}}{BV_{oc,ref}}\right) - 1\right]\right\} + \Delta I(t), \tag{8.3}$$

$$V_{pv}(t) = V_{mp,ref}\left[1 + 0.0539\log\left(\frac{G_{in}(t)}{G_{st}}\right)\right] + \mu\Delta T(t), \tag{8.4}$$

$$A = \left(1 - \frac{I_{mp,ref}}{I_{sc,ref}}\right) \exp\left(-\frac{V_{mp,ref}}{BV_{mp,ref}}\right), \tag{8.5}$$

$$B = \frac{\left(\frac{V_{mp,ref}}{V_{oc,ref}} - 1\right)}{\ln\left(1 - \frac{I_{mp,ref}}{I_{sc,ref}}\right)}, \tag{8.6}$$

$$\Delta I(t) = \gamma \frac{G_{in}(t)}{G_{st}} \Delta T(t) + \frac{G_{in}(t)}{G_{st} - 1} I_{sc,ref}, \tag{8.7}$$

$$\Delta T(t) = T_{amb} + 0.02 G_{in}(t), \tag{8.8}$$

where $I_{sc,ref}$ is the module short-circuit current, $I_{mp,ref}$ is the module maximum power current, $V_{mp,ref}$ is the module maximum power voltage, $V_{oc,ref}$ is the module open circuit voltage, G_{in} is the total irradiation incident, G_{st} is the standard light intensity, μ is the module voltage temperature coefficient, and γ is the module current temperature coefficient.

The hourly output voltage and power of the PV array are given as follows:

$$V_{pv,out}(t) = \beta_{pv,s} V_{pv}(t), \tag{8.9}$$

$$u_{pv}(t) = \beta_{pv,s}\beta_{pv,p} V_{pv}(t) I_{pv}(t) \zeta_{loss}, \tag{8.10}$$

where $\beta_{pv,s}$ is the serial connection number of PV modules, $\beta_{pv,p}$ is the parallel connection number of PV module strings, and ζ_{loss} is the connection loss and further loss caused by other factors.

3.4. Energy Storage System Dynamics

The MG is equipped with an ESS that aims to achieve a balance between electric loads and RES power generation. Therefore any local shortage in supplying loads to consumers should be met either by discharge of the ESS or by the purchasing of power from one of the MG neighbors or from the main electric grid. The function of the ESS is to participate in the balance of the electric loads and RES power generation. The ESS capacity must be sized on the number of residences connected to the MG.

3.5. Loads

The total load in each MG includes controllable and uncontrollable loads. Load forecasting techniques performed by each EMU can help to make important decisions for the network, such as purchasing power from or selling power to a particular MG, charging/discharging the ESS, and purchasing power from or selling power to the main grid. To provide accurate predictions for MGs, appropriate weather data (such as temperature and humidity) and historical data on energy consumption are required.

3.6. Energy Management Unit

The MG power system supports the power in grid-connected and/or MG-connected mode. The main function of the EMU is to optimize the MG operation in the case of autonomous operation, or alternatively to act as an interface between the MG components and the other MGs. In both cases, the goal of the EMU is to receive and to send a control signal such as the one for the ESS. The EMU can also disconnect the MG from other MGs and/or from the main grid in the case of network failure. It uses data gathered from different sensors available on-site to compute the predicted amount of power generated from RESs and the electric load demands for a few seconds, minutes, or hours ahead.

3.7. Demand-Side Management and Demand Response

Demand-side management (DSM) is a set of programs which enable customers to plan and monitor activities related to the use of electricity. In particular, it helps them play an important role in shifting their own demands for electricity during peak periods to consume less power and/or shift their demands to off-peak hours to smooth the demand curve. In fact, by reducing the overall load on an electricity network, DSM can mitigate electrical system emergencies, increasing system reliability, reducing the number of blackouts and regulating the voltage profiles, and balancing the grid. The concept of DSM was initially introduced in the late 1970s in response to the rise in energy costs and the rationale for energy conservation [40]. The primary DSM programs were implemented on the basis of one-way data communication and were directly operated by utilities, aggregators, and third-party control companies. In this case, power consumption control is achieved by direct load control (DLC) with which users allow utilities involuntarily to disconnect selected appliances when needed [40]. For example, utilities may control lighting, thermal comfort equipment (i.e., heating, ventilating, and air conditioning), refrigerators, and pumps. However, problems related to the high number of appliances and customers' privacy have constituted important barriers for the implementation of DLC. A solution for DLC is decentralized or autonomous DSM, where

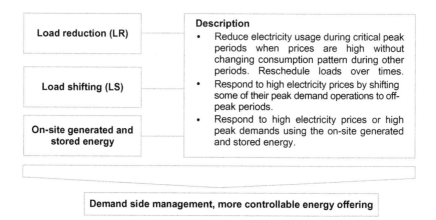

Fig. 8.3 Schemes for demand-side management.

users (homeowners, commercial/industrial buildings) are encouraged to individually and voluntarily manage their loads by reducing or shifting their consumption at peak hours [41, 42].

DR is the reduction of electric energy consumption by DSM from the normal consumption in response to programs related to an increase in the price of electricity, or to programs related to incentive payments designed to induce lower consumption of electric energy [43]. One of the main objectives of DR is to reduce the peak load and the ability to control consumption according to generation [44]. The price-based programs use dynamic electricity pricing rates that would reflect the price and availability of electricity in real time. Time-of-use (TOU) rates and real-time pricing (RTP) are among the main pricing-based programs [45]. DR includes all intentional modifications to patterns of consumption of electricity by customers that are intended to alter the timing, level of instantaneous demand, or the total electricity consumption [46]. Fig. 8.3 shows three schemes that can be used in DSM.

In the SG and MG environment, the intermittent nature of RESs, and especially that of solar energy, has led to much concern over the flexibility, stability, and reliability of grid-connected MGs. Therefore the MGs need to be flexible, and their EMS needs to respond to real-time fluctuations and uncertainties in energy generation and demand. DSM is among the solutions in balancing supply and demand with penetration of RESs. In particular, DR in a grid-connected MG can be realized through the application of other sources such as elastic loads and electric vehicles, which may help to balance the main grid power supplied against demand in real time.

DR programs have many potential benefits for SGs. They can support power markets to set efficient energy prices, mitigate market power, improve economic efficiency, and increase security [47]. In particular, depending on the customers' classification (commercial, industrial, residential, or individual) as well as the technologies used and the whole structure of the SG, the benefits of DR can be classified as follows [48]:

- Financial benefits include cost savings for customers (electricity bill reduction) due to the use of less energy when prices are high or due to their shifting their electricity consumption to lower-priced hours.
- Reliability benefits include customers' benefits perceived from the reduced probability of supply being involuntarily curtailed in a blackout.
- Market performance: DR prevents the exercise of market power by electric power producers.
- System security: system operators are endowed with more flexible means to meet contingencies.
- Improved choice: customers have more options to manage their electricity costs.

3.8. Price-Based Demand Response Programs

Price-based DR programs refer to changes in use by customers in response to changes in the prices they pay, and include RTP, critical peak pricing, and TOU rates [49]. The main ideas for all pricing schemes are (1) to allow retail prices to reflect fluctuating wholesale prices so the end users pay what the electricity is worth at different times of the day and (2) to encourage users to shift their use of high-load household appliances to off-peak hours to not only reduce their electricity costs but also to help to reduce the peak-to-average ratio (balance the supply and demand or grid balancing) in load demand [50–52].

- TOU rates: TOU is a service that gives different electricity prices for different periods; it consists of peak and off-peak rates [53]. A TOU program sets prices months in advance and therefore it does not provide incentives for critical days with actual power system conditions and unusual wholesale prices. The TOU price is the pricing scheme used by the Ontario Hydro Company in Toronto, Canada [54].
- RTP: In this case, DR, makes use of real-time schemes to allow users to modify their load demand patterns according to the real costs of their

energy consumption [55]. RTP reflects the changes in the wholesale price of electricity by changing the price hourly or more often retail prices. The main advantage of RTP is to enable users to reduce their energy consumption when the price is high, and hence to lower their electricity bills.

- Critical peak pricing: The critical peak pricing program sets high prices during a restricted number of days or hours in the case of high wholesale prices or contingencies to discourage electricity consumption during those times [56]. Critical peak pricing programs usually start with a TOU rate structure, but then they add one more rate that applies to critical peak hours, which the utility can call on at short notice.

3.9. Control Strategies for MGs

During grid-connected operation, it is possible to dispatch DG active power and reactive power essentially according to economic criteria, as voltage and frequency control are performed at the level of the main system that the MG is connected to (main grid) [57]. During this mode of operation, the primary function of the MG is to satisfy all of its load requirements and contractual obligations with the main grid. In grid-connected mode, the point of common coupling voltage is dominantly determined by the main grid, and the main role of the MG is to accommodate the real or reactive power generated by the DG units and the load demand. Reactive power injection by a DG unit can be used for power factor correction, reactive power supply, or voltage control [58].

During islanded operation, it is necessary to control local generators to ensure voltage and frequency stability [57]. In islanded mode, the MG operates as an independent entity and must provide voltage and frequency control, as well as real and reactive power balance. For example, if the net load demand is less than the total generation, the MG central controller should decrease the net generated power. This is done by assignment of new set points to the DG units. On the other hand, if the power generated within the MG cannot meet the load demand, either noncritical load shedding or activation of storage units must be considered [58].

This means that during islanded operation, the major concern is the MG primary and secondary frequency regulation. In this case the most interesting DR services are related to the provision of primary and secondary reserve. This type of service requires fast responses to transients, usually being implemented on the basis of the system frequency [59]. In general, in islanded operation, the MG will face the following challenges:

- Voltage and frequency management: The voltage and frequency are established by the grid when the MG is connected, however, when the MG is islanded, one or more primary or intermediate energy sources should form the grid by establishing its voltage and frequency, otherwise the MG will collapse. Both voltage and frequency should be regulated within acceptable limits.
- Balance between supply and demand: Sudden islanding of the MG may cause high imbalances between local load and generation. If the MG is exchanging power with the grid before disconnection, then secondary control actions should be applied to balance generation and consumption in island mode to ensure initial balance after a sudden fluctuation in load or generation. Because of the slow response of microsources to control signals, the power balance needs to be ensured by fast-acting storage units with enough capacity, such as flywheels, complemented by load shedding schemes to ensure balance. In addition, programs such as DR should be implemented if the load exceeds the available generation.
- Power quality: The MG should maintain an acceptable power quality while in islanded operation. There should be an adequate supply of reactive energy to mitigate voltage sags. The energy storage device should be capable of reacting quickly to frequency and voltage deviations and inject or absorb large amounts of real or reactive power. Finally, the MG should be able to supply the harmonics required by nonlinear loads.
- Microsource issues: The DG units in a MG operated in islanded mode should share power between each other in an appropriate ratio to prevent circulating current and thermally overstressing or damaging components.
- Communication among MG components: The MG should have plug-and-play architecture for microsources to rely on locally available information to control their generated power.

3.10. Hierarchical Control of MGs

The hierarchical control in a MG can be decomposed into two levels:
- Low-level control, which is decentralized and is performed by the local controllers of the programmable generators (primary and secondary droop control of voltage and frequency) [57].

- High-level control, which has to centrally perform an economic optimization of the operation of the MG as a whole. This is obtained by implementation of innovative optimal power flow as well as an appropriate storage system management [57].

Consequently, three levels of control can be identified for a MG that can be centrally or locally performed: primary, secondary, and tertiary. The main differences among them lie in the speed of response, the time frame in which they operate, and the infrastructure requirements. The primary and secondary droop controls can be implemented in a MG on the grounds of local measurements, thus reducing the need for communication between generators and a central controller so as to speed up the system response to perturbations while keeping the voltage and frequency within the allowed ranges. However, to obtain good operation also from the economical viewpoint, a higher-level coordinated control action, performed by a hierarchical control system, is required to optimally dispatch the generators after primary and secondary droop controls have occurred [57].

- Primary control: Primary control is the first control level in the control hierarchy and has the fastest response. Primary control responds to system dynamics and ensures that the system variables (e.g., voltage and frequency) track their set points. Primary control mostly employs conventional linear control methods and is performed locally, on the basis of locally measured signals. Because of their speed implications, islanding detection and the subsequent change of controller modes lie at this control level [58].
- Secondary control: Secondary control is the next level of control and is responsible for ensuring power quality and mitigating long-term voltage and frequency deviations by determining the set points for the primary control. In the central controller, the role of secondary control is to maintain a stable voltage and frequency during load change or power variation of RESs. Secondary control operates on a slower time scale than primary control (e.g., a settling time on the order of 1 min in a conventional grid), so the primary control loop reaches its steady state before the secondary controller updates the set point [58].
- Tertiary control: Tertiary control is the highest/top level of control and sets the steady-state set points depending on the requirements for an optimal power flow (e.g., on the basis of the information received about the status of the distributed energy resource units, market signals, and other system requirements) [58].

4. POWER CONTROL STRATEGY FOR THE NETWORK OF MGS

Optimal control theory provides a modern, direct, and systematic approach to a large variety of control design problems, including constrained optimization with interconnected variables. The optimal control theory provides many mathematical formulations that can be exploited directly to reduce the multistage resolution and the complexity of constrained performance optimization, such as a genetic algorithm, mixed-integer programming, or an evolutionary algorithm.

The optimal control problem is formulated with use of the PMP, in terms of dual variables (Hamiltonian costate and multipliers of constraints) with use of the Euler-Lagrange approach. Specifically, it proposes an original formulation based on the PMP that may be viewed as an approach to model and control the exchange of power in a MG network. The main innovation is the use and the exchange of information and the forecast of power production and consumption on the whole set of MGs to improve the overall quality of the power management and energy storage. In addition, the PMP decreases the computational loads as the number of nonlinear second-order differential equations increases in a linear manner with the dimension of the state variables.

4.1. Problem Formulation

The MG network is modeled as a directed connected graph $G = \{\mathcal{N}, \mathcal{L}\}$, where $\mathcal{N} = \{1 \ldots N\}$ is the vertex set representing both the MGs and the main electric grid. As a convention, the Nth vertex is associated with the main electric grid. $\mathcal{L} = 1 \ldots L$ is the set of undirected power links defining the network topology and playing the role of a data link. The MG has power connections with other MGs, and at least for one MG with the main electric grid. The objective is to minimize the power exchanged among the grids, and to keep each local ESS working around a proper optimal value for all MGs in the network.

The aim is to find the optimal control $u_i^*(t)$ and the state $S_i^*(t)$ (where $(*)$ designates the optimal value) that minimizes the performance measure (cost function), assuming that there is perfect knowledge of the state of each ESS, and under a cooperative strategy among the MGs. The cooperative strategy aims to maintain an optimal level of energy in the distributed ESS, as well as to achieve low power flow among the MGs. The objective function to be minimized is expressed as follows:

$$J(S, u) = \int_0^{t_f} f(S(t), u(t)) dt, \tag{8.11}$$

$$J(S, u) = S(t_f)' M_{t_f} S(t_f) + \int_0^{t_f} S(t)' MS(t) + u(t)' Qu(t) dt, \tag{8.12}$$

subject to

$$\begin{cases} \dot{S}(t) = AS(t) + Bu(t) + \mu(t), \\ S(t_0) = S_0, \end{cases} \tag{8.13}$$

with

$$S_{\min} \leq S(t) \leq S_{\max}, \tag{8.14}$$

where $S_{\min} \in \mathbb{R}^N$ and $S_{\max} \in \mathbb{R}^N$ are respectively the minimum and the maximum limits of the state vector related to the ESSs, B is the $N \times L$ incidence matrix, such that each element $b_{i,j} = -1$ if there is a link exiting the ith MG, $b_{i,j} = 1$ if there is a link entering the ith MG, and $b_{i,j} = 0$ otherwise, A is an $N \times N$ diagonal matrix related to the efficiency of the ESS in the MG network, M ($M \succ 0$) is an $N \times N$ matrix related to the cost of excess/lack of energy stored in each energy storage device, and M_{t_f} ($M_{t_f} \succ 0$) has the same definition as M but it is only defined for instant $t = t_f$, for the following reason.

While $M_{t_f} = M$ may be chosen in general, by choice of a proper M_{t_f}, the decision maker may decide to enhance the minimization of the final state, with the goal of leaving the state closer to the desired value at the end of the optimization horizon. Q is a $L \times L$ matrix ($Q \succ 0$) related to the cost of the power sent on each edge of the network.

4.2. Necessary Conditions for Global Optimality

It is worth mentioning that the conditions given by the PMP are necessary only for optimality but are not sufficient. This means that the solution might not be a global optimum. Three approaches are available to verify that the necessary conditions from the PMP become sufficient for global optimal control: (1) the optimal trajectory obtained from the PMP must satisfy the necessary and boundary conditions; (2) the optimal trajectory obtained from the PMP must be a unique trajectory; and (3) the absolute optimality is, mathematically, proven by clear propositions.

4.3. State Variable Constraints

In the context of optimal control problems, inequality constraints related to the state require more consideration than constraints related to the control inputs. This is because the constraints on the control could be handled at the end of the resolution, by removal of values of the control variables that are not within the range of the inequality constraints. In the case of state constraints, the mathematical formalization of the optimal control problem must take into account the constraints before the necessary conditions for the optimality are developed. The state constraint can be converted to two equality constraints. The reason for such a transformation is to take into account the variation of the state in the Hamiltonian. The inequality constraints related to the state can be written as follows:

$$g_1(t) = S_{\max} - S(t), \qquad (8.15)$$

$$g_2(t) = S(t) - S_{\min}. \qquad (8.16)$$

4.4. The Optimal Control Strategy of the Network of MGs

The problem statement can be considered as an optimal control of energy in a network of MGs, which can be explained as a transfer of the state of the system from any initial state that is different from 0 (St_0) to a certain value at the final time (St_f) that is known. In addition, the problem takes into account the minimization of the energy transfer and storage costs, while considering the state constraint equations.

For the problem, defined in previous sections, the optimal control problem can be solved with use of the PMP. The optimal solution of the problem is given by

$$u^*(t) = -CB'p^*(t), \qquad (8.17)$$

where $C = [Q + Q']^{-1}$ and p^* is the optimal value of the Lagrange multiplier vector (called also costate vector).

4.5. Example

The following example can be taken as an application of the control scheme introduced in the previous sections. It is assumed that a network of MGs is installed in a region, where four MGs ($N = 4, L = 5$) having a given storage system can cooperate by exchanging the power produced from RESs to supply their electric loads. The four MGs are supposed to be cooperative and connected from a power exchange viewpoint. Q, M, and M_T are taken as the identity matrices.

In general, the topology of the network adopted has a major effect on the management and the control of the network. Each MG can have several connections with the other MGs, which may influence the control results. In the proposed case study, a network of four MGs is considered, which may lead to many configurations of the network. So the control of the power exchanges among the four MGs will depend on the configuration. In this simulation a network with five links is assumed. The computation of the optimal control using the PMP is performed for a period of $T = 72$ hours. The predictions of the local energy balance are supposed to be available. Furthermore, it is assumed that the costate variables are taken as constant. In addition to simplifying the computation, the main reason behind the adoption of this assumption is that the optimal control based on the PMP is ensured to be a global optimal control as reported in several literature studies.

The optimal control of the state of the energy storage device available in one MG is shown in Fig. 8.4. The optimal control of the energy exchanged through one link is reported in Fig. 8.5.

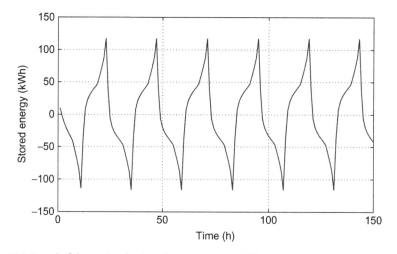

Fig. 8.4 Trend of the optimal values for the state variable.

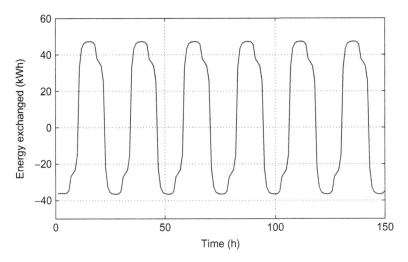

Fig. 8.5 Trend of the optimal values for the control variable.

5. LQG-BASED OPTIMAL CONTROL OF POWER FLOWS IN A SMART NETWORK OF MGS

5.1. Problem Formulation

In this section a centralized control model for optimal management and operation of a smart network of MGs is designed. The proposed control strategy considers grid interconnections for additional power exchanges. The proposed model is based on an original LQG problem definition for the optimal control of power flows in a smart network of MGs. The control strategy incorporates storage devices, various distributed energy resources, and loads. The objective function aims to minimize the power exchanges among MGs, and to make each local ESS in a MG work around a proper optimal value.

The optimal control module receives information from different MGs about the current state (i.e., the local stored energy) and the produced/consumed forecasted power. In a computational and transmission time that is negligible with respect to the duration of the time interval, the optimal control module computes and sends u_t to all the MG controllers; that is the flow control optimal strategy for each time interval.

The main decisions are whether to store instantaneous excess energy produced or to send it to some of the grid connections, or, alternatively, in the case of lack of energy, whether it is convenient to acquire energy from some other grids or to use (if any) the energy stored in the local ESS. The central controller of the smart network of MGs needs to evaluate how far the introduction of the possibility of directly exchanging power among the MGs according to the network can help to improve the overall performance of the system; for example, by reducing the overall power load on the main grid as well as limiting the variations of the energy stocked in the local storage units.

The optimization function to be minimized is

$$J(z, u) = E \left\{ z'_T M_T z_T + \sum_{t=0}^{T-1} (z'_t M z_t + u'_t R u_t) \right\}. \tag{8.18}$$

Each MG is supposed to be subject to the following discrete-time state equation:

$$z_{t+1} = A z_t + B u_t + \mu_t + \omega_t, \tag{8.19}$$

$$z_0 = z0, \tag{8.20}$$

$$\mu_t = \eta_t + (A - I) z^*, \tag{8.21}$$

where z_t, the state variable, is the vector of the energy storage device inventory at instant t, and A is a diagonal matrix describing, in each diagonal element α^i, the efficiency of the energy storage technology in the ith grid. In this respect, it holds that $0 \prec \alpha^i \leq 1$, and μ_t is a known sequence of deterministic values.

5.2. LQG Optimal Control Strategy

The optimal control of the problem defined in the previous section is given by the following equation:

$$u_t^* = K_t(z_t - z_t^{d2}) + K_t^g g_{t+1}, \tag{8.22}$$

where K_t is a matrix given by

$$K_t = -(R + B' P_{t+1} B)^{-1} (B' P_{t+1} A), \tag{8.23}$$

where P_t is a matrix given by the discrete-time algebraic Riccati equations,

$$P_t = M + A' P_{t+1} (I + BR^{-1} B' P_{t+1})^{-1} A, \qquad (8.24)$$

$$P_T = M_T, \qquad (8.25)$$

and K_t^g is a matrix given by

$$K_t^g = (R + B' P_{t+1} B)^{-1} B'. \qquad (8.26)$$

The vector z_t^{d2} is given by

$$z_{t+1}^{d2} = A z_t^{d2} + \mu_t, \qquad (8.27)$$

$$z_0^{d2} = z0. \qquad (8.28)$$

The vector g_t is given by

$$g_t = (A' - A' P_{t+1} (I + BR^{-1} B' P_{t+1})^{-1} BR^{-1} B') - M z_t^{d2}. \qquad (8.29)$$

$$g_T = M_T z_T^{d2}. \qquad (8.30)$$

The behavior of the smart network of MGs and the related degree of cooperation are dependent on the definition of matrices M, M_T, and R. An extreme problem formulation—whose solutions can be taken into account as a reference for other formulations and that can be achieved with less computational effort—is to weight power flows among the grids in a significantly less relevant way with respect to the system state. Under this assumption the optimization of the control variable can be ignored (i.e., $R \longmapsto 0$), resulting in a "weak control" optimization.

Under the hypothesis of weak control (i.e., when R tends to the null matrix), the optimal control for the system is

$$u_t^* = -Q^{-1} B' M [A z_t + \mu_t], \qquad (8.31)$$

where

$$Q = B' MB. \qquad (8.32)$$

5.3. Example

The proposed example is performed on four MGs supposed to be cooperative and interconnected. The variations of the state variables that represent the level of the ESS available in one MG under a cooperative strategy are reported in Fig. 8.6. The trend of the energy exchanged in a link is shown in Fig. 8.7.

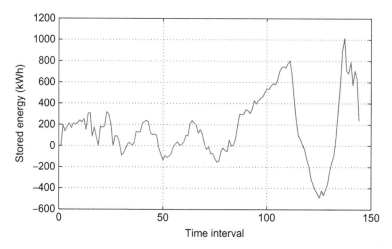

Fig. 8.6 Optimal values for the state variable.

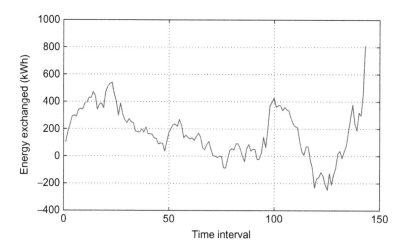

Fig. 8.7 Optimal values for the control variable.

6. MODEL PREDICTIVE CONTROL-BASED POWER SCHEDULING IN A NETWORK OF MGS

An MPC method for the optimal power exchanges in a smart network of power MGs is presented. The main purpose is to present an innovative control strategy for a cluster of interconnected MGs to maximize the global benefits. An MPC-based algorithm is used to determine the scheduling of power exchanges among MGs and the charge/discharge in each local storage system. The MPC algorithm requires information on power prices, power generation, and load forecasts.

6.1. Model Predictive Control-Based Power Scheduling

The basic theory of the MPC-based power scheduling strategy is that at each time step (t), a finite-horizon (N_c) optimal control sequence is computed for the ESS state, power exchanges among MGs, and power exchanges with the main electric grid for the whole network of MGs. However, only the first step of control actions is applied. For example, $\bar{u}_m^*(t + k|t)$ will denote the vector of optimal power exchanged between MGs at time ($t + k$) predicted at time t. The method operates following a rolling-horizon scheme, which means that at the next time step ($t+1$), new measurements of renewable resources (wind, solar), loads, and prices are available, giving updated information into the future. With these updates, the optimal control routine is recalculated for the next N_c periods. The MPC-based algorithm is implemented with use of the following steps:

- At $t = 1$, initialize with the actual current state of the MGs (i.e., storage systems, loads, renewable energy power generation, energy price predictions).
- Compute an optimal control sequence for the selected rolling optimization horizon (N_c) on the basis of loads, renewable energy, and energy price predictions for the next prediction periods (N_p).
- Implement the first control period operation of the scheduling problem for all MGs.
- Update the information available in each MG for the next period (i.e., ESS state, loads, renewable energy power generation, and energy price predictions). Then move to the next sampling instant, and repeat the algorithm.

The high-level control generates optimal set points for all DG units, ESSs and power exchanges so as to maximize the total network profit and to meet the loads in each MG. The EMU available in each MG must guarantee that the system tracks the power reference values delivered by the global central controller. The voltage and frequency stabilities are supposed to be controlled by a lower-level controller in each MG. In particular, in this study, each MG is connected to the main electric grid, so the frequency of each MG is maintained within some limits by the main electric grid. The EMU is in charge of the forecasts for the energy prices, renewable power generation, and loads by appropriate models. It transfers the forecasts to the global central controller, which computes the optimal system operation and applies the first control input set points to all MGs. The EMUs update the parameters of the prediction model with variations to reduce the errors and they send their updates to the global central controller.

6.2. Optimization Problem Formulation

In a network of MGs the primary objective is to maximize the benefits, while satisfying power balance, power generation, ESS, and energy exchange constraints. The first two terms in the objective function are related to the cost of the power sold to the distribution network operator and the other MGs, while the second two terms are related to the cost of the power purchased from the main electric grid and the MGs. The objective function to be maximized at each time step (t) can be formulated as follows:

$$J = \sum_{k=1}^{N_c} \sum_{j=1}^{M} \phi(k) \cdot \bar{u}_{g,s}(j, t+k) \cdot \tilde{C}_{g,s}(j, t+k)$$

$$+ \sum_{k=1}^{N_c} \sum_{j=1}^{M} \sum_{i,i\neq j}^{M} \tilde{\phi}_j(k) \cdot \bar{u}_{m,s}(j, i, +k) \cdot \tilde{C}_{m,s}(j, t+k)$$

$$- \sum_{k=1}^{N_c} \sum_{j=1}^{M} \psi(k) \cdot \bar{u}_{g,p}(j, t+k) \cdot \tilde{C}_{g,p}(j, t+k)$$

$$- \sum_{k=1}^{N_c} \sum_{j=1}^{M} \sum_{i,i\neq j}^{M} \tilde{\psi}_j(k) \cdot \bar{u}_{m,p}(j, i, +k) \cdot \tilde{C}_{m,p}(i, t+k). \quad (8.33)$$

The energy storage state equation for each MG is given by

$$\bar{x}(i, t+k) = \bar{x}(i, t+k-1) + \beta_{Char,i} u_{Char}(i, t+k) - \beta_{Dis,i} u_{Dis}(i, t+k). \quad (8.34)$$

The stored energy in each ESS is constrained by upper and lower bounds:

$$x_{i,\min} \leq x(i, t+k) \leq x_{i,\max}. \quad (8.35)$$

The power charged/discharged needs to be lower than certain maximum charging/discharging power limits:

$$0 \leq u_{Char}(i, t+k) \leq u_{Char,i,\max}. \tag{8.36}$$

$$0 \leq u_{Dis}(i, t+k) \leq u_{Dis,i,\max}. \tag{8.37}$$

The predicted powers sold to and purchased from the main electric grid are constrained by upper and lower bounds:

$$u_{g,s,\min} \prec u_{g,s}(i, t+k) \prec u_{g,s,\max}, \tag{8.38}$$

$$u_{g,p,\min} \prec u_{g,p}(i, t+k) \prec u_{g,p,\max}. \tag{8.39}$$

The predicted power balance $\Delta \tilde{u}_{bal}(j, t+k)$ in the jth MG and at instant $(t+k)$ is given by the following equation:

$$\Delta \tilde{u}_{bal}(j, t+k) = \tilde{u}_{wt}(j, t+k) + \tilde{u}_{pv}(j, t+k) - \tilde{D}(j, t+k) \tag{8.40}$$

$$= \sum_{i=1,ij}^{M} \bar{u}_{m,s}(j, i, t+k) + \bar{u}_{g,s}(j, t+k)$$

$$+ \beta_{Char,j} \bar{u}_{Char}(j, t+k)$$

$$- \beta_{Dis,j} \bar{u}_{Dis}(j, t+k)$$

$$- \sum_{i=1,ij}^{M} \bar{u}_{m,p}(j, i, t+k) - \bar{u}_{g,p}(j, t+k).$$

6.3. Example

A cooperative network of five MGs is considered, where each one is interconnected with four adjacent MGs and with a main electric grid. It is assumed that all MGs are connected to the same main electric grid and the power exchange can take place in both directions. The MGs are assumed to be equipped with renewable generators (wind turbine and PV modules), ESSs, and inelastic loads. The lengths of the prediction horizon N_p and the control horizon N_c are set equal to 24 hours, and the control interval is 1 hour.

As an illustrative example, the optimal state of an ESS is shown in Fig. 8.8. The charged/discharged energy is reported in Fig. 8.9. Furthermore, the energy exchanged with the other MGs is shown in Fig. 8.10.

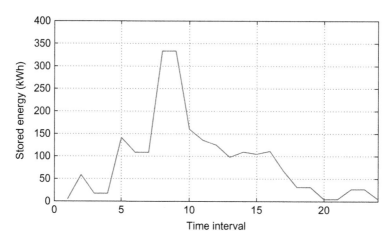

Fig. 8.8 The optimal state of an ESS.

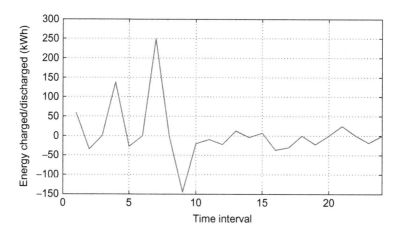

Fig. 8.9 The charged/discharged energy.

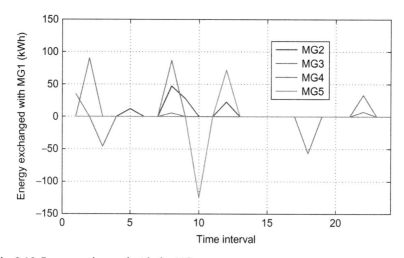

Fig. 8.10 Energy exchanged with the MGs.

7. CONCLUSIONS

This chapter had the modest ambition of showing how a network of MGs—under an original modeling approach—may be controlled with conventional optimal control theory. In the literature, work on the control of power MG networks is increasing, so the objective was to show that traditional approaches, under a simplifying modeling hypothesis, may work as well. The aim of this work was to introduce a view of an emergent problem that may represent a promising field for control researchers. In conclusion, it has been demonstrated that the cooperation among grids has significant advantages and benefits for each single grid operation in terms of integrating a common strategy to face shortage or excess of power production due to the uncontrollable RES behavior.

REFERENCES

[1] M. Hashmi, S. Hanninen, K. Maki, Survey of smart grid concepts, architectures, and technological demonstrations worldwide, in: Proceedings of IEEE PES Conference on Innovative Smart Grid Technologies, 2011.

[2] P. Acharjee, J. Gunda, Development prospect of smart grid in India, in: Proceedings of IEEE International Conference on Power and Energy, 2010, pp. 953–957.

[3] Q. Li, G. Cao, Multicast authentication in the smart grid with one-time signature, IEEE Trans. Smart Grid 2 (4) (2011) 686–696.

[4] F. Pilo, G. Celli, S. Mocci, G.G. Soma, Multi-objective programming for optimal DG integration in active distribution systems, in: Power and Energy Society General Meeting IEEE 2010, Minneapolis, 2010, pp. 1–7.

[5] J. McDonald, Adaptive intelligent power systems: active distribution networks, Energy Policy 36 (2008) 4346–4351.

[6] P.M. Costa, M.A. Matos, Assessing the contribution of microgrids to the reliability of distribution networks, Electr. Power Syst. Res. 79 (2009) 382–389.

[7] A. Borghetti, M. Bosetti, S. Grillo, S. Massucco, C.A. Nucci, M. Paolone, et al., Short-term scheduling and control of active distribution systems with high penetration of renewable resources, IEEE Syst. J. 4 (2010) 313–322.

[8] Y. Zhang, N. Gatsis, G.B. Giannakis, Robust energy management for microgrids with high penetration renewables, IEEE Trans. Sustain. Energy 4 (4) (2013) 944–953.

[9] E. Dall'Anese, H. Zhu, G.B. Giannakis, Distributed optimal power flow for smart microgrids, IEEE Trans. Smart Grid 4 (3) (2013) 1464–1475.

[10] S.A. Arefifar, Y.A.R.I. Mohamed, T.H.M. El-Fouly, Supply-adequacy-based optimal construction of microgrids in smart distribution systems, IEEE Trans. Smart Grid 3 (3) (2012).

[11] C.H. Lo, N. Ansari, Decentralized controls and communications for autonomous distribution networks in smart grid, IEEE Trans. Smart Grid 4 (1) (2013).

[12] L. Zhu, F.R. Yu, B. Ning, T. Tang, Optimal charging control for electric vehicles in smart microgrids with renewable energy sources, in: IEEE 75th Vehicular Technology Conference (VTC Spring), 2012.

[13] T. Logenthiran, D. Srinivasan, A.M. Khambadkone, T. Sundar Raj, Optimal sizing of distributed energy resources for integrated microgrids using evolutionary strategy, in: WCCI 2012 IEEE World Congress on Computational Intelligence, 2012.

[14] S. Bolognani, S. Zampieri, Distributed control for optimal reactive power compensation in smart microgrids, in: 2011 50th IEEE Conference on Decision and Control and European Control Conference (CDC-ECC), 12–15 December, Orlando, FL, USA, 2011.

[15] D. Zhu, R. Yang, G. Hug-Glanzmann, Managing microgrids with intermittent resources: a two-layer multi-step optimal control approach, in: North American Power Symposium (NAPS), 2010.

[16] A. Khodaei, M. Shahidehpour, Microgrid-based co-optimization of generation and transmission planning in power systems, IEEE Trans. Power Syst. 28 (2) (2013).

[17] L. Valverde, C. Bordons, F. Rosa, Power management using model predictive control in a hydrogen-based microgrid, in: IECON 2012—38th Annual Conference on IEEE Industrial Electronics Society, 2012.

[18] D.E. Olivares, C.A.C. nizares, M. Kazerani, A centralized optimal energy management system for microgrids, in: Power and Energy Society General Meeting, 2011.

[19] N.C. Ekneligoda, W.W. Weaver, Game-theoretic communication structures in microgrids, IEEE Trans. Power Deliv. 27 (4) (2012).

[20] A. Molderink, V. Bakker, M.G.C. Bosman, J.L. Hurink, G.J.M. Smit, Management and control of domestic smart grid technology, IEEE Trans. Smart Grid 1 (2010) 109–118.

[21] P.K. Naraharisetti, I.A. Karimi, A. Anand, D.-Y. Lee, A linear diversity constraint e application to scheduling in microgrids, Energy 36 (2011) 4235–4243.

[22] E.R. Sanseverino, M.L. Di Silvestre, M.G. Ippolito, A. De Paola, G. Lo Re, An execution, monitoring and replanning approach for optimal energy management in microgrids, Energy 36 (2011) 3429–3436.

[23] H. Dagdougui, R. Minciardi, A. Ouammi, M. Robba, R. Sacile, A dynamic decision model for the real time control of hybrid renewable energy production systems, IEEE Syst. J. 4 (2010) 323–333.

[24] Q. Zhao, K. Sun, D.Z. Zheng, J. Ma, Q. Lu, A study of system splitting strategies for island operation of power system: a two-phase method based on OBDDS, IEEE Trans. Power Syst. 18 (2003) 1556–1565.

[25] J. Li, C.C. Liu, K.P. Schneider, Controlled partitioning of a power network considering real and reactive power balance, IEEE Trans. Smart Grid 1 (2010) 261–269.

[26] D.E. Olivares, A. Mehrizi-Sani, A.H. Etemadi, C.A. Canizares, R. Iravani, M. Kazerani, et al., Trends in microgrid control, IEEE Trans. Smart Grid 5 (2014) 1905–1919.

[27] M. Barnes, A. Engler, C. Fitzer, N. Hatziargyriou, C. Jones, S. Papathanassiou, et al., MicroGrid laboratory facility, in: IEEE International Conference on Future Power Systems, 2005.

[28] K. Huang, S. Srivastava, D. Cartes, Decentralized reconfiguration for power systems using multi agent system, in: IEEE Systems Conference, 2007.

[29] K. De Brabandere, K. Vanthournout, J. Driesen, G. Deconinck, R. Belmans, Control of microgrids, in: IEEE PES General Meeting, 2007, pp. 1–7.

[30] N.C. Ekneligoda, W.W. Weaver, Game-theoretic communication structures in microgrids, IEEE Trans. Power Deliv. 27 (4) (2012) 2334–2341.

[31] N. Hatziargyriou, A. Dimeas, A. Tsikalakis, Centralized and decentralized control of microgrids, Int. J. Dist. Energy Resour. 1 (3) (2005) 197–212.

[32] H. Dagdougui, R. Sacile, Decentralized control of the power flows in a network of smart microgrids modeled as a team of cooperative agents, IEEE Trans. Control Syst. Technol. 22 (2) (2014) 510–519.

[33] A. Ouammi, H. Dagdougui, R. Sacile, Optimal control of power flows and energy local storages in a network of microgrids modeled as a system of systems, IEEE Trans. Control Syst. Technol. 23 (2015) 128–138.

[34] H. Dagdougui, A. Ouammi, R. Sacile, Optimal control of a network of power microgrids using the Pontryagin's minimum principle, IEEE Trans. Control Syst. Technol. 22 (2014) 1942–1948.

[35] M. Fathi, H. Bevrani, Statistical cooperative power dispatching in interconnected microgrids, IEEE Trans. Sustain. Energy 4 (2013) 586–593.

[36] G. Carpinelli, G. Celli, S. Mocci, F. Mottola, F. Pilo, D. Proto, Optimal integration of distributed energy storage devices in smart grids, IEEE Trans. Smart Grid 4 (2013) 985–995.

[37] A. Ouammi, H. Dagdougui, L. Dessaint, R. Sacile, Coordinated model predictive-based power flows control in a cooperative network of smart microgrids, IEEE Trans. Smart Grid 6 (5) (2015) 2233–2244.

[38] E. Camponogara, H.F. Scherer, Distributed optimization for model predictive control of linear dynamic networks with control-input and output constraints, IEEE Trans. Autom. Sci. Eng. 8 (1) (2011) 233–242.

[39] B. Ai, H. Yang, H. Shen, X. Liao, Computer-aided design of PV/wind hybrid system, Renew. Energy 28 (2003) 1491–1512.

[40] G.T. Costanzo, G. Zhu, M.F. Anjos, G. Savard, A system architecture for autonomous demand side load management in smart buildings, IEEE Trans. Smart Grid 3 (4) (2012).

[41] K. Herter, Residential implementation of critical-peak pricing of electricity, Energy Policy 35 (2007) 2121–2130.

[42] C. Triki, A. Violi, Dynamic pricing of electricity in retail markets, J. Oper. Res. 7 (1) (2009) 21–36.

[43] M.C. Vlot, J.D. Knigge, J.G. Slootweg, Economical regulation power through load shifting with smart energy appliances, IEEE Trans. Smart Grid 4 (3) (2013).

[44] P. Palensky, D. Dietrich, Demand side management: demand response, intelligent energy systems, and smart loads, Ind. Inform. 7 (3) (2011) 381–388.

[45] L. Gelazanskas, K.A.A. Gamage, Demand side management in smart grid: a review and proposals for future direction, Sustain. Cities Soc. 11 (2014) 22–30.

[46] M.H. Albadi, E.F. El-Saadany, Demand response in electricity markets: an overview, in: Power Engineering Society General Meeting, 2007.

[47] X. Guan, Z. Xu, Q. Jia, Energy-efficient buildings facilitated by microgrid, IEEE Trans. Smart Grid 1 (3) (2010) 243–252.

[48] P. Siano, Demand response and smart grids-a survey, Renew. Sustain. Energy Rev. 30 (2014) 461–478.

[49] S. Borenstein, M. Jaske, A. Rosenfeld, Dynamic pricing, advanced metering and demand response in electricity markets, Center for the Study of Energy Markets, University of California, Berkeley, 2002.

[50] A. Ipakchi, F. Albuyeh, Grid of the future, IEEE Power Energy Mag. 8 (4) (2009) 52–62.

[51] B. Ramanathan, V. Vittal, A framework for evaluation of advanced direct load control with minimum disruption, IEEE Trans. Power Syst. 23 (4) (2008) 1681–1688.

[52] M.A.A. Pedrasa, T.D. Spooner, I.F. MaxGill, Scheduling of demand side resources using binary particle swarm optimization, IEEE Trans. Power Syst. 24 (3) (2009) 1173–1181.

[53] C.W. Gellings, J. Chamberlin, Demand Side Management: Concepts and Methods, The Fairmont Press Inc., Lilburn, 1988.

[54] Smarter electricity pricing coming to Ontario: McGuinty government rolls out time-of-use rates, Ontario Newsroom For Residents, May 14, 2009.

[55] S. Salinas, M. Li, P. Li, Multi-objective optimal energy consumption scheduling in smart grids, IEEE Trans. Smart Grid 4 (1) (2013).

[56] J. Aghaei, M.I. Alizadeh, Critical peak pricing with load control demand response program in unit commitment problem, IET Generat. Transm. Distrib. 7 (7) (2013) 681–690.

[57] S. Conti, R. Nicolosi, S.A. Rizzo, H.H. Zeineldin, Optimal dispatching of distributed generators and storage systems for MV islanded microgrids, IEEE Trans. Power Deliv. 27 (3) (2012).

[58] A. Mehrizi-Sani, R. Iravani, Potential-function based control of a microgrid in islanded and grid-connected models, IEEE Trans. Power Syst. 25 (2010) 1883–1891.

[59] C. Gouveia, J. Moreira, C.L. Moreira, J.A.P. Lopes, Coordinating storage and demand response for microgrid emergency operation, IEEE Trans. Smart Grid 4 (4) (2013).

CHAPTER 9

Power Electronics for Microgrids: Concepts and Future Trends

T. Dragičević, F. Blaabjerg
Aalborg University, Aalborg, Denmark

1. STATE OF THE ART IN DC MICROGRID TECHNOLOGY

While remarkable progress has been made in improving the performance of AC microgrids (MGs) in recent years [1–7], DC MGs have also been recognized as an attractive technology for many applications because of higher efficiency, more natural interface to many types of renewable energy sources and energy storage systems (ESS), better compliance with consumer electronics, etc. [8–13]. Besides, when components are coupled around a DC bus, there are no issues with reactive power flow, power quality, and frequency regulation, resulting in a notably less complex control system. This chapter aims to reflect the state of the art in DC MGs in terms of architectures, control, power electronic interfaces, and protections issues. Also, it will point out some future research trends in this field.

A typical DC MG with the corresponding communication infrastructure is depicted in Fig. 9.1. While a DC MG can be conceived as an extension of historically important DC applications such as distributed power systems, telecom and data center stations, and vehicular power systems, development in this area was also largely influenced by its older counterpart, the AC MG. In that view, it is shown that a DC MG is actually a hybrid mixture of these technologies. A common DC bus is the backbone of the entire system, with all of the source and load subsystems being interfaced with it either directly or through the power electronic converter interface.

Today, DC power distribution is typically used in critical industrial applications such as spacecraft, hybrid electric vehicles (HEVs), consumer electronics, data centers, and telecom systems, where all its qualities in relation to AC power distribution are very much needed. In the next section, the typical architecture used in the context of applications of DC MGs is briefly described. In the subsequent sections, control issues as well

Microgrid
http://dx.doi.org/10.1016/B978-0-08-101753-1.00009-7

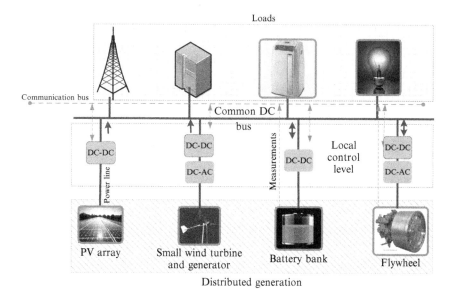

Fig. 9.1 Typical DC MG.

some power electronic topologies are reviewed. Lastly, issues associated with the design of protection systems are discussed.

2. DC MG APPLICATIONS

Even though AC power systems have established their dominant position in large-scale electricity distribution, DC has been constantly used in many stand-alone industrial applications. The common feature of all of these applications is the requirement for high reliability and efficiency. Some meaningful examples of installations that use DC distribution are the electrical distribution of consumer electronics, electric vehicles, naval ships, spacecraft, submarines, telecom systems, and rural areas. Stand-alone operation of these systems implies the use of a centralized ESS that serves as the source of power during periods without an external supply, whether it is from the grid or other sources. Batteries of various kinds are the most employed storage technology. To that extent, one of the battery types that are available on the market is chosen depending on concrete application requirements for cycle life, power, and energy density. The resource coordination strategies that have been deployed within this chapter can be used in any of these systems, as they are custom designed so as to respect the efficient operation of multiple batteries within the system. Specific characteristics of these types of systems are analyzed and the reasons to deploy flexible MGs are explained in that context. Moreover, the way of extending these conventional DC distributions to more controllable and flexible power systems such as DC MGs is given in the next section, where some flexible DC MG control principles are discussed in more detail.

2.1. Battery-Based DC Architectures

In applications such as HEVs, remote telecom stations, and consumer electronics, batteries are the heart of the system. Fig. 9.2 shows a typical architecture that can be considered as representative of this kind of system. Consequently, the entire respective electrical distribution system is built around the DC bus formed by its terminals. It consists of several power converter stages that transform the battery voltage to levels appropriate for other fundamental elements. On some occasions, the voltage difference can be significant and conventional DC-DC converter topologies cannot be used. In those cases, it is necessary to convert the DC link voltage to a high-frequency AC voltage and transform it to the appropriate voltage level with use of an isolation transformer. In this regard, a single-phase dual active bridge (DAB) is the most commonly used topology, but many others have been proposed in the literature with improvements such as reduction of the number of switches, and minimization of DC link voltage ripple. Some of these topologies are presented later in this chapter.

Regarding the prominent battery technologies, the lithium battery prevails in more demanding applications as it is currently able to provide the best performance in terms of power and energy density. In particular,

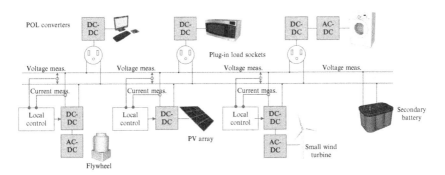

Fig. 9.2 Conventional DC MG architecture.

lithium-ion polymer battery cells are a common choice today because of their especially high energy density and convenient property that allows for flexibility in the design. On the other hand, valve-regulated lead-acid batteries are typically used for less demanding applications such as telecom stations. Unlike in case of the batteries from previous generations, shortening of the lifetime or even hazardous conditions will likely occur if a lithium-based battery is overcharged or overdischarged. Therefore a battery management system whose main functions is state of charge monitoring and recharge control has become an integral part of these systems. Direct connection of battery stacks is also the principle in remote telecom stations, where it is used to achieve high reliability in the telecom industry. It has a dual role in that case: backup power and filtering. The capacity of the battery is usually high enough to filter out the load and generation transients, provide backup power, and ensure stability. It also limits the voltage variation, which is determined by the number of series cells, the battery technology, and the management regime.

2.2. Flexible DC MG Architecture

Although very robust and reliable, the architecture with direct connection of the battery stack has many limitations. The most prominent one is its inability to regulate the common DC bus voltage, which does not allow its use in any multiterminal topologies. Moreover, the battery management system is not unified, leading to the problems of circulating current between distributed converters. Insertion of a regulated converter between the battery and the common DC bus, as shown in Fig. 9.3, solves these problems, but opens a whole new field of control possibilities. Some of them are discussed in the next section.

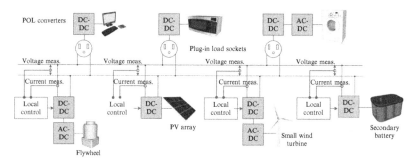

Fig. 9.3 DC MG with flexible architecture.

3. DC MG CONTROL

The configuration of a single-terminal DC MG is shown in Fig. 9.3. It consists of a number of sources and loads. Here, the bus in the system is fully controllable as there are no ESSs directly connected to any of them. Normally, only the sources are responsible for the control over the common bus bar voltage, although some loads allow a direct load control over them. In Fig. 9.3 one can note a local control level and a central controller that serves as a data collection station with which a wide range of possibilities are achievable depending on the bandwidth of the communication interfaces between it and other units. If all units are wired together, the local control level unites the whole system and the power sharing control can be as fast as the converters allow. However, as the wiring can become unacceptably extensive with an increase in the number of units, this kind of strategy applies only to extremely small, specific systems. On the other hand, a more realistic scenario for interconnection of several dispersed generators is through the use of a low-bandwidth communication interface on the upper control layer. Then, the DC MG can adopt a structure similar to the one of full-scale electricity distribution systems by the installation of secondary, tertiary, or supervisory control on top of the primary control level. The reason for their application is to achieve a higher quality of service through restoration of voltage deviations (also frequency deviations in the case of AC), to obtain full control over the power exchanges and ancillary services, to be able to deploy energy management system (EMS), etc. One way or another, the primary controllers still retain the high-bandwidth local control based on direct voltage and current measurements. The kinds of schemes that have a primary control system distributed to a number of dispersed generators that are independent of one another from a control point of view are often referred to as *distributed control strategies*. A special case is decentralized control, where even the low-bandwidth communication link is disabled and the local control level is responsible for all the functionalities. Specific control issues for a DC MG are discussed in the following section, where a distributed, droop-based, primary control is presented at first. Then, its extension with higher-level loops to accomplish secondary and tertiary control capabilities is explained. Finally, proposals for an EMS that have recently appeared in the literature are reexamined in the last paragraph of that section. The energy management capabilities of these respective strategies are realized by means of concepts ranging from fully centralized to fully decentralized. However, a strategy that is somewhere in the middle of them, a so-called *distributed control*, prevails in the recent approaches.

3.1. Hierarchical Layers

As mentioned before, the tightest control can be achieved if fast inter-communication links between the sources are available. However, with an increasing number of units and/or their spatial diffusion, wiring hardware becomes a serious limitation. Moreover, physical differences between converters and lines can trigger circulating current problems. To overcome these constraints, a droop control method, taken from traditional power system control, has emerged as the most promising solution for obtaining distributed control in MGs. In this way, the automatic primary control of the system can be realized by implementation of the droop control law to some of the paralleled units.

Apart from the droop control law, two other laws can be implemented on top of it, leading to a kind of a hierarchical control structure (see Fig. 9.4) that typically comprises the three layers:

- Primary control is responsible for individual converter power and voltage regulation. As already mentioned, this typically comprises a droop (which corresponds to virtual impedance) strategy that operates on top of inner voltage and current loops.
- The main function of secondary control is to simultaneously shift the droop characteristics of associated converters so as to perform the restoration of voltage levels to nominal values or to values that ensure proper power exchange between different DC buses.
- Tertiary control is in charge of regulating power exchange with the external grid and/or with other MGs. This layer of tertiary control is considered as part of secondary control in Fig. 9.4. Tertiary control can also include advanced functions related to efficiency and economic enhancements that constitute a higher management level, commonly referred to as the *EMS* [14, 15]. These functionalities are discussed further later.

The control bandwidth is gradually decreased as we climb up the hierarchy. Besides, unlike the secondary and tertiary layers, all the functions of the primary layer are by definition achieved without use of digital communication technologies. As primary control is very well established, the following subsection will provide information about the family of control strategies that have received quite high interest from the academic community recently time (i.e., distributed control), and particularly the part

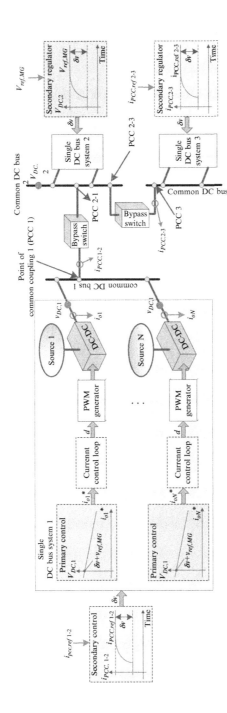

Fig. 9.4 Hierarchical control layers in MGs.

responsible for load coordination activities. Therefore, the next subsection will focus on this particular aspect of MG control.

3.2. Distributed Control Principles

A principle of distributed control for MGs is depicted in Fig. 9.5. It shows how different converters in the system actually communicate only with each other rather than with the centralized supervisory controller. Distributed control is typically based on consensus-based algorithms, and all the nodes of the network are activated synchronously at each time step and update their current state with respect to local information and that gathered from their neighbors. These algorithms have a long history in the field of computer science and distributed computation as well as in many applications involving multiagent systems, where groups of agents need to agree on certain quantities of interest [16, 17]. They rely on the principle of dynamic averaging of certain signals between only neighboring agents. The use of these algorithms for secondary control of both AC and DC MGs has been considered recently [18–21]. In contrast, gossip algorithms are asynchronous, meaning that only one random node chooses another node (or more) to exchange their estimates and update them to the global information (e.g., the average value at each time step). Gossip algorithms are attractive because they are robust to unreliable wireless network conditions, and they have no bottleneck or single point of failure [22]. Many applications of these algorithms in MGs have been reported recently. Information is discovered in each local controller by use of a dynamic consensus algorithm assisting the realization of local decision making. Similarly, in [23] the essential information is obtained in each local controller for optimization and optimal sharing of compensating efforts

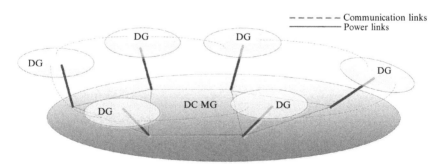

Fig. 9.5 Distributed control of MGs with physical and communication connections.

among distributed generators in an islanded AC MG. Xu and Liu [24] propose a fully distributed agent-based load restoration algorithm. As a consensus algorithm is used, the communication links are established only between neighboring nodes. The information state vector of each agent includes three parameters: the total net power, the indexes of the buses, and the loads that are ready to be restored. The global information is discovered by each agent knowing the current state of the generation and consumption in all the buses. Local decisions for restoration of loads can be made according to the priority of the loads. Furthermore, to ensure the fast and stable convergence of the consensus algorithm, Xu and Liu [24] also propose a distributed adaptive weights setting method and compares this method with existing algorithms. In the simulation it is demonstrated that this approach can be applied to MGs with any kind of topology, while guaranteeing better convergence of the global information discovery. A similar approach was applied in [25] for balancing the generation and consumption by coordinating the operation of doubly fed induction generators.

Considering the localized generation, storage, and consumption feature of the future smart grid, a wireless communication network is generally accepted as a highly flexible and low-cost way of facilitating the control and monitoring in MGs. Accordingly, a consensus theory-based distributed coordination scheme for a MG via wireless communication is proposed in [26] for coordinating the generation, storage, and consumption. Two types of information states are included: status and performance. The status information indicates the binary state of apparatus, such as fault/normal and available/unavailable. The performance information provides the measurement values (e.g., instance power generation, energy storage). By using the consensus algorithm, all the agents discover the status and performance of the apparatus in the MGs, and take decisions locally. In addition, the practical issues regarding wireless communications are also considered, such as the communication rates and range.

Meng et al. [23, 27, 28] propose consensus algorithm-based distributed hierarchical control for both DC and AC MGs. In those cases the primary power sharing control, secondary voltage/frequency restoration, and power quality compensation as well as tertiary optimization and decision making are all implemented in locally distributed controllers, facilitating the flexible operation of distributed generation units. Two study cases are established to verify the proposed method. In [28], the optimal load power sharing is achieved among several paralleled converters with enhanced overall system efficiency.

4. CONVERTERS IN DC MGS

Most of the publications concerned with design, modeling, and control of DC MGs use conventional and well-known topologies for power converters (e.g., nonisolated DC-DC converters). However, alternative topologies are required for many specific MG applications. For instance, in the case of bipolar MGs, as proposed by Kakigano et al. [29], a voltage balancer is required ensure that smoothing capacitors over each pole are equally charged (see Fig. 9.6). Later, three-level balancers were proposed to reduce the voltage stress on the devices (see Fig 9.7). The topology is very similar to that of neutral point-clamped converters but it has an inductor that is connected between the two neutral points of the series capacitors and the series switches. This inductor has two important roles: achieving the three-level states and transmitting the unbalanced power to the possible unbalanced load.

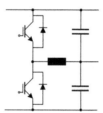

Fig. 9.6 Voltage balancing circuit for bipolar DC MGs: two-level balancer.

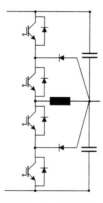

Fig. 9.7 Voltage balancing circuit for bipolar DC MGs: three-level balancer.

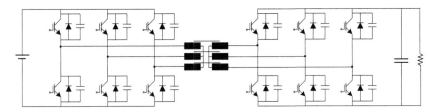

Fig. 9.8 Three phase dual active bridge converter.

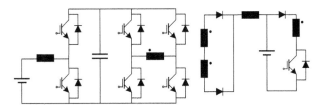

Fig. 9.9 Integrated full-bridge-forward DC-DC converter.

In Fig. 9.8, one can see a three-phase DAB converter that is often used for DC MGs where a high voltage gain is needed. A typical example is HEVs where a battery with a couple of hundred volts is interfaced with a distribution that operates at less than 100 V. Another example is a 380-V DC MG interfaced with a medium-voltage network. A single-phase DAB is also a prominent topology for lower-power applications but is not discussed here for lack of space.

Another example, shown in Fig. 9.9, is an alternative to the classical single-phase DAB. Comparing with the DAB, it has a lower number of devices and also low input and output current ripples. Therefore it would be natural to expect higher reliability and at the same time higher lifetime of batteries and other energy storage components in the system.

5. PROTECTION SYSTEMS FOR DC MGS

Protection is a field where power electronics will play an increasingly important role within MGs in the coming years. This will be specifically pronounced in the case of DC MGs, which have naturally resistive lines that lead to fast response. Moreover, DC does not provide natural extinguishment of the arc. To that end, despite many advantages that DC systems bring in relation to AC systems, the design of their protection presents a much more challenging task and also there is still a general lack of experience with operational issues of low-voltage DC systems [30].

The aim of this section is to provide the current status of the DC MG protection area, addressing the two main aspects: namely, protection devices, and design of protection systems.

5.1. Protection Devices

The commonest protection devices for DC systems are fuses and circuit breakers (CBs) [31]. However, both of them introduce large time constants and long time delays between the fault and the activation respectively. In addition, interruption of the current in both cases is accompanied by the appearance of the arc, which is not extinguished naturally as in the case of AC systems. Arc occurrence is a dangerous condition not only from the safety point of view but also causes contact erosion in a CB and consequently a short lifetime and high maintenance costs [30, 32].

Fuses operate on the principle of melting of the fuse in a heat-absorbing material. They are cheap and robust protection solutions that are equally applicable to AC and DC systems. Standard molded-case circuit breakers (MCCBs) that consist of a contractor, a quenching chamber, and a tripping device can also be applied. They can have either a thermal-magnetic or an electronic tripping device. In either case, sufficient voltage blocking capability can be achieved by the connection of contractors in series [31]. The problem with using MCCBs in power electronic intensive systems is that short-circuit currents are supplied mainly by filter capacitors that quickly discharge with short-circuit current of very high magnitude. However, it lasts for a very short time and it is therefore questionable if enough force can be generated to open the contacts of MCCBs [30].

Because of inherent tripping time delays and other problems associated with fuses and CBs, protection that uses only these devices has performance limitations. To cope with that, the used of power electronics-based protection devices has been proposed [33]. Several semiconductor devices such as gate turn-off thyristors (GTOs), insulated-gate bipolar transistors (IGBTs), and insulated-gate commutated thyristors (IGCTs) have been used in this context, each of them having some specific advantages and drawbacks. Meyer et al. [34] summarize and compare a number of different switch topologies from technical and economic aspects. It is demonstrated that GTOs and IGCTs have much lower on losses than IGBTs. Also it is estimated that an IGCT in a rectifier configuration has by far the highest reliability.

The performance of solid-state switches in terms of selectivity can be enhanced with current limiting circuits. Better selectivity comes from the fact that if an overload current is not interrupted immediately, as is the case with the sole CB, there is still time for the most selective CB to act and only the corrupted part of the grid is disconnected. Current limiters are usually installed in parallel with the solid-state switches and comprise one or several reactive elements [34]. Proper combination of inductors and capacitors can ensure a good compromise between the magnitude of short-circuit current and overvoltage before the fault. Damping resistors are also commonly added to avoid resonances.

An insightful example of the selection of appropriate protection devices for different components of a DC MG was presented in [31]. According to the specific fault-withstanding capability of each associated unit, it was concluded that converter protection must be very fast to limit the fault current through IGBT's antiparallel diodes, while battery protection can allow slower reaction times as it can withstand a considerably higher fault current without damage. For that matter, an ultrafast hybrid CB was proposed for converter protection and regular CBs were proposed for battery protection. On the other hand, fuses and MCCBs were used for feeder protection. In particular, it was claimed that it is better to use MCCBs closer to the loads since they can simultaneously trip both poles, while fuses are more appropriated for installation closer to the bus because of their magnetic sensing that gives good selectivity.

5.2. Design of Protection Systems

The objective of a protection system is to detect, locate, and isolate any fault so that the uncorrupted part of the DC MG can continue operating. To achieve this, overall system needs to be analyzed before the fault so as to select the most appropriate protection devices and protection method for a complete protection system. Desirable characteristics of such a system are reliability, selectivity, speed, cost, and simplicity, and it is often the case that these objectives are in conflict with each other [30]. In the remainder of this subsection, several protection approaches are summarized.

Protection devices were selected independently for each unit and their coordination was studied on an exemplary DC MG in [31]. A number of possible problems have been identified for this configuration, and a fully decentralized method was proposed as a remedy (i.e., use of DC link voltage information together with fault current). A handshake method was used in [35] to split the DC system into three possible zones and isolate the possible fault on any of the buses. The drawback of this method is that it needs to completely shut down the MG following the fault, which may be unnecessary in most conditions. An improvement on that solution was suggested in [36], where a loop-type DC bus was used together with protection that is able to detect the fault and separate only the faulty part, so that the rest of the system keeps operating. The method was realized by use of a communication system where one master unit receives current measurements from two slave units and gives commands to control the bus switches depending on the difference between the two measured currents.

A protection scheme that relies on a solid-state converter with fast-acting current-limiting capability was proposed in [33]. The respective converter uses two emitter turn off devices in an antiparallel configuration to realize both switching and protection. Furthermore, a relay coordination and backup protection schemes were addressed [33]. Coordination was realized in a fully decentralized fashion, by use of appropriate devices such as fuses and diodes.

Consideration of the dynamic characteristics of fault impedances is addressed in detail in [37], giving motivation for the use of a unit protection of DC MGs. For that matter, a flexible design framework for unit protection of DC MGs that achieves high sensitivity while taking into account the total cost of the system by use of commercial of-the-shelf technologies is proposed. Moreover, the practical limitations of communication technology are discussed, and it is shown that better discrimination of faults can be achieved by the proposed method.

5.3. Conclusions and Future Trends

Although the definition of a MG was established in early years of this century, there has been a lot of debate recently to reconsider the definition. There are several reasons for that. Firstly, the voltage and power levels that distinguish MGs from other power systems and power electronics applications are not clearly defined. Secondly, the power architectures deployed for different MG applications as reported in the literature have covered virtually all possible configurations, going from the highest to the lowest voltages (and powers). For instance, multiple DC MG clusters are controlled in essentially the same way as multiterminal DC systems used in high-voltage DC networks for the collection of offshore wind energy or medium-voltage DC networks used in all electric ships. Another example is the architecture used for conventional remote telecom stations which is structurally the same as the one used in modern HEVs. It is again interesting to notice that the very same architecture has been analyzed in the literature under the term *MG*. Finally, in some cases even the whole power distribution system has been considered as a connection of a number of MGs that can be reconfigured online according to different operating scenarios.

For all these reasons, it can be said that it is very difficult to isolate the MGs from many other applications and therefore the future research trends can, in the authors' opinion, be stated in only a generic way. In that sense, three major trends can be singled out:

- Implementation of advanced control techniques: MG power electronic converters should achieve the same generic functionalities as conventionally controlled systems but with better transient and steady-state performance.
- Development of new protection devices and their coordinated integration in the overall protection philosophy.
- Reliability oriented design of power electronic system-intensive MGs: Assessment of the system-level reliability is critical for new applications that include ESSs.

REFERENCES

[1] N. Pogaku, M. Prodanovic, T.C. Green, Modeling, analysis and testing of autonomous operation of an inverter-based microgrid, IEEE Trans. Power Electron. 22 (2) (2007) 613–625.

[2] R. Majumder, B. Chaudhuri, A. Ghosh, G. Ledwich, F. Zare, Improvement of stability and load sharing in an autonomous microgrid using supplementary droop control loop, IEEE Trans. Power Syst. 25 (2) (2010) 796–808.

[3] K. De Brabandere, B. Bolsens, J. den Keybus, A. Woyte, J. Driesen, R. Belmans, A voltage and frequency droop control method for parallel inverters, IEEE Trans. Power Electron. 22 (4) (2007) 1107–1115.

[4] J.A. Pecas Lopes, C.L. Moreira, A.G. Madureira, Defining control strategies for microgrids islanded operation, IEEE Trans. Power Syst. 21 (2) (2006) 916–924.

[5] J. Rocabert, A. Luna, F. Blaabjerg, P. Rodriguez, Control of power converters in AC microgrids, IEEE Trans. Power Electron. 27 (11) (2012) 4734–4749.

[6] Y.W. Li, C.-N. Kao, An accurate power control strategy for power-electronics-interfaced distributed generation units operating in a low-voltage multibus microgrid, IEEE Trans. Power Electron. 24 (12) (2009) 2977–2988.

[7] Q.-C. Zhong, Robust droop controller for accurate proportional load sharing among inverters operated in parallel, IEEE Trans. Ind. Electron. 60 (4) (2013) 1281–1290.

[8] D. Salomonsson, L. Soder, A. Sannino, An adaptive control system for a DC microgrid for data centers, IEEE Trans. Ind. Appl. 44 (6) (2008) 1910–1917.

[9] H. Kakigano, Y. Miura, T. Ise, Distributed voltage control for DC microgrids using fuzzy control and gain-scheduling technique, IEEE Trans. Power Electron. 28 (5) (2013) 2246–2258.

[10] K. Sun, L. Zhang, Y. Xing, J.M. Guerrero, A distributed control strategy based on DC bus signaling for modular photovoltaic generation systems with battery energy storage, IEEE Trans. Power Electron. 26 (10) (2011) 3032–3045.

[11] J. Schonberger, R. Duke, S.D. Round, DC-bus signaling: a distributed control strategy for a hybrid renewable nanogrid, IEEE Trans. Ind. Electron. 53 (5) (2006) 1453–1460.

[12] P. Karlsson, J. Svensson, DC bus voltage control for a distributed power system, IEEE Trans. Power Electron. 18 (6) (2003) 1405–1412.

[13] R.S. Balog, P.T. Krein, Bus selection in multibus DC microgrids, IEEE Trans. Power Electron. 26 (3) (2011) 860–867.

[14] E. Barklund, N. Pogaku, M. Prodanovic, C. Hernandez-Aramburo, T.C. Green, Energy management in autonomous microgrid using stability-constrained droop control of inverters, IEEE Trans. Power Electron. 23 (5) (2008) 2346–2352.

[15] J. Lagorse, M.G. Simoes, A. Miraoui, A multiagent fuzzy-logic-based energy management of hybrid Systems, IEEE Trans. Ind. Appl. 45 (6) (2009) 2123–2129.

[16] R. Olfati-Saber, J.A. Fax, R.M. Murray, Consensus and cooperation in networked multi-agent systems, Proc. IEEE 95 (1) (2007) 215–233.

[17] R. Olfati-Saber, R.M. Murray, Consensus problems in networks of agents with switching topology and time-delays, IEEE Trans. Autom. Control 49 (9) (2004) 1520–1533.

[18] A. Bidram, A. Davoudi, F.L. Lewis, J.M. Guerrero, Distributed cooperative secondary control of microgrids using feedback linearization, IEEE Trans. Power Syst. 28 (3) (2013) 3462–3470.

[19] F.L. Lewis, Z. Qu, A. Davoudi, A. Bidram, Secondary control of microgrids based on distributed cooperative control of multi-agent systems, IET Gener. Transm. Distrib. 7 (8) (2013) 822–831.

[20] J.W. Simpson-Porco, F. Dorfler, F. Bullo, Q. Shafiee, J.M. Guerrero, Stability, power sharing, and distributed secondary control in droop-controlled microgrids, in: Proceedings of the IEEE International Conference on Smart Grid Communications (SmartGridComm), 2013, pp. 672–677.

[21] V. Nasirian, S. Moayedi, A. Davoudi, F. Lewis, Distributed cooperative control of DC microgrids, IEEE Trans. Power Electron. PP (99) (2014) 1–11.

[22] A.G. Dimakis, S. Kar, J.M.F. Moura, M.G. Rabbat, Gossip algorithms for distributed signal processing, Proc. IEEE 98 (11) (2010) 1847–1864.

[23] L. Meng, T. Dragicevic, J.M. Guerrero, J. Vasquez, M. Savaghebi, F. Tang, Agent-based distributed unbalance compensation for optimal power quality in islanded microgrids, in: IEEE International Symposium on Industrial Electronics (ISIE 2014), 2014.

[24] Y. Xu, W. Liu, Novel multiagent based load restoration algorithm for microgrids, IEEE Trans. Smart Grid 2 (2011) 140–149.

[25] W. Zhang, Y. Xu, W. Liu, F. Ferrese, L. Liu, Fully distributed coordination of multiple DFIGs in a microgrid for load sharing, IEEE Trans. Smart Grid 4 (2) (2013) 806–815.

[26] H. Liang, B. Choi, W. Zhuang, X. Shen, A.A. Awad, A. Abdr, Multiagent co-ordination in microgrids via wireless networks, IEEE Wirel. Commun. 19 (2012) 14–22.

[27] L. Meng, J.C. Vasquez, J.M. Guerrero, T. Dragicevic, Agent-based distributed hierarchical control of DC microgrid systems, in: Proceedings of the ElectrIMACS Conference, 2014.

[28] L. Meng, T. Dragicevic, J.M. Guerrero, J.C. Vasquez, Dynamic consensus algorithm based distributed global efficiency optimization of a droop controlled DC microgrid, in: Proceedings of the IEEE International Energy Conference (EnergyCon2014), 2014.

[29] H. Kakigano, Y. Miura, T. Ise, R. Uchida, DC microgrid for super high quality distribution—system configuration and control of distributed generations and energy storage devices, in: Proceedings of the 37th IEEE Power Electronics Specialists Conference (PESC'06), 2006, pp. 1–7.

[30] R.M. Cuzner, G. Venkataramanan, The status of DC microgrid protection, in: Proceedings of the IEEE Industry Applications Society Annual Meeting (IAS'08), 2008, pp. 1–8.

[31] D. Salomonsson, L. Soder, A. Sannino, Protection of low-voltage DC microgrids, IEEE Trans. Power Deliv. 24 (3) (2009) 1045–1053.

[32] J.-M. Meyer, A. Rufer, A DC hybrid circuit breaker with ultra-fast contact opening and integrated gate-commutated thyristors (IGCTs), IEEE Trans. Power Deliv. 21 (2) (2006) 646–651.

[33] M.E. Baran, N.R. Mahajan, Overcurrent protection on voltage-source-converter-based multi-terminal DC distribution Systems, IEEE Trans. Power Deliv. 22 (1) (2007) 406–412.

[34] C. Meyer, S. Schroder, R.W. De Doncker, Solid-state circuit breakers and current limiters for medium-voltage systems having distributed power systems, IEEE Trans. Power Electron. 19 (5) (2004) 1333–1340.

[35] L. Tang, B.-T. Ooi, Locating and isolating DC faults in multi-terminal DC systems, IEEE Trans. Power Deliv. 22 (3) (2007) 1877–1884.

[36] J.-D. Park, J. Candelaria, Fault detection and isolation in low-voltage DC-bus microgrid system, IEEE Trans. Power Deliv. 28 (2) (2013) 779–787.

[37] S.D.A. Fletcher, P.J. Norman, S.J. Galloway, P. Crolla, G.M. Burt, Optimizing the roles of unit and non-unit protection methods within DC microgrids, IEEE Trans. Smart Grid 3 (4) (2012) 2079–2087.

CHAPTER 10

Power Electronic Converters in Microgrid Applications

M. Shahbazi[*], A. Khorsandi[†]

[*]School of Engineering and Computing Sciences, Durham University, Durham, United Kingdom
[†]Science and Research Branch, Islamic Azad University, Tehran, Iran

1. INTRODUCTION

Power electronics (PE) enables conversion from one form of electrical energy to another form, as shown in Fig. 10.1. Each PE interface includes a power converter that consists of power semiconductor switches and primary electronic elements (resistors, capacitors, inductors, transformers, diodes, etc.), and a control unit that manages the flow of power and conversion of voltages and currents.

The voltage levels, frequencies, and the voltage and current forms (AC or DC) of the two sides of the converter may be different. In some cases the power converter must make the bidirectional flow of power possible (e.g., battery interface) but in other cases the power might flow from one side (source) to the other (load).

2. POWER SEMICONDUCTOR SWITCHES

The heart of a PE converter is its semiconductor power switches. A power switch is a controlled electronic device that can switch between "on" and "off" states, and is used in PE converters to manipulate and shape the output voltage and currents. In an ideal case a power switch is switched immediately, has no resistance in its on state, and has infinite resistance in the off state. However, in reality, power switches are not ideal. Switching between on and off states takes time and the on state resistance has a positive value. Also, real switches have limitations in terms of the on state current and off state voltage that they can tolerate. Figs. 10.2 and 10.3 show approximate waveforms of voltage and current in an ideal and a real switch respectively, which in turn illustrates that in reality every switching leads to power loss in the switches [1]. Therefore even though PE converters offer

Microgrid
http://dx.doi.org/10.1016/B978-0-08-101753-1.00010-3

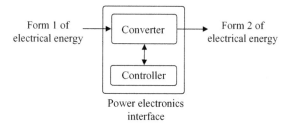

Fig. 10.1 General representation of a power electronics interface.

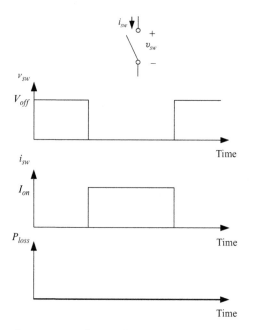

Fig. 10.2 Voltage and current waveforms of a power switch for an ideal switch.

considerably higher efficiency than other conversion circuits (e.g., linear regulators) and typically have efficiencies higher that 85% up to close to 100%, switching loss has to be considered in their design and is a limiting factor for the switching frequency of the converter [1].

Two types of power switches are normally used in power converters, especially for low to medium voltage and power applications such as in microgrids: metal-oxide-semiconductor field-effect transistors (MOSFETs)

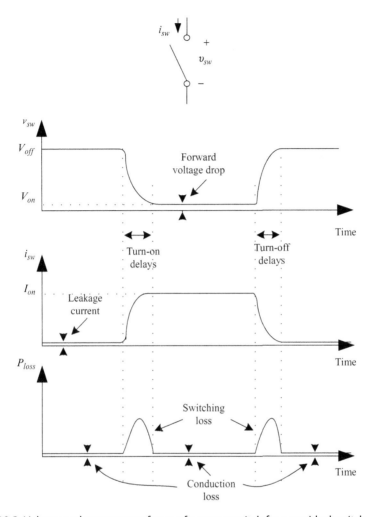

Fig. 10.3 Voltage and current waveforms of a power switch for a nonideal switch.

and insulated-gate bipolar transistors (IGBTs). IGBTs that can tolerate very large voltages (up to 6.5 kV) are available, and high-current IGBT modules are used in many applications with currents up to 3 kA [2]. Therefore IGBTs can be used in a wide power range of up to several megawatts [3]. MOSFETs, however, have a voltage rating of a maximum of a few hundred volts, and their power handling capability is limited to 100 kVA [4]. As the fastest switching devices, they can reach switching frequencies of more than 1 MHz [1]. The maximum switching frequency of IGBTs is, however,

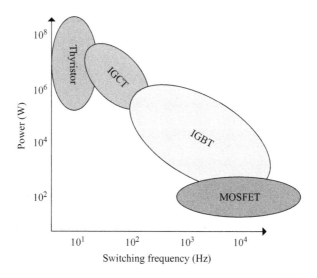

Fig. 10.4 Comparison of power switches in terms of power and frequency.

limited to a few tens of kilohertz. While there are other types of power semiconductor switches such as thyristors, gate turn-off thyristors and integrated gate-controlled thyristors (IGCTs), they are mostly used for very high power applications, and therefore their use in microgrids is not as common as that of IGBTs and MOSFETs. Fig. 10.4 shows how different power switches compare on the basis of their power ratings and switching capabilities [2].

3. CLASSIFICATION OF POWER CONVERTERS

On the basis of the input and output types of the voltages and currents shown in Fig. 10.1, a PE converter can be classified in general as follows [5]:

- DC–DC converters (see Fig. 10.5A): They get DC voltages and currents in the input, and generate controlled DC voltages and currents in the output. Although they have huge application (e.g., in consumer electronics), their use in microgrids is mostly limited to DC microgrids. However, they are also used in multistage power converters (e.g., when a change in the amplitude of the generated DC voltage of a photovoltaic (PV) module is needed before it is fed to the DC–AC converter).

- DC–AC converters (see Fig. 10.5B): These are also called *inverters*, and they produce AC outputs with controllable phase, frequency, and

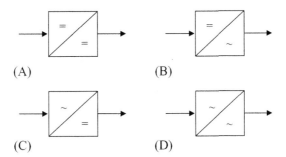

Fig. 10.5 General classification of power electronic converters: (A) DC-DC converter; (B) DC-AC converter; (C) AC-DC converter; (D) AC-AC converter.

magnitude. They are the central part of electric drive systems for control of the speed and torque of electric motors, and therefore they play an important role in many industrial applications, as well as in traction systems (including electric vehicles) and wind energy applications. Moreover, PVs, batteries, and fuel cells all produce DC voltages that must be converted to AC if they are to be used in an AC system.

- AC-DC converters (see Fig. 10.5C): These converters transform AC to DC with controllable voltage, and are also known as *rectifiers*. In its simplest form, a rectifier can consist of diode circuits, but controlled converters based on semiconductor switches can also be used to offer higher degrees of controllability and bidirectional power transfer.
- AC-AC converters (see Fig. 10.5D): These can transform AC to another AC with controllable phase, magnitude, and frequency. In most cases they include two stages of conversion from AC to DC and back to AC, and therefore normally have a DC link in between. However, some structures such as matrix converters can convert AC to AC without an intermediary DC link, but they normally have disadvantages such as higher complexity and cost. Most wind energy systems produce variable-frequency AC and therefore need an AC-AC converter for them to be connected to the grid.

Controlled AC-DC and DC-AC conversion can be done with a simple structure known as a *two-level converter*. This structure is also widely used in AC-DC-AC converters such as those used in some wind energy conversion systems (WECS). In the following, the structure, essential operation, modulation, and modeling of a two-level converter are explained.

4. CONVENTIONAL TWO-LEVEL CONVERTER

The three-phase two-level converter consists of three legs, each made of two switches and two antiparallel diodes connected to them. Fig. 10.6 shows the structure of a two-level converter. The switches are normally IGBTs or MOSFETs, as previously mentioned, but it is possible to use IGCTs to achieve higher powers. All three legs are connected to a DC link that includes a capacitor and provides constant DC voltage in inverter mode. When this structure is used in rectifier mode, the capacitor voltage is controlled to be constant. An L or LCL filter is typically used in the output to filter the voltage harmonics and allow connection to the grid in the case of grid-connected operation (e.g., in wind energy systems).

The converter is controlled by the appropriate switching on and off of its six switches. Each switch has a driver that gets the gating signals from the controller and turns the switch on/off. If two switches of the leg are switched on at the same time, the DC-link capacitor will be short-circuited. The gate signals of the two switches of a leg are therefore complementary, and are shown by T_x for leg x. If $T_x = 1$, this means that the upper switch of the leg is commanded to be turned on and the lower switch is to be in the off state. Conversely, $T_x = 0$ means that the upper switch is turned off and the lower switch is commanded to be in the on state.

Fig. 10.7 shows how the output voltage of a single leg will vary on the basis of its switching state. The output voltage of legs will be $\frac{+V_{DC}}{2}$ when the upper switch is on and $\frac{-V_{DC}}{2}$ when the lower switch is on. The output current can pass through the conducting switch or its antiparallel diode in each case, depending on the current direction. This also shows the importance of antiparallel diodes, considering that they provide a path for currents to pass, which is necessary because of the series filter inductances.

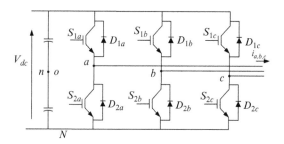

Fig. 10.6 Two-level three-phase converter.

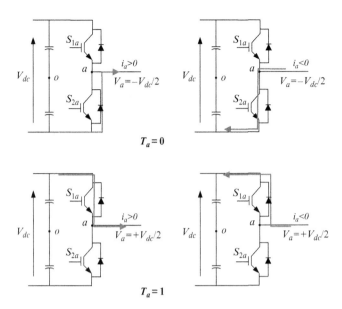

Fig. 10.7 Current flow paths in different switching states and phase current polarity.

Also, loads in a three-phase inverter normally have inductive parts and therefore need to always have a path for their currents.

By modulation of the two output values of each leg, it is possible to control its effective output voltage, and hence the output voltage of the converter. This is called *pulse width modulation* (PWM) and can be done in different ways, as explained in the following.

4.1. Pulse Width Modulation

PWM is used to control the average value of a waveform over a switching period by controlling the pulse width. The PWM generation methods can be classified into carrier-based and space vector modulation (SVM) methods. Both of these methods provide high-quality output voltages and currents and good transient response.

4.2. Carrier-Based Pulse Width Modulation

This is the most widely used PWM method mostly because of its ease of implementation. Many modern controller devices (microcontrollers and control ICs) have dedicated hardware for this type of PWM method. In this method a set of carrier and reference waveforms are compared to generate the PWM signals. Fig. 10.8 shows an example where a fixed reference value

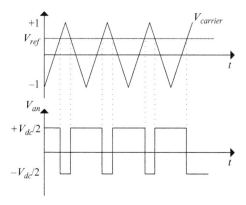

Fig. 10.8 Carrier and reference signals *(top)* and the resulting output voltage *(bottom)*.

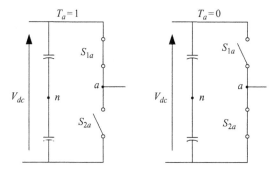

Fig. 10.9 The two switching cases of a single leg.

is compared with a carrier, and if the reference value is higher than that of the carrier, the upper switch of the controlled leg is switched on; otherwise it is switched off (and the lower switch is turned on). For AC systems, the carrier is typically a triangular signal with peak values of ±1. Fig. 10.9 shows the two switching cases. The frequency of the carrier determines the frequency at which the leg switches will be turned on and off. The resulting converter leg's output voltage is also shown in Fig. 10.8. The average voltage can be calculated as

$$V_{an,avg} = \frac{V_{dc}}{2}(V_{a,ref}).$$ (10.1)

Therefore if we choose an appropriate value for V_{ref}, the average value of the output voltage can be controlled. Normally, three reference voltages are used for the three legs of the converter in a three-phase system. In the case of a sinusoidal PWM (SPWM), these references are given as

$$V_{a,ref} = M \cos \theta, \tag{10.2}$$

$$V_{b,ref} = M \cos \left(\theta - \frac{2\pi}{3} \right), \tag{10.3}$$

$$V_{c,ref} = M \cos \left(\theta + \frac{2\pi}{3} \right), \tag{10.4}$$

where M is the modulation index and affects the magnitude of the output phase voltages and θ is the phase angle offset. Fig. 10.10 shows the voltage reference, carrier, and PWM voltages for a sample SPWM. Fig. 10.11 shows the harmonic content of the output line voltage. High-order harmonics are centered around multiples of the switching frequency and can be filtered to have the desired sinusoidal output voltage. Use of higher frequencies will result in higher-frequency harmonics, which are easier to filter, but this comes at the expense of higher switching losses and lower efficiency. On the other hand, as the power rating of converters and devices increases, lower switching frequencies are used. Therefore, as will be shown later, to prevent the need for irregularly large (and therefore expensive) filters in high-power applications, multilevel converter topologies are normally used [6]. Use of multilevel converters may lead to considerably fewer harmonics, which in turn allows easier and smaller filter design, as well as lower switching frequencies.

4.3. Zero Sequence Injection

Use of three sinusoidal voltage references in SPWM will result in sinusoidal phase and line voltages (after the filtering of higher-order harmonics). It is however possible to add a zero sequence signal (ZSS) to these reference values to form new modulation signals. Addition of the same ZSS to all three reference voltages does not change the output line-to-line and phase voltages; therefore it is used as a degree of freedom to reduce the current harmonics or improve the DC-bus utilization. In the case of the three sinusoidal references shown in Eqs. (10.2)–(10.4), the maximum value of M is 1, and higher values lead to overmodulation, which in turn leads to low-frequency voltage harmonics and is considered undesirable. For SPWM with three phase-shifted voltage references as shown in Eqs. (10.2)–(10.4),

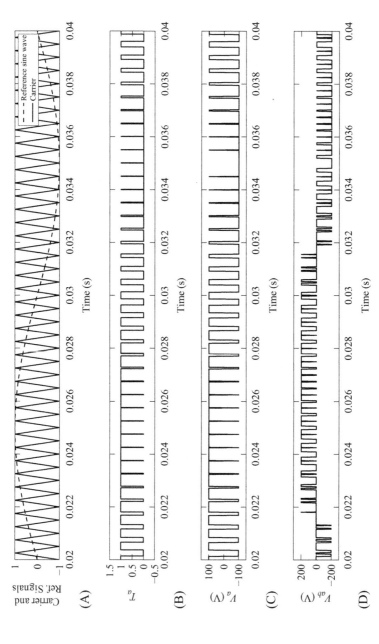

Fig. 10.10 Pulse width modulation signals for a two-level converter with 200 V in DC link, switching frequency of 2 kHz, and fundamental frequency of 50 Hz: (A) carrier and reference signals for phase a; (B) gate signal of phase a; (C) output voltage of phase a; (D) line voltage V_{ab}.

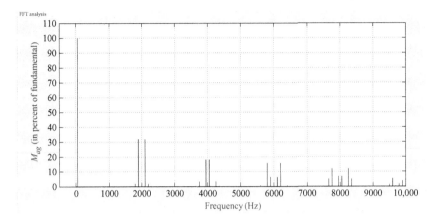

Fig. 10.11 Fast Fourier transform analysis of the output line voltage.

third-harmonic injection of the following form is a classic example of ZSS injection [6]:

$$V_{zss,3h} = -\frac{1}{6}\cos(3\theta). \tag{10.5}$$

In a more general case, the most widely used ZSS for a three-phase system with any type of voltage references is calculated as follows [7]:

$$V_{zss} = -\frac{1}{2}(\max(V_{a,ref}+V_{b,ref}+V_{c,ref})+\min(V_{a,ref}+V_{b,ref}+V_{c,ref})). \tag{10.6}$$

Fig. 10.12 shows the modified voltage references in the case of SPWM using the ZSS generation method of Eq. (10.6). In this case, the modulation index may be increased further without overmodulation resulting. It can be shown that the maximum modulation index can be increased in this way to [1]

$$M_{\max} = \frac{2}{\sqrt{3}}. \tag{10.7}$$

Therefore with the same DC-link voltage, larger sinusoidal output voltages can be constructed and therefore DC-link utilization can be improved.

4.4. Space Vector Modulation

For a three-leg two-level converter, each leg has two possible switching states; therefore the converter has $2^3 = 8$ total switching states. These states

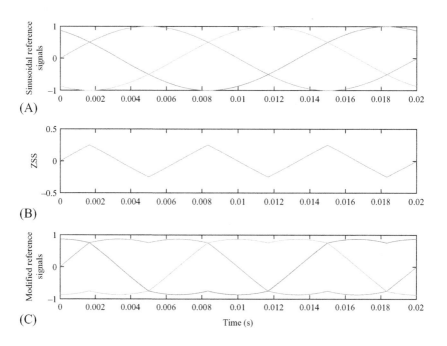

Fig. 10.12 (A) Sinusoidal reference voltages, (B) zero sequence signal, and (C) modified reference voltages with lower magnitudes.

are shown in Fig. 10.13. With use of space vector theory, the three-phase output voltages of each of these states can be represented by a complex vector [6]:

$$V = 2\sqrt{3}(V_a + V_b e^{\frac{j2\pi}{3}} + V_c e^{-\frac{j2\pi}{3}}). \tag{10.8}$$

Table 10.1 shows all eight possible switching states, as well as their corresponding output voltages and voltage space vectors, and Fig. 10.14 shows these space vectors. To realize a specific set of three-phase voltages, their space vector representation V is first calculated with use of Eq. (10.8). A number of voltage vectors of the converter $V_1, V_2, \ldots, V_n$ are then used with duty cycles of $d_1, d_2, \ldots, d_n$ according to the following equation so that their average resulting vector is equal to V:

$$V = d_1 V_1 + d_2 V_2 + \cdots + d_n V_n, \tag{10.9}$$

where $d_1 + d_2 + \cdots + d_n = 1$.

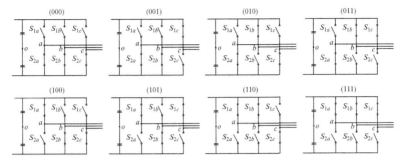

Fig. 10.13 Switching states of a two-level converter.

Table 10.1 Line voltages and space vectors of a two-level converter [6]

State (T_a, T_b, T_c)	Line voltages (V_{ab}, V_{bc}, V_{ca})	Space vector
000	0,0,0	0
100	$V_{dc},0,-V_{dc}$	$\frac{2}{\sqrt{3}}V_{dc}$
110	$0,V_{dc},-V_{dc}$	$\frac{2}{\sqrt{3}}e^{\frac{j\pi}{3}}V_{dc}$
010	$0,-V_{dc},V_{dc}$	$\frac{2}{\sqrt{3}}e^{\frac{2j\pi}{3}}V_{dc}$
011	$-V_{dc},0,V_{dc}$	$\frac{2}{\sqrt{3}}e^{j\pi}V_{dc}$
001	$0,-V_{dc},V_{dc}$	$\frac{2}{\sqrt{3}}e^{\frac{j4\pi}{3}}V_{dc}$
101	$V_{dc},-V_{dc},0$	$\frac{2}{\sqrt{3}}e^{\frac{j5\pi}{3}}V_{dc}$
111	0,0,0	0

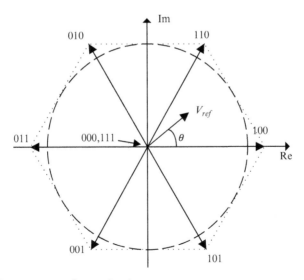

Fig. 10.14 Space vectors of a two-level converter.

In the simplest form of SVM, the two adjacent voltage vectors of V in the complex plane and zero voltage vectors are chosen and modulated.

4.5. Modeling

In leg x, where $x = a, b$, or c, when the upper switch (S_{1x}) is on, the output voltage of the leg (V_{an}) is independent of the output current direction and is equal to $\frac{V_{DC}}{2}$. In this state, if the output current is positive, it will flow through (S_{1x}); otherwise it will flow through (D_{1x}). When the lower switch of the xth leg is on, if the output current is negative, it will flow through (S_{2x}); otherwise it will flow through (D_{2x}). The output voltage of leg x is given by the following equations:

$$V_{xN} = T_x V_{dc}, \tag{10.10}$$

$$V_{xn} = \left(T_x - \frac{1}{2} \right) V_{dc}. \tag{10.11}$$

The output line-line voltage is given by

$$V_{xy} = T_x V_{dc} - T_y V_{dc}, \tag{10.12}$$

where $x, y = a, b$, or c and $x \neq y$. $T_y = 1$ if the upper switch of the yth leg is on, and $T_y = 0$ if the lower switch of the yth leg is on.

Eqs. (10.10)–(10.12) show the switched model for the two-level converter in Fig. 10.6, which includes high-frequency and low-frequency components. Generally, the compensators and filters of the control system have a low-pass behavior and almost eliminate high-frequency components. Therefore, for control design purposes, there is no need to know about the high-frequency components, and an average model is used [8]. The average output voltage of the xth leg is [8]

$$V_{xn,avg} = \left(d_x - \frac{1}{2} \right) V_{dc}, \tag{10.13}$$

where d_x is the duty ratio of the upper switch and is equal to the average value of T_x ($d_x = \overline{T}_x$). As the reference voltage ($V_{x,ref}$) varies between -1 and 1, the duty ratio (d_x) varies linearly between 0 and 1 and the output voltage varies between $\frac{-V_{dc}}{2}$ and $\frac{V_{dc}}{2}$ (see Fig. 10.8) which means

$$V_{x,ref} = 2d_x - 1. \tag{10.14}$$

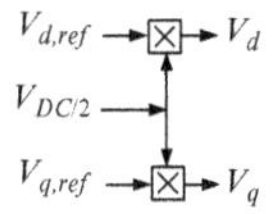

Fig. 10.15 Model of the two-level converter in the dq-frame.

Eqs. (10.13) and (10.14) result in

$$V_{xn,avg} = V_{x,ref} \frac{V_{dc}}{2}. \tag{10.15}$$

The dq-frame representation of the converter is shown in Fig. 10.15 and is given by [8]

$$V_{d,avg} = \frac{V_{dc}}{2} V_{ref,d}, \tag{10.16}$$

$$V_{q,avg} = \frac{V_{dc}}{2} V_{ref,q}, \tag{10.17}$$

where $V_{d,ref}$ and $V_{q,ref}$ are the components of V_{ref} on the d-axis and q-axis, respectively.

5. THREE-LEVEL NEUTRAL POINT-CLAMPED INVERTERS

A neutral point-clamped (NPC) inverter is a multilevel inverter used in microgrid applications. In this inverter the voltage across switches is halved in comparison with that in a two-level inverter. Moreover, lower total harmonic distortion in the output voltage is obtained. Fig. 10.16 shows a three-level NPC inverter. As can be seen, this inverter is composed of three legs, and each leg is composed of four switches (S_{1x}, S_{2x}, S_{3x}, S_{4x}, and $x = a, b$, and c) and their antiparallel diodes (D_{1x}, D_{2x}, D_{3x}, D_{4x}) and two additional diodes called *clamp diodes* (D_{5x} and D_{6x}).

To describe the principles of operation, leg a of the inverter is considered. To produce a positive output voltage (i.e., $v_a = \frac{V_{dc}}{2}$), S_{1a} and S_{2a} are turned on. In this case, if i_a is positive, S_{1a} and S_{2a} conduct. When i_a is negative, D_{1a} and D_{2a} conduct as shown in Fig. 10.17.

To generate a zero output voltage (i.e., $v_a = 0$), S_{2a} and S_{3a} are turned on. In this switching state, if i_a is positive, D_{5a} and S_{2a} conduct. When i_a is negative, D_{6a} and S_{3a} conduct as shown in Fig. 10.18.

To produce a negative output voltage (i.e., $v_a = \frac{-V_{dc}}{2}$), S_{3a} and S_{4a} are turned on. When i_a is positive, D_{3a} and D_{4a} conduct. If i_a is negative, S_{3a} and S_{4a} conduct as shown in Fig. 10.19.

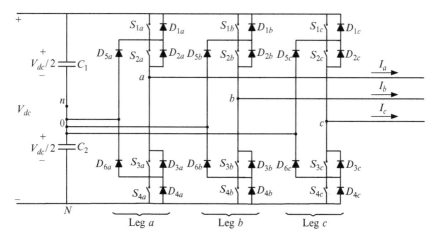

Fig. 10.16 Three-level neutral point-clamped inverter.

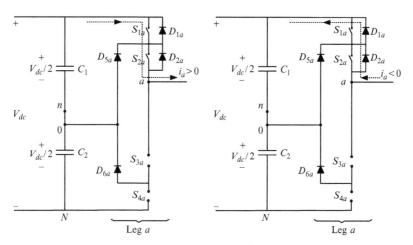

Fig. 10.17 Switching state of the three-level neutral point-clamped inverter when $v_a = \frac{V_{dc}}{2}$ (state 1).

According to the above explanations, three different switching states are possible:

- State 1: S_{1a} and S_{2a} are turned on (see Fig. 10.17) and $v_a = \frac{V_{dc}}{2}$.
- State 2: S_{2a} and S_{3a} are turned on (see Fig. 10.18) and $v_a = 0$.
- State 3: S_{3a} and S_{4a} are turned on (see Fig. 10.19) and $v_a = \frac{-V_{dc}}{2}$.

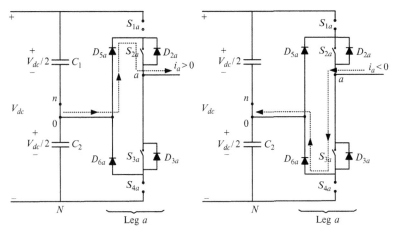

Fig. 10.18 Switching state of the three-level neutral point-clamped inverter when $v_a = 0$ (state 2).

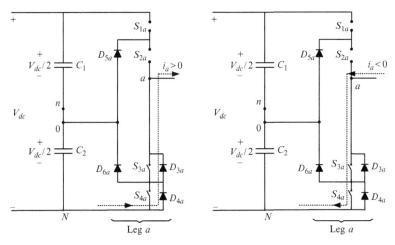

Fig. 10.19 Switching state of the three-level neutral point-clamped inverter when $v_a = \frac{-V_{dc}}{2}$ (state 3).

The instantaneous output voltage of the converter can be calculated as

$$V_a = (1 - T_a)\frac{V_{dc}}{2}, \tag{10.18}$$

where $T_a = 0$ if the switching state is in state 1, $T_a = 1$ if the switching state is in state 2, and $T_a = 2$ if the switching state is in state 3.

The average model of the NPC converter is similar to that of the two-level inverter given in Eq. (10.15).

6. DIFFERENT MODES OF OPERATION OF POWER CONVERTERS

Typically, power converters interfacing with distributed generation resources or storage systems are connected in parallel to each other and to the grid in grid-connected operation mode of the microgrid. However, they may also be used in islanded operation mode of the microgrid. In grid-connected mode the voltage at the point of common coupling is determined by the grid itself, and the converters are not allowed to regulate or oppose the voltage. The converters are typically used in a current or power control mode. There are also limits on the total harmonic distortion of the current injected into the grid [6].

When the grid connection or grid power is lost and the microgrid is working in islanded mode, the converters are in charge of controlling the voltage and frequency of the microgrid. If several converters are working in parallel, the load should be shared between them proportional to their nominal capacity. Many power management schemes are proposed in the literature for this case, most of them based on droop control methods [9–14].

7. POWER CONVERTER TOPOLOGIES FOR RENEWABLE AND DISTRIBUTED ENERGY SYSTEMS

Some level of power conversion is necessary between renewable and distributed energy generation systems and the rest of power system to convert the generated power to utility-compatible form. In this section, some of the PE topology interfaces in these applications are briefly reviewed.

7.1. PV Systems

PV modules produce DC currents and voltages and therefore need an inverter for them to be connected to the microgrid (or utility). Many different inverters are proposed for PV applications in comparison with other renewable systems such as wind energy systems. This is mostly because PV systems cover a wider range of power, and are also regularly installed on a residential scale, where efficiency and personal safety are of high importance.

In centralized PV systems (solar farms), PV modules can be connected to each other in series and parallel to get to the required current and voltage

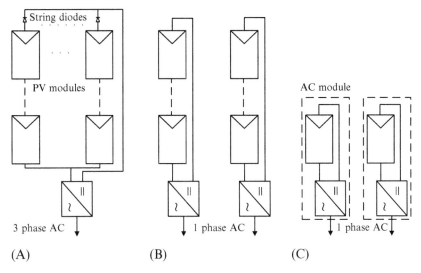

Fig. 10.20 Different arrangements of PV modules and inverters: (A) centralized; (B) string configuration; (C) AC module technology.

level. The output of the PV array is then converted to AC by means of a three-phase inverter. Fig. 10.20A shows how PV modules are connected to the inverter in such a configuration [15]. Fig. 10.21 shows as example of the inverter and connection to the utility [5]. A filter capacitor is used in the input, and an LC filter may be used in the output of the inverter. A

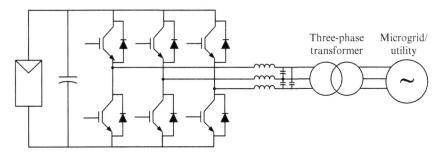

Fig. 10.21 Single-stage PV power converter.

three-phase transformer boosts the voltage level and connects the inverter to the microgrid/utility, while providing galvanic isolation.

Putting solar modules in such a series and parallel configuration has its own problems such as inflexible design and power loss due to centralized maximum power point tracking (MPPT) [15]. In newer designs, a string of PV modules are connected in series and then the DC output is converted to AC by means of single-phase inverters. If enough modules are put in series, the resulting voltage may be large enough that no voltage amplification will be necessary, otherwise a DC-DC converter or a line-frequency transformer may be used for voltage amplification. Use of this configuration leads to higher efficiency compared with a central scheme. This configuration is depicted in Fig. 10.20B.

On the other hand, low-frequency transformers are bulky, expensive, and have low efficiency, and therefore are considered as poor components. To avoid their use in PV applications, multistage conversion systems are used, especially in residential-scale applications. Fig. 10.22 shows an example of a dual-stage conversion system [5]. A DC-DC converter is used to boost the voltage of the PV array and to perform the MPPT. Normally a high-frequency transformer is used in the DC-DC converter to provide galvanic isolation as well as voltage boost. The three-phase inverter then converts the resulting DC voltage to AC voltage and controls the power flow and power factor at the point of coupling.

In residential applications, module-integrated converters have been the focus of research in recent years. Each PV module has a grid-connected inverter, constructing a so-called *AC module*, which allows a high degree of modularity and plug-and-play capability. Since each individual module is controlled by MPPT, the overall energy-harvesting capability is improved. This is the configuration shown in Fig. 10.20C.

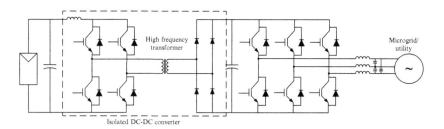

Fig. 10.22 Dual-stage PV power converter.

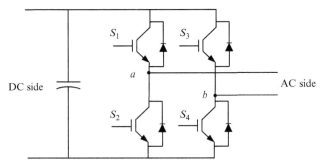

Fig. 10.23 Full-bridge inverter.

The AC module consists of a PV module and an inverter, and supports individual MPPT. Because of the modularity offered, the system can be enlarged easily, and also plug-and-play capability will be inherent in such a system. As shown in Fig. 10.20, these modules normally use single-phase inverters to connect to the microgrid/utility.

The most basic single-phase inverter is the full-bridge (or H-bridge) converter, which is the simplified and single-phase version of the three-phase two-level inverter in Fig. 10.6. Fig. 10.23 shows the full-bridge inverter. However, the full-bridge inverter itself is normally not used for PV applications, and more complicated topologies based on this converter are employed for inverters of AC modules. Many of these topologies are proposed in the literature [15–22]. In general, like three-phase inverters, single-phase inverters can also be based on single-stage conversion with a low-frequency transformer or double-stage conversion with a high-frequency transformer at the DC side. In recent years, many transformerless topologies have been proposed for AC modules. They offer more efficiency compared with structures with low-frequency or high-frequency transformers, and are also lighter, less costly, and less bulky [20]. Transformerless structures can be divided into three categories: two-stage, pseudo-DC-link, and single-stage topologies.

Two-stage topologies have a DC-DC converter that amplifies the voltage and performs MPPT, as well as a DC-DC inverter stage, similarly to what is presented in Fig. 10.22 but without a transformer and with a single-phase inverter. Fig. 10.24 shows an example structure, consisting of a boost DC-DC converter placed in series with a full-bridge converter [20]. Many alternative topologies have been presented [20], including a soft switching, buck-boost converter as the DC-DC converter, and half-bridge or neutral-point converters as the inverter.

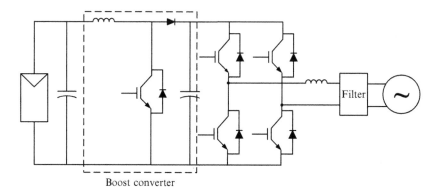

Boost converter

Fig. 10.24 Two-stage transformerless converter (boost converter and full-bridge inverter).

Since each level of power conversion produces losses, two-level conversion can decrease the overall efficiency of the system. Reducing the number of power conversion stages can lead to an increase in the overall efficiency; therefore single-stage topologies are also being studied, and several topologies have been proposed in the literature. They can also offer higher reliability and lower cost. Fig. 10.25 shows an example where a boost converter is integrated with the full-bridge converter [20]. Several other topologies based on integration of boost or buck-boost converters or based on a Z-source inverter are also available [20]. However, single-stage topologies generally need a capacitor in parallel with the PV module with values higher than that of the two-stage capacitors, where it is placed in the DC link. Bulky electrolyte capacitors that are used in these structures may be a limiting factor for the lifetime and reliability of the converter. Their control is complex, and despite there being a single stage, it may be difficult to reach higher efficiencies in practice.

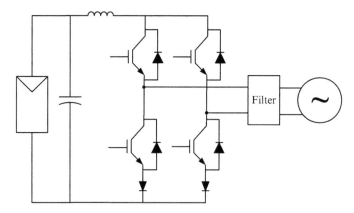

Fig. 10.25 Single-stage transformerless converter with integrated boost inverter.

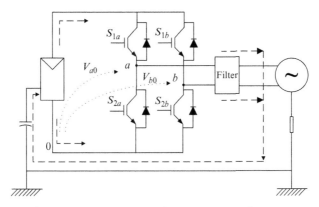

Fig. 10.26 Path of leakage current in a transformerless PV configuration.

Another very important aspect in transformerless grid-connected systems is that the leakage current generated by the PV parasitic capacitors and the grounding should be carefully dealt with. Because of the parasitic capacitance between the PV cells and the metal frame of the panel (which is usually grounded), a common mode current can flow if there is no galvanic isolation. This will result in greater electromagnetic interference and also loss of electrical safety and possibly disconnection from the grid by the protection devices on the basis of standards such as DIN VDE 0126-1-1 [18]. Fig. 10.26 shows the path of current in a simple transformerless system [18, 23, 24].

In a single-phase full-bridge converter, bipolar and unipolar PWM methods can be used, but in most cases a unipolar PWM is used to improve the output quality of the converter. In the case of bipolar PWM, the switching command of phase a is calculated similarly to that in the case of a three-phase inverter by comparison of V_{ref} with the carrier signal. The switching commands of the other leg (leg b) are complementary of those of the first leg, meaning that the gate commands for S_1 and S_4 are the same, and similarly the same gate signal is applied to both S_2 and S_3. Fig. 10.27 shows the carrier and reference voltages, as well as the output voltage of each leg and the resulting output voltage of the inverter.

In the case of unipolar PWM, however, the switching commands for the two legs of the inverter are calculated by comparison of V_{ref} and $-V_{ref}$ with the carrier signal. Fig. 10.28. shows as example of unipolar waveforms, and shows that this PWM results in three-level output voltage with twice the switching frequency. It can be verified that bipolar PWM results in a

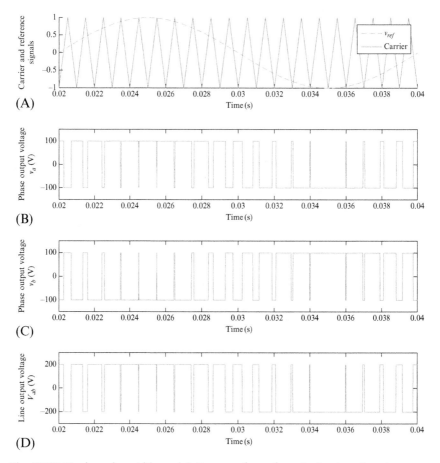

Fig. 10.27 Bipolar pulse width modulation waveforms for a single-phase inverter.

two-level output voltage with higher total harmonic distortion and more stress on the output filter. That is generally why unipolar PWM is used.

When S_1 and S_4 are on, the common mode voltage $v_{cm} = (V_{a0} + V_{b0})/2$ is equal to V_{dc}, as it is when S_2 and S_3 are on. On the other hand, in the "freewheeling" interval when S_1 and S_3 or S_2 and S_4 are on, the common mode voltage will be $+V_{dc}$ or $-V_{dc}$ respectively, giving a high-frequency common mode voltage, which in turn will result in high leakage current. One can minimize the common mode current by keeping the common mode voltage constant.

Two solutions to reduce the leakage current by the disconnecting of the output of the inverter from the input during the freewheeling period

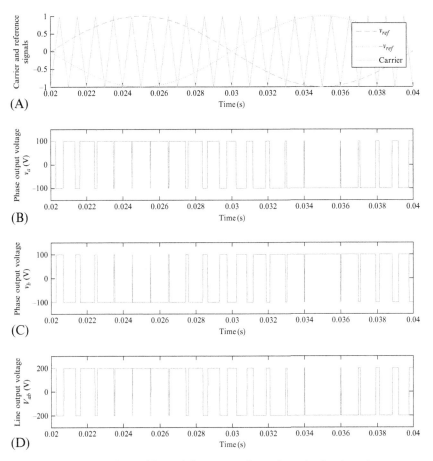

Fig. 10.28 Unipolar pulse width modulation waveforms for a single-phase inverter.

are shown in Fig. 10.29 [18]. An additional switch can be added at the DC side to disconnect the two sides during freewheeling, as shown in Fig. 10.29A. This topology is called *H5* and is used in SMA inverters. It is also possible to use two switches at the AC side to decouple the AC and DC sides (and therefore decoupling of the grid and PV system), as shown in Fig. 10.29B. This topology is called the highly efficient reliable inverter concept (HERIC) and is used in Sunways inverters [18]. Both these topologies can reduce the leakage current. Several other topologies are also available that use additional switches to clamp the load voltages to half of the DC bus voltage during freewheeling, therefore keeping the common mode voltage constant and the leakage current at a minimum [16–18, 25].

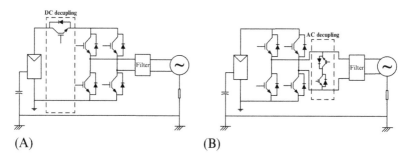

Fig. 10.29 AC and DC decoupling for a full-bridge inverter.

7.2. Wind Energy Conversion Systems (WECS)

Modern variable-speed wind power systems use PE to adjust the rotor speed of the generator and to control the flow of power to the grid. They can be divided into two categories: WECS with partially rated converters and systems with fully rated converters.

The partially rated WECS is based on a doubly fed induction generator. A back-to-back converter is commonly used in the rotor circuit, while the stator is connected to the grid, as shown in Fig. 10.30A [5]. Only a fraction of the total power (less than 30%) flows through the rotor and the PE converter; therefore the converter power rating is less than that for WECS with fully rated power converters. The machine's torque and reactive power can be controlled by means of the rotor-side converter. The grid-side converter is in charge of controlling the DC-link voltage, as well as its own reactive power. The generator speed can be controlled around its synchronous speed by means of the power converter.

In another configuration, it is possible to use a conventional or permanent magnet synchronous or induction machine as the generator, and use a full-scale PE converter to convert the generated power to acceptable voltage and frequency and also to control the torque and the speed of the generator. Fig. 10.30B shows a WECS with a fully rated PE interface [5]. An AC-DC-AC converter is used to convert variable-frequency and variable-amplitude AC to constant-voltage DC and then to constant-voltage and constant-frequency AC of the grid. The whole generated power flows through the converter in this case. The converter also performs reactive power compensation at the grid side. The rectifier

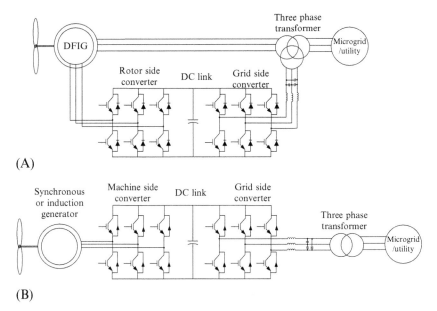

Fig. 10.30 Power converters for WECS: (A) partially rated for DFIG; (B) series fully rated for a synchronous/induction generator.

may be a controlled two-level converter, but in the case of a permanent magnet synchronous generator can be also a simple diode-based rectifier to reduce the cost. In this case a boost DC-DC converter may be placed in the DC link to smooth the DC voltage and therefore avoid low-frequency pulsations of the diode rectifier. For higher powers it is possible to use multilevel converters such as NPC or cascaded H-bridge converters instead of conventional two-level ones [26].

7.3. Storage System Converters

Batteries produce DC voltages and need a PE interface for their connection to the grid. Since they can be either in a charging or a discharging state, the PE interface should be bidirectional. Normally several battery cells are placed in a series and parallel combination to provide the required voltage and current level. A two-level converter, such as that in Fig. 10.21, is inherently bidirectional. Therefore it can be used as a simple form of

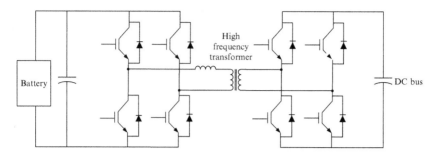

Fig. 10.31 Battery charge/discharge control with a bidirectional DC-DC power converter.

a battery interface for three-phase systems. To add galvanic isolation, a transformer can be placed either as a low-frequency transformer in the output or as a high-frequency transformer in the DC part, similarly to what is shown in Fig 10.22 for PV systems. Fig. 10.31 shows a PE interface for a battery system with galvanic isolation [5]. For single-phase applications, similar topologies with single-phase inverters can be used [5].

8. CONCLUSION

In this chapter, power semiconductor switches as the building block of converters were reviewed. Then the conventional two-level converter as one of the most widely used converters in microgrids was examined and different switching strategies (i.e., PWM, SPWM, ZSS injection, and SVM) for this converter were introduced. Next, an NPC converter was analyzed as a multilevel inverter that can be used in higher-power applications. Moreover, the switching model and the average model of the two converters were presented. Finally, some of the main PE interfaces for renewable and distributed generation systems were reviewed.

REFERENCES

[1] M.H. Rashid, Power Electronics Handbook: Devices, Circuits and Applications, Academic Press, New York, 2010.
[2] M. Rahimo, Future trends in high power MOS controlled power semiconductors, in: IEEE ISPS, 2012.
[3] N. Mohan, T.M. Undeland, W.P. Robbins, Power Electronics: Converters, Applications, and Design, 3rd ed., John Wiley & Sons, New York, 2007.
[4] R.M. Strzelecki, Power Electronics in Smart Electrical Energy Networks, Springer Science+Business Media, Berlin, 2008.
[5] S. Chakraborty, M.G. Simões, W.E. Kramer, Power electronics for renewable and distributed energy systems, in: A Source Book of Topologies, Control and Integration, Springer, London, 2013.

[6] S.M. Sharkh, M.A. Abusara, G.I. Orfanoudakis, B. Hussain, Power Electronic Converters for Microgrids, John Wiley & Sons, New York, 2014.

[7] B.M. Wilamowski, J.D. Irwin, Power Electronics and Motor Drives, CRC Press, Boca Raton, 2016.

[8] A. Yazdani, R. Iravani, Voltage-Sourced Converters in Power systems: Modeling, Control, and Applications, John Wiley & Sons, New York, 2010.

[9] A. Khorsandi, M. Ashourloo, H. Mokhtari, R. Iravani, Automatic droop control for a low voltage DC microgrid, IET Gener. Transm. Distrib. 10 (1) (2016) 41–47.

[10] J.M. Guerrero, J.C. Vasquez, J. Matas, L.G. de Vicuna, M. Castilla, Hierarchical control of droop-controlled AC and DC microgrids: a general approach toward standardization, IEEE Trans. Ind. Electron. 58 (1) (2011) 158–172.

[11] S. Augustine, M.K. Mishra, N. Lakshminarasamma, Adaptive droop control strategy for load sharing and circulating current minimization in low-voltage standalone DC microgrid, IEEE Trans. Sustain. Energy 6 (1) (2015) 132–141.

[12] R. Eriksson, J. Beerten, M. Ghandhari, R. Belmans, Optimizing DC voltage droop settings for AC/DC system interactions, IEEE Trans. Power Deliv. 29 (1) (2014) 362–369.

[13] X. Lu, J.M. Guerrero, K. Sun, J.C. Vasquez, An improved droop control method for DC microgrids based on low bandwidth communication with DC bus voltage restoration and enhanced current sharing accuracy, IEEE Trans. Power Electron. 29 (4) (2014) 1800–1812.

[14] A. Khorsandi, M. Ashourloo, H. Mokhtari, A decentralized control method for a low-voltage DC microgrid, IEEE Trans. Energy Convers. 29 (4) (2014) 793–801.

[15] S.B. Kjaer, J.K. Pedersen, F. Blaabjerg, A review of single-phase grid-connected inverters for photovoltaic modules, IEEE Trans. Ind. Appl. 41 (5) (2005) 1292–1306.

[16] T. Kerekes, R. Teodorescu, P. Rodríguez, G. Vázquez, E. Aldabas, A new high-efficiency single-phase transformer-less PV inverter topology, IEEE Trans. Ind. Electron. 58 (1) (2011) 184–191.

[17] W. Li, Y. Gu, H. Luo, W. Cui, X. He, C. Xia, Topology review and derivation methodology of single-phase transformer-less photovoltaic inverters for leakage current suppression, IEEE Trans. Ind. Electron. 62 (7) (2015) 4537–4551.

[18] G. Buticchi, D. Barater, E. Lorenzani, G. Franceschini, Digital control of actual grid-connected converters for ground leakage current reduction in PV transformer-less systems, IEEE Trans. Ind. Inform. 8 (3) (2012) 563–572.

[19] Q. Li, P. Wolfs, A review of the single phase photovoltaic module integrated converter topologies with three different DC link configurations, IEEE Trans. Power Electron. 23 (3) (2008) 1320–1333.

[20] D. Meneses, F. Blaabjerg, O. Garcia, J.A. Cobos, Review and comparison of step-up transformerless topologies for photovoltaic AC-module application, IEEE Trans. Power Electron. 28 (6) (2013) 2649–2663.

[21] M. Amirabadi, A. Balakrishnan, H.A. Toliyat, W.C. Alexander, High-frequency AC-link PV inverter, IEEE Trans. Ind. Electron. 61 (1) (2014) 281–291.

[22] M. Islam, S. Mekhilef, M. Hasan, Single phase transformer-less inverter topologies for grid-tied photovoltaic system: a review, Renew. Sustain. Energy Rev. 45 (2015) 69–86.

[23] B. Yang, W. Li, Y. Gu, W. Cui, X. He, Improved transformer-less inverter with common-mode leakage current elimination for a photovoltaic grid-connected power system, IEEE Trans. Power Electron. 27 (2) (2012) 752–762.

[24] H. Xiao, S. Xie, Leakage current analytical model and application in single-phase transformer-less photovoltaic grid-connected inverter, IEEE Trans. Electromagn. Compat. 52 (4) (2010) 902–913.

[25] H. Xiao, S. Xie, Y. Chen, R. Huang, An optimized transformer-less photovoltaic grid-connected inverter, IEEE Trans. Ind. Electron. 58 (5) (2011) 1887–1895.

[26] F. Blaabjerg, M. Liserre, K. Ma, Power electronics converters for wind turbine systems, IEEE Trans. Ind. Appl. 48 (2) (2012) 708–719.

CHAPTER 11

Power Talk: Communication in a DC Microgrid Through Modulation of the Power Electronics Components

M. Angjelichinoski, C. Stefanovic, P. Popovski, F. Blaabjerg
Aalborg University, Aalborg, Denmark

1. INTRODUCTION

In a typical microgrid setting the distributed generators (DGs) interface with the buses through flexible *power electronic converters* that support digital signal processing; the high penetration of flexible and programmable power electronic interfaces is a distinguishing characteristic of small microgrid systems, not commonly encountered in traditional power grids [1–4].

The converters implement a set of control mechanisms to regulate the power flow in the microgrid and to balance the power supply and demand [5]. Microgrid control at different levels has different requirements when it comes to communication support [5–8]. Traditionally, an external communication network, such as wireless or power-line communication, is used to support the secondary/tertiary levels [9].

However, because of the small scale and simple architecture, the control messages are rather sporadic and short, requiring small bandwidth but high reliability, for which the installation of external communication hardware, including a dedicated power-line communication modem, might not prove to be cost-efficient.

Another important argument against making the microgrid system reliant on an external communication system is the accordance with the principle of self-sustainability, which foresees that the system is able to provide (optimized) services by its own resources whenever possible. Finally, the reliability of the microgrid control architecture, when coupled with an external communication network, can be calculated as the product of the

Microgrid
http://dx.doi.org/10.1016/B978-0-08-101753-1.00011-5

separate reliabilities of the microgrid and the communication network; thus the overall reliability of the control algorithms is inherently reduced when an external communication system is used.

Recent work suggests exploitation of the potential residing in the power electronic converters enabled with digital signal processing to provide communication capabilities over bus lines [10–13]. Motivated by this idea, as well as the requirement for low communication bandwidth, we present a novel, inexpensive, and reliable ultra narrowband power-line communication solution designed specifically for DC microgrids. We refer to it as *power talk*, as the information transfer is achieved through subtle deviations of the power supplied by control units [14–16].

In particular, power talk exploits the flexibility of the electronic converters and modulates the information in the parameters of the primary loops that regulate the common bus voltage level, without using additional communication hardware. The general architecture is illustrated in Fig. 11.1.

The higher control layers "embed" the information that they need to exchange with peer entities into subtle deviation of the parameters of the primary control loops. In this way the control units *create* communication channels, reflected in the relation that characterizes the inputs (i.e., modulated control parameters) and the outputs (i.e., the observed bus voltage). In this respect, power talk is envisioned as an advanced tool to upgrade the control functionality of power electronic converters with low-bandwidth communication capability. Its reliability/availability draws on the reliability/availability of the microgrid bus, providing complete

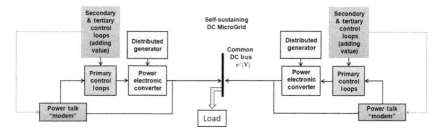

Fig. 11.1 General architecture of a DC microgrid system using power talk. The local upper layer control entities "embed" the information that is necessary to communicate for the respective control applications into modulated deviations of the primary control parameters, thereby creating communication channel over the microgrid DC bus.

coverage over the microgrid system (as all control units measure the bus voltage) without using additional communication hardware. Moreover, unlike the approaches proposed in [10–13], power talk offers a general digital interface with a potential to meet the communication requirements of the secondary/tertiary control and can be used in a variety of applications (i.e., it is not confined to a prespecified application framework).

The main objective of this chapter is to introduce and illustrate set of communication and signal processing techniques that foster reliable power talk communication. In particular, Section 2 describes the motivation for the development of power talk and highlights its application potential. Section 3 discusses the general power talk communication model, which exhibits some nonstandard properties in communication terms. Section 4 focuses on the design of communication strategies that enable reliable power talk communication amid the nonstandard features of the microgrid as a communication channel. Section 5 discusses and illustrates the effect of microgrid constraints on power talk, and, Section 6 concludes the chapter.

2. CONTROL IN DC MICROGRIDS

To describe the motivation for the development of power talk, we review the basic principles of control in DC microgrids. Leveraging on the advantages of both centralized and distributed strategies, the control mechanisms in a microgrid are usually a combination of both and are organized in a layered, hierarchical structure [5, 6]. The hierarchy comprises three levels: *primary*, *secondary*, and *tertiary*. The distinction between the levels is based on the frequency of each of the loops (i.e., the control bandwidth) as well as the set of responsibilities they have.

In summary, the primary level is traditionally designed in a distributed manner, where each unit uses only locally available measurements and avoids the use of communication. This approach makes the basic system operation independent of an external communication network and provides a strong self-sustaining capability. However, primary control based only on locally observable quantities produces a strictly *suboptimal* operating point. Therefore centralized or distributed secondary/tertiary control algorithms are designed to provide *optimal* references for the primary level, requiring exchange of local information among units.

In the following section, we briefly review the details of each control level.

2.1. Primary Control

The primary control level defines the mechanisms for stable operation of different DGs in the microgrid and enables load-dependent power sharing among them. This is the fastest control level, and its mechanisms are essential for proper operation of the microgrid system. The general trend is to make the primary control as robust as possible by use of decentralized mechanisms that do not require external communication infrastructure and rely only on local measurements at each DG. This way, the microgrid can provide its basic services to end consumers even when the system operates in stand-alone mode and the upper-layer controllers might be inactive. Depending on the configuration of the primary control loops in DC microgrids, the power electronics-controlled DGs fall into two general categories: voltage source converters (VSCs), which perform voltage regulation, and current source converters (CSCs), which do not participate in voltage regulation.

Fig. 11.2 shows the primary control diagram of a DG m that operates as a VSC. The primary control consists of two sets of (nested) control loops: fast inner and slower outer (i.e., power sharing) control loops. The inner loop comprises the current and voltage control loops. These loops are usually very fast (on the order of kilohertz depending on the switching frequency of the converter), forcing the output current I_m and the output voltage V_m^* to follow the predefined references, as shown in Fig. 11.2. The bus voltage reference signal V_m^* that is fed to the inner voltage control loop of the unit m is generated by the outer, power sharing loop and standard set according to the *droop* law [5, 6]:

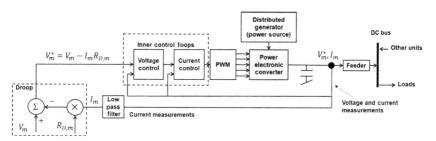

Fig. 11.2 Primary control of a power electronically controlled distributed generator configured as a voltage source converter connected to a DC bus. The inner control loops force the output voltage and current to follow the reference provided by the outer, power sharing control loop, which uses droop control to generate the voltage reference.

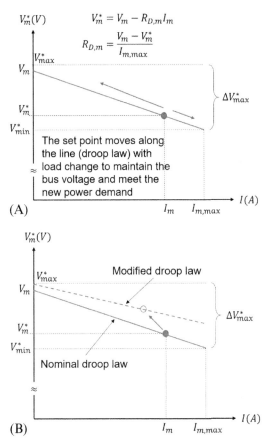

Fig. 11.3 Droop control in DC microgrids. (A) The (nominal) droop control law that regulates the bus voltage by generating a voltage reference for the inner loops, responding to the load changes in the system. (B) Modification of the droop control law as the basic idea of power talk (i.e., changing the droop control parameters in a controlled manner results in deviations of the operating set point that can be used for information transfer).

$$V_m^* = V_m - I_m R_{D,m}, \tag{11.1}$$

where V_m and $R_{D,m}$ are the droop control parameters—namely the *reference voltage* and the *virtual resistance* (also referred to as *droop slope*)—and I_m is the output current of the unit. The droop law is depicted in Fig. 11.3A. Evidently, the operational set point (V_m^*, I_m) moves along the droop line as the loads change, maintaining the bus voltage within some predefined limits, denoted as V_{min}^* and V_{max}^*, meeting the new demand. V_m and $R_{D,m}$ are controllable; usually their values are set by the upper (tertiary) control layers in an optimized manner. For example, the reference voltage V_m

represents the operating point for the voltage of bus m and is set by the upper control layers (i.e., the secondary/tertiary controllers) usually in an optimized manner (often the reference voltage is an outcome of a global optimization procedure that produces the optimal operating points of the microgrid given the conditions of the system). Moreover, the virtual resistance $R_{D,m}$ is set to enable proportional power sharing among the units on the basis of their current rating $I_{m,\max}$. In the steady state the VSC units can be represented by equivalent voltage sources with parameters V_m and $R_{D,m}$.

On the other hand, CSC units are usually constant-power units that either consume or supply power [17]. They do not implement the inner voltage control loop; instead, the reference for the inner current control loop is provided by upper control layers and is determined by the amount of power that the unit is able to supply/consume. Typically, renewable power sources, extracting the maximum possible power, are operated as CSC units, with the inner current control loop reference generated by the maximum power point tracking algorithm. In principle, a unit can switch seamlessly between VSC mode and CSC mode. As discussed later, in power talk we assume that each unit operates in VSC mode; nevertheless, we note that the concepts can be easily extended to the general case, including CSC units with proper modifications.

The basic idea of power talk is to change (i.e., modulate) V_m and $R_{D,m}$ in a controlled manner, thereby inducing variations in the output voltage V_m^* and current I_m that can be detected by other units in the system, thus leading to information exchange [14–16]. The principle is illustrated in Fig. 11.3B.

Effectively, by modulating the droop control parameters that maintain and regulate the state of the system, the VSC units *create* a communication channel over the microgrid with some unique properties from a communication point of view; the physical information conveyor in this channel is the steady state of the system itself. We defer the discussion of the basic properties of the power talk channel to Section 3. In the rest of this section, we review the upper layer control applications in microgrids and discuss their information exchange requirements; the observations regarding the characteristics of the data traffic needed to support the upper control levels represent the main motivation for the development of power talk.

2.2. Upper Control Layers: Secondary and Tertiary Control

Primary control, when implemented in a distributed manner with use of the droop method, has some drawbacks. In particular, as the load in the system varies, the output current of the unit also varies, causing bus voltage deviations; see Eq. (11.1). To counter this effect, the secondary control level has been defined with the main focus on restoring the bus variables to their rated values and thus improving the power quality of the microgrid [2, 5, 6]. The secondary control level is closely tied to the primary control, feeding it with appropriate voltage reference correction signals that help restore the bus voltage deviations introduced by the droop control:

$$V_m^* = V_m - I_m R_{D,m} + dV_m^*. \tag{11.2}$$

The reference voltage correction term dV_m^* is usually generated through a proportional-integral controller using the average value of the output voltage observations of all other units in the system. Secondary control is optional, as many microgrid applications do not require tight voltage regulation and can handle the voltage drops introduced by droop control.

The tertiary control level deals with microgrid optimization and related processing tasks. This is the highest control level and is also active when the microgrid operates in grid-connected mode, enabling the transfer of power between the microgrid and the main grid, determining the microgrid operation set points (i.e., the bus voltages V_m^*) and providing the optimized control parameters for the primary controllers—namely, the droop control parameters (i.e., the reference voltage V_m and/or virtual resistances $R_{D,m}$). Because of the nature of the microgrid system and the requirement to have an overview of the complete microgrid system, the tertiary control mechanisms are traditionally implemented in a centralized manner, but in some cases, depending on the control objective, they can also be implemented as decentralized algorithms. Although also optional, tertiary control may be necessary to improve the overall efficiency and reliability of the system. Standard control applications on the tertiary level include, but are not limited to [2, 5, 6, 8], (1) unit commitment, which corresponds to the determination of the current role of a specific unit in the system given the momentary power balance and the expected future power demand, usually formulated as an optimal scheduling problem over a finite horizon of time periods (e.g., assigning charging/discharging/off tokens to batteries, sending on/off signals to backup distributed energy resources, notifying the renewable energy sources to reduce/increase their power output, etc.), (2) optimal dispatch (i.e., long-term determination of optimal operating points given the available power of the sources, including the energy storage systems, the cost per kilowatt hour, and the momentary power demand of the system), and (3) determination of the optimal power flow. In addition to these, there are several supporting techniques that are usually implemented on the tertiary level, such as (1) determining the equivalent system parameters as seen from each unit and (2) topology discovery and determining the presence of specific nodes in the system.

2.3. Communications for Microgrid Control

As already discussed, the primary control level works only with local voltage and current measurements of the output current and voltage (i.e., no communication is necessary for the basic operation of the microgrid system). In contrast, the secondary and the tertiary controls require information about the status of other/all units to operate properly. The secondary layer requires only the averages of the local measurements of other units in the system, while the tertiary control level, depending on the specific application, might demand more comprehensive information, such as generation capacities, battery levels, power generation prices, and power demands.

There is extensive literature on defining the tertiary control mechanisms and optimization algorithms, including the type and the format of the information messages that should be exchanged among the units and (possibly) with a central controller. An interesting result is presented in [8]; there, besides showing how to implement the optimal dispatch with a linear cost function and ordered generation costs in a distributed manner, Liang et al. also show that the sufficient information for the optimal assignment of the duties to each distributed energy resource is the average values of the disposable power, the power demand, and the power cost in the system. Moreover, the control messages can be exchanged infrequently; as an example, the period for which certain dispatch is valid is usually on the scale of tens of minutes or even more [8]. In light of these facts, we suggest the use of power talk as a candidate solution for reliable and inexpensive microgrid communications, which meets the communication needs of the upper control layers and avoids the use of external hardware.

3. POWER TALK FOR DC MICROGRIDS: THE FOUNDATIONS

In this section we elaborate the pivotal ideas of power talk. In particular, we introduce the general model of a low-voltage DC microgrid, in which we "embed" the power talk communication channel and discuss its properties from the communication point of view.

3.1. Model of a Low-Voltage DC Microgrid

We model the low-voltage DC microgrid with M buses, indexed with $m = 1, \ldots, M$. For completeness, we assume that bus m hosts a DG and a generic load, comprising a constant-resistance part, $R_{CR,m}$, a constant-current part, $I_{CC,m}$, and a constant-power component $P_{CP,m}$. To keep the exposition simple, in this and later sections we use the notation R_{eq} to denote the "equivalent" load (i.e., R_{eq} comprises the values of all loads in the system). The buses are interconnected with low-voltage DC distribution lines. The subset of buses that bus m is connected to is denoted by $\mathcal{N}_m$; the impedance of the line connecting bus m with bus n is denoted by $R_{m,n} n \in \mathcal{N}_m$. Each bus is characterized by a steady-state voltage level, denoted by $V_m^*, m = 1, \ldots, M$. The values of the steady-state voltages are determined by (1) the primary control configuration of the DGs, (2) the types and the instantaneous values of the loads, and (3) the impedances of the distribution lines. We assume that in power talk, all DGs are operated as droop-controlled VSCs. The droop control parameters (i.e., the reference voltage and the virtual resistance) are denoted by V_m and $R_{D,m}$. The architectural configuration of the bus is summarized in Fig. 11.4. All voltages, currents, and impedances in DC systems are real numbers.

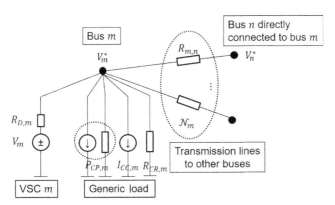

Fig. 11.4 General architecture of a DC microgrid bus. The bus hosts a single DG operated as a VSC. In addition to the power source, the bus also can host a local, generic load consisting of resistive, constant-current, and constant-power parts. The bus is connected to other buses through distribution lines.

The steady-state behavior of the system is governed by Ohm's law and Kirchoff's law, leading to the following power balance equation for bus m:

$$\underbrace{\frac{(V_m - V_m^*)V_m^*}{R_{D,m}}}_{\text{power supplied by the VSC}} = \underbrace{\frac{(V_m^*)^2}{R_{CR,m}} + I_{CC,m}V_m^* + P_{CP,m}}_{\text{power consumed by the local loads}} + \underbrace{\sum_{n \in \mathcal{N}_m} \frac{(V_m^* - Y_n)V_m^*}{R_{m,n}}}_{\text{power loses}}.$$

(11.3)

Using Eq. (11.3), we arrive at the following quadratic equation for the steady-state voltage of bus m:

$$(V_m^*)^2 \left(\frac{1}{R_{D,m}} + \frac{1}{R_{CR,m}} + \sum_{n \in \mathcal{N}_m} \frac{1}{R_{m,n}} \right)$$

$$- V_m^* \left(\frac{V_m}{R_{D,m}} - I_{CC,m} + \sum_{n \in \mathcal{N}_m} \frac{Y_n}{R_{m,n}} \right) + P_{CP,m} = 0, \qquad (11.4)$$

from which the steady-state voltage of bus m can be written as

$$V_m^* = \frac{\frac{V_m}{R_{D,m}} + \sum_n \frac{Y_n}{R_{m,n}} - I_{CC,m} + \sqrt{\left(\frac{V_m}{R_{D,m}} + \sum_n \frac{Y_n}{R_{m,n}} - I_{CC,m} \right)^2 - 4P_{CP,m} \left(\frac{1}{R_{D,m}} + \frac{1}{R_{CR,m}} + \sum_n \frac{1}{R_{m,n}} \right)}}{2 \left(\frac{1}{R_{D,m}} + \frac{1}{R_{CR,m}} + \sum_n \frac{1}{R_{m,n}} \right)}.$$

(11.5)

Eq. (11.5) describes the bus voltages in the steady state for given droop control parameters of the VSC units and given load values.

3.2. The General Power Talk Multiple-Access Channel

In power talk the droop control parameters V_m and $R_{D,m}$ deviate from their nominal values, causing changes in the bus voltages V_m^* for any $m = 1, \ldots, M$, so as to transmit local upper layer control information among different, remote units in the system [14–16]. Thus V_m and $R_{D,m}$ can be interpreted as inputs to the power talk channel, while V_m^* is the output, observed by VSC m in some observation noise. In this context, Eq. (11.5) gives the general input-output relation of the power talk channel, established over the primary control level of the low-voltage DC microgrid. Obviously, this is a multiple-access channel, in which all the inputs simultaneously determine the output. This calls for the use of multiple-access communication techniques; some candidate solutions are presented in Section 4. An equivalent representation of the channel output is the power $P_m = V_m^* I_m$ that VSC m supplies to the bus, as the output current I_m is uniquely determined by the bus voltage and its own droop parameters; this is the motivation for the use of the term *power talk*. From Eq. (11.5), we can directly observe the *three* major challenging aspects of the power talk channel:

1. In general, the power talk channel is nonlinear.
2. The output depends on the specific physical configuration of the system; from Eq. (11.5) we can see that the functional dependence of the output observed by VSC m (i.e., the bus voltage V_m^*) on the inputs of VSCs connected to other buses is not immediately clear unless the complete configuration of the system, including the impedances of the lines, is known. This type of knowledge is typically unavailable at the primary control level.
3. The output is determined by the instantaneous values of the loads in the system. Since they change sporadically, they constitute an arbitrary random component in the channel, besides the observation noise that will always be added to the output voltage measurement, and represent the main communication impairment of the power talk.

To address the challenges outlined previously, we adopt a system-configuration agnostic reception method based on locally constructed *detection spaces*, comprising the set of all possible combinations of bus voltages and output currents that can be observed at each unit under power talk [15, 16]. The detection space can be viewed as a representation of the rest of the system with the Thevenin equivalent seen from each unit. With use of the equivalent representation, the rest of the system, seen from VSC m, is reduced to two parameters: the *equivalent voltage* G_m and the *equivalent resistance* H_m, whose values are functions of the loads and the droop control combinations of the peer VSCs $(V_j, R_{D,j})$, $j = 1, \ldots, M$, $j \neq m$, as well as the interconnecting line impedances. In this case, Eq. (11.5) can be simplified to the following form:

$$V_m^* = \frac{\frac{V_m}{R_{D,m}} + \frac{G_m}{H_m}}{\frac{1}{R_{D,m}} + \frac{1}{H_m}}. \tag{11.6}$$

Obviously, via Eq. (11.6), VSC m can learn only the equivalent aggregate representation of the inputs of the peer VSCs (e.g., G_m and H_m). This is an alternative view of the multiple-access feature, and we rely on this representation to design effective power talk signaling strategies, where the process of constructing the local detection spaces effectively amounts to estimating the Thevenin equivalent for any input combination of the transmitting VSCs, given the instantaneous values of the loads R_{eq}. To construct the detection spaces (i.e., to learn all the potential values of G_m and H_m, for all m), we employ a solution based on training sequences used in a predefined training phase.

A possible alternative approach assumes that the deviations of the droop control parameters ΔV_m and $\Delta R_{D,m}$ are relatively small compared with their nominal values, allowing us to linearize Eq. (11.5) around the operating point. While the solution based on detection spaces uses all the available (and allowable) dynamic range of the droop control parameters, the second approach is confined only to small input signals. Nevertheless, the linearized channel lends itself to a number of modulation and coding strategies developed for linear communication channels and it is therefore useful to consider it, especially when the system imposes strict constraints on the allowable droop control parameter deviation.

4. COMMUNICATION WITH DETECTION SPACES

Before formally introducing the concept and ideas of power talk communication with detection spaces, we briefly discuss the assumptions and related terminology regarding the operation of the system in power talk mode. We refer to the operating mode of the system when it is not "power talking" as the nominal mode, and denote the droop control parameters by V_m^n and $R_{D,m}^n$.

In power talking mode, we assume that the time axis is slotted and the VSC units are slot-synchronized (e.g., ensured with GPS-based clocks or by use of distributed synchronization strategies [18]). Fig. 11.5 depicts an illustration of a bus voltage waveform as observed by VSC m. Each slot is of duration T_S. The communicating units change their droop parameters at the beginning of the slot. After a transient period of duration $\tau \ll T_S$, a steady state is established and the receiving units measure their output voltages and currents.

The steady-state voltage and output current (see Fig. 11.5) are sampled by each receiving VSC, with frequency f_S (the sampling frequency of the converter, also known as the *switching frequency*, is on the order of a couple of tens of kilohertz [2, 3]); the operating values of the output voltage and current used for subsequent demodulation of information bits are obtained by the averaging of these samples. Since the establishment of a steady state is

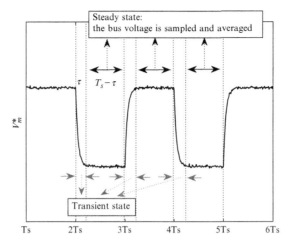

Fig. 11.5 Example waveform of bus voltage as observed by voltage source converter m for $T_S = 50$ ms. The bus voltage and output current are sampled and averaged after the steady state has been reached; the averages of the bus voltage and output current are used for demodulation of the information sent by the transmitting units after the end of the power talk slot. The duration of the slot depends on the control bandwidth of the primary controller, which in turn determines the duration of the transient periods. In most practical configurations, the control bandwidth of the power electronic converters is strongly dominated by the output capacitor and the bus capacitance.

mandatory, the duration of T_S has to be on the order of milliseconds, which makes power talk inherently a communication technique with a low rate.

We note that each bus voltage/output current sample is subject to measurement noise, inherently induced by the power electronic equipment. After averaging has been performed over the steady-state period of duration $T_S - \tau$, the noise component of the averaged values can be assumed to be additive and Gaussian in nature, owing to the law of large numbers [19–22]. Moreover, investigations have shown that the noise term can be successfully suppressed by proper choice of the duration of the averaging period (i.e., power talk slot T_S), making its effect negligible [16]. Usually, if $T_S \geq 50$ ms in typical DC microgrids, the effect of the observation noise can be ignored.

Next we illustrate the basic idea of power talk communication using detection spaces. We denote the droop parameters of VSC m as the input $\mathbf{x}_m$:

$$\mathbf{x}_m = (V_m, R_{D,m}), \quad m = 1, \dots, M. \tag{11.7}$$

We focus on binary power talk [16], where each VSC unit when power talking chooses between two different combinations to represent the value of the transmitted bit $b_m \in \{0, 1\}$:

$$\mathbf{x}_m^0 = (V_m^0, R_{D,m}^0) \equiv (V^0, R_D^0) \leftrightarrow 0, \tag{11.8}$$

$$\mathbf{x}_m^1 = (V_m^1, R_{D,m}^1) \equiv (V^1, R_D^1) \leftrightarrow 1, \tag{11.9}$$

for $m = 1, \ldots, M$. This is the simplest case to deal with, and we note that the case for modulations of higher orders can be straightforwardly accommodated. Finally, we assume that the symbols $\mathbf{x}$ from the binary input constellation have been chosen to satisfy the following condition:

$$P_m(\mathbf{x}^1) > P_m(\mathbf{x}^n) > P_m(\mathbf{x}^0); \tag{11.10}$$

that is, the logical 1/0 when inserted by VSC m, $m = 1, \ldots, M$, corresponds to a higher/lower output power compared with the nominal output power when all VSC units operate in nominal mode $\mathbf{x}_m^n$, $m = 1, \ldots, M$. An example of a power talk symbol constellation satisfying the above power condition is as follows: Keep the virtual resistances fixed at their nominal values (i.e., $R_{D,m} = R_{D,m}^n$, $m = 1, \ldots, M$) and use the following reference voltage constellation:

$$V^0 = V_m^n - \frac{\Delta V_m}{2},$$

$$V^1 = V_m^n + \frac{\Delta V_m}{2};$$

that is, the reference voltage constellation is antipodal with respect to V_m^n. The above example is denoted as the *fixed virtual resistance power talk constellation*, and we will use it extensively throughout the rest of the chapter to illustrate the basic mechanisms of establishing reliable power talk communication.

The locally observable voltage and current at VSC m (i.e., the output symbol) is denoted by

$$\mathbf{s}_m = (V_m^*, I_m), \quad m = 1, \ldots, M, \tag{11.11}$$

where, as discussed earlier, the output voltage V_m^* and current I_m are obtained as averages over a single power talk slot of duration $T_S - \tau$ (see Fig. 11.5). The output $\mathbf{s}_m$ is inevitably corrupted by observation noise and converter uncertainties; however, as discussed earlier, we assume that the duration of the power talk slot T_S is properly chosen to enable suppression of the noise and minimize its impact on the performance of power talk. The set of all possible output symbols $\mathbf{s}_m$ of VSC m for any input combination of the transmitting VSCs and for given values of the loads in the system is referred to as the *detection space*. As already noted, each value in the detection space is associated with output power $P_m = V_m^* I_m$.

In principle, a receiving VSC unit can learn all output symbols in its local detection space that correspond to specific input combinations of the transmitting VSCs through a training phase. However, in a general multibus DC microgrid, the number of all possible output symbols that a single unit can observe grows exponentially with the number of simultaneously transmitting units since, in general, every input combination maps to a different output symbol. As an example, in a DC microgrid with M VSC units where all units transmit and receive at the same time (communication denoted as *full duplex*, FD) using binary power talk, the number of output symbols that each VSC as a receiver can observe in power talk mode is $2^{(M-1)}$ since there are $M - 1$ transmitting units as seen by each VSC. To simplify the analysis and to foster the development of effective power talk communication strategies, we review two different multiple-access communication strategies:

1. A time division multiple-access (TDMA) strategy, where each unit transmits at a time. We develop and analyze this strategy for a general low-voltage DC microgrid system consisting of multiple buses.
2. An FD strategy, where all VSCs transmit and receive at the same time. This strategy is developed specifically for single-bus systems, where the effects of the transmission network are negligible.

To evaluate the efficiency of the proposed solutions, we use the net transmission rate per unit η (i.e., the average number of information bits transmitted per unit per time slot).

4.1. Time Division Multiple Access for Single-Bus and Multibus Systems

In TDMA power talk, only a single unit transmits at a time (e.g., VSC m), while all other VSC units $j, j = 1, \ldots, M, j \neq m$, listen. The scheduling of transmission is presumed to be done and fixed before power talk communication (e.g., in a simple polling manner). Assuming fair scheduling, the net transmission rate when the load does not change is simply

$$\eta^{\mathrm{S}}_{\mathrm{TDMA}} = \frac{1}{M}. \tag{11.12}$$

When listening in TDMA mode, the VSC employs the nominal droop combination. Fig. 11.6 depicts an example of detection spaces of the transmitting unit, VSC m, and the receiving unit, VSC j, $j \neq m$. When the

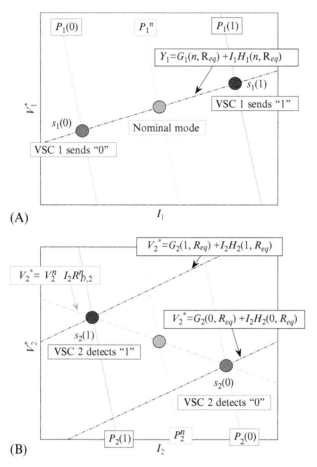

(A)

(B)

Fig. 11.6 Power talk communication using detection spaces in time division multiple-access mode. (A) Local $V^* - I$ diagram of the transmitting unit. The transmitter sends either 0 or 1. (B) Local $V^* - I$ or detection space of the receiving unit. The outputs corresponding to the inputs sent by the transmitter locally at the receiver lie at the intersections of the local droop law of the receiver and the law describing the system through the Thevenin equivalent as seen from the receiver. The results were obtained with PLECS simulation of a single-bus system with two voltage source converter (VSC) units with use of the following parameters: $V^0 = 399\,\mathrm{V}$, $V^1 = 401\,\mathrm{V}$, $R_D^0 = R_D^1 = 2\,\Omega$, $V_1^n = V_2^n = 400\,\mathrm{V}$, $R_{D,1}^n = R_{D,2}^n = 2\,\Omega$, $R_{eq} = 100\,\Omega$. VSC 1 is the transmitter and VSC 2 is the receiver.

transmitting unit inserts $\mathbf{x}^1$ to signal 1, it produces greater output power $P_m(1) > P_m^n$, which locally at the receiving unit is reflected in reduced output power $P_j(1) < P_j^n$ compared with the nominal mode of operation. The same reasoning applies for the case when the transmitting unit inserts $\mathbf{x}^0$ to signal 0.

Thus the receiving unit can determine the transmitted bit by comparing its output power level with respect to the nominal level. The TDMA strategy as illustrated here can be applied to arbitrary constellation sizes and an arbitrary number of units.

The previous discussion exposes two major aspects of power talk communication with detection spaces. First, the output symbols have to be discovered before the power talk communication commences, which can be easily done in a training phase. For the TDMA strategy, each VSC has to learn $M-1$ separate detection spaces, one for each transmitting unit. The detection spaces after construction are stored in memory and invoked whenever the corresponding VSC transmits. For binary power talk, each detection space for each transmitting unit has two symbols. A possible strategy would be for each VSC to send predefined training sequences consisting of $\mathbf{x}^1$ and $\mathbf{x}^0$, in a sequential manner. Assuming that the single point is learned in K time slots and, since there are M VSC units in the system, the duration of the training period for TDMA-based binary power, in number of slots, talk would be

$$L_{\text{TDMA}} = 2MK. \tag{11.13}$$

Second, the positions of the output symbols in the detection space also depend on the value of the loads R_{eq}. In general, when the load changes, then the output symbols will take new positions and the complete layout of the detection space will change, as illustrated in Fig. 11.7.

The new detection space can be easily reconstructed by use of the training sequence as discussed earlier. Fortunately, the loads in typical microgrid settings change infrequently compared with the target power talk rates. In this sense, the power talk channel behaves similarly as the slow fading wireless channel, where the channel state (also known as *channel gain*) stays fixed during multiple symbols. Therefore, the training sequences in power talk that enable the updating of the detection spaces and, thus the updating of the knowledge of the channel, play a role similar to that of the channel estimation phases in slow fading wireless channels.

The detection space, as already discussed, provides an alternative view of the multiple-access feature of the general power talk channel. Namely,

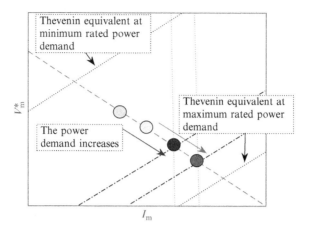

Fig. 11.7 The effect of load change on the detection space. The output symbols move along the local droop law used by the receiving unit, leading to an outdated detection space. The results were obtained with PLECS simulation of a single-bus system with two voltage source converter (VSC) units with use of the following parameters: $V^0 = 399\,V$, $V^1 = 401\,V$, $R_D^0 = R_D^1 = 2\,\Omega$, $V_1^n = V_2^n = 400\,V$, $R_{D,1}^n = R_{D,2}^n = 2\,\Omega$. VSC 1 is the transmitter and VSC 2 is the receiver. The load R_{eq} is purely resistive and changes from 100 to 60 Ω.

learning all possible points in the detection space, locally at each VSC, amounts to estimating the parameters of the Thevenin equivalent as seen by each unit, for a given input symbol by the transmitting unit. This is illustrated in Fig. 11.6A. The general layout of the detection space for the TDMA power talk communication strategy can be summarized as follows: All outputs, observed by receiving VSC j for transmitting VSC m, $\mathbf{s}_j$ lie on the droop control line $V_j^* = V_j - R_{D,j}I_j$, where $\mathbf{x}_j = (V_j, R_{D,j})$ is the symbol VSC j is inserting. For TDMA power talk, since VSC j is receiving, $\mathbf{x}_j = \mathbf{x}_j^n = (V_j^n, R_{D,j}^n)$ and the outputs lie on a single line. The outputs $\mathbf{s}_j$ also depend on the symbol the transmitting unit is inserting. Expressed through the Thevenin equivalent, $\mathbf{s}_j$ must also lie on the line $V_j^* = G_j + H_jI_j$. Therefore the loci of $\mathbf{s}_j$ are at the *intersections* of the droop control line $V_j^* = V_j - R_{D,j}I_j$ and $V_j^* = G_j + H_jI_j$. Evidently, for TDMA power talk, when VSC m is transmitting, $G_j = G_j(b_m, R_{eq})$ and $H_j = H_j(b_m, R_{eq})$; that is, there are two possible intersections in the detection space of VSC j for each transmitting unit $m = 1, \ldots, M, m \neq j$. As the loads in the system vary, the points $\mathbf{s}_j$ in the detection space of receiving VSC j shift along the droop line $V_j^* = V_j - R_{D,j}I_j$, since H_j and G_k also vary with the loads (see Fig. 11.7). Evidently, to learn the position of the output symbols in the detection space, each unit needs only to estimate the equivalent Thevenin circuit parameters for every possible input combination, which is a usual and widespread technique employed in power systems and can be easily adapted for power talk purposes.

4.2. Full Duplex for Single-Bus Systems

4.2.1. Full-Duplex Systems

Full duplex (FD) is a communication strategy where each unit transmits and receives at the same time. In principle, FD can be applied to a general multibus DC microgrid. Here, we restrict ourselves to the single-bus system where the effect of the transmission network can be assumed to be negligible and all VSC units connected to the bus observe the same output voltage. Then Eq. (11.5) can be rewritten as follows:

$$V^* = \frac{\sum_{m=1}^{M} \frac{V_m}{R_{D,m}} - I_{CC} + \sqrt{\left(\sum_{m=1}^{M} \frac{V_m}{R_{D,m}} - I_{CC}\right)^2 - 4P_{CP}\left(\frac{1}{R_{CR}} + \sum_{m=1}^{M} \frac{1}{R_{D,m}}\right)}}{2\left(\frac{1}{R_{CR}} + \sum_{m=1}^{M} \frac{1}{R_{D,m}}\right)},$$

(11.14)

where R_{CR}, I_{CC}, and P_{CP} represent the components of the aggregated load. In addition, to simplify the exposition, we also assume that the units employ the fixed virtual resistance constellation with the same virtual resistances; that is,

$$R_{D,m} = R_D, \quad m = 1, \ldots, M.$$

(11.15)

As already mentioned, in FD power talk all units transmit and receive at the same time. Fig. 11.8A depicts examples of the layout of the detection spaces of a receiving unit in simple single-bus systems with $M = 2$ and $M = 3$ VSC units that communicate in FD mode. These two cases expose the main principles of FD power talk communication in single-bus systems. To begin with, observe the case with two VSC units. The same detection principle as in the TDMA case is applied: by observing its local output, the receiving VSC detects the input of the transmitting VSC. The difference from the TDMA case is that the local output depends on the input combination of both VSCs: Consider the case when VSC 1 inserts $\mathbf{x}^0$ (i.e., signals 0). Then, depending on the symbol inserted by VSC 2, VSC 1 outputs different power P_1. In particular, for $b_2 = 0$, VSC 2 outputs less power than nominally, while for $b_2 = 1$, VSC 2 outputs more power than

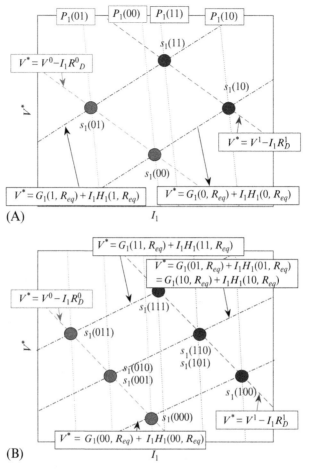

Fig. 11.8 Power talk communication using detection spaces in full-duplex mode. (A) The detection space of voltage source converter (VSC) 1 is a single-bus system with two VSC units as an example of two-way power talk communication. (B) The detection space of VSC 1 is a single-bus system with three VSC units. The cases when VSC 2 and VSC 3 send either 0 and 1, or $1''$ and 0, respectively, map in the same output in the detection space and cannot be distinguished without multiple-access coding. The results were obtained with PLECS simulation of a single-bus system with two or three VSC units with use of the following parameters: $V^0 = 399\,V$, $V^1 = 401\,V$, $R_D^0 = R_D^1 = 2\,\Omega$, $V_1^n = V_2^n = V_3^n = 400\,V$, $R_{D,1}^n = R_{D,2}^n = R_{D,3}^n = 2\,\Omega$, $R_{eq} = 100\,\Omega$.

nominally; correspondingly, P_1 increases or decreases. The same reasoning applies when VSC 1 inserts $\mathbf{x}^1$, as well as for the detection of b_1 at VSC 2.

Now consider the case with $M = 3$ VSC units (see Fig. 11.8B for illustration of the detection space of VSC 1). In this case the additional challenge for the FD power talk in single-bus systems is as follows: the receiving VSC cannot distinguish between the cases in which the sum of the bits of the other units is the same, as then the sum of their output powers is the same and, thus, the local output power is the same. This is a major difference between TDMA power talk and FD power talk for single-bus systems, as in the FD case the individual values of bits of other VSCs cannot always be directly detected when $M \geq 3$. It is easy to verify that the position of the specific output symbol within the detection space depends on the local input bit and the integer sum of the bits inserted by the other units in the time slot. Thus in FD power talk, employed in single-bus systems, the receiving VSC observes the same output symbol for input bit combinations of the transmitting units with equal integer sums of logical 1's. In this case, each receiving VSC constructs a detection space for joint detection of the transmitted inputs by all transmitting units. The number of points in the detection space of each unit is $2M$ and the total number of points to be learned by each receiving unit is $2M^2$. Then assuming that each VSC constructs its detection space separately and K slots are devoted to each point, the length of the training phase for FD power talk is

$$L_{FD} = 2KM^2. \tag{11.16}$$

Evidently, FD power talk requires M times more time slots than the TDMA variant.

Fig. 11.8 illustrates in more detail the general outline of the detection space for specific receiving unit VSC j in FD power talk mode for single-bus systems. As in the TDMA case, all outputs $\mathbf{s}_j$ of VSC j lie on the local droop line $V_j^* = V_j - R_{D,j}I_j$. For FD power talk, VSC j inserts either $\mathbf{x}^0$ or $\mathbf{x}^1$ (as it also transmits at the same time while receiving), and there are two lines on which the outputs may lie. Expressed through the Thevenin equivalent, $\mathbf{s}_j$ must also lie on the line $V_j^* = G_j + H_j I_j$, and the locations of the output symbols can be found at the intersections of the lines $V_j^* = V_j - R_{D,j}I_j$ and $V_j^* = G_j + H_j I_j$. For FD power talk, G_j and H_j are functions of the integer sum of the bit inputs of all other units and, therefore, there are M possible values of G_j and H_j, one for each value of the sum $\sum_{j \neq m} b_j$, where b_j denotes the input bit of unit j. Thus there are M possible intersections for each local input. As the loads vary, the points $\mathbf{s}_j$ shift along $V_j^* = V_j - R_{D,j}I_j$, since H_j and G_j also vary with the load.

We conclude by noting that, seen from each VSC locally and given the value of the local input, the FD power talk channel in single-bus systems can be equivalently represented by a multiple-access adder channel with binary inputs [23–25]. The property of addition in the binary domain in FD power talk arises due to the condition in Eq. (11.15). However, it can be easily shown that the same phenomenon can be observed for general symbol constellations as long as they satisfy $V_m/R_{D,m} = $ const for all $m = 1, \ldots, M$. To obtain individual bit streams from the observations, one can use coding methods developed for a multiple-access adder channel with binary inputs. Chang and Weldon [23] proposed a coding solution for this type of multiple access that enables unique decodability of user codewords (i.e., input streams) and that is asymptotically optimal. For the specific implementation of the Chang-Weldon code construction and subsequent decoding, as well the achievable rates of the code, we refer the interested reader to [23]. It can be shown that the net transmission rate when Chang-Weldon code is used and when the load does not change is strictly greater than the net rate for the TDMA strategy:

$$\eta_{FD}^{S} > \eta_{TDMA}^{S} = \frac{1}{M}. \tag{11.17}$$

We also note that the decoding complexity of the Chang-Weldon code if one relies on the fact that the number of units in practice is expected to be on the order $M \approx 10$, implying short codewords and enabling use of lookup tables.

4.3. The Main Communication Impairment: Dealing With Load Variations

As discussed in the previous sections, all communicating units in power talk have to construct and maintain up-to-date detection spaces so as to foster reliable communication. The construction and reconstruction of the detection spaces is achieved in a specifically dedicated training phase, in which, depending on the strategy, each VSC inserts a predefined sequence of power talk input symbols in a coordinated manner. The receiving units either observe the possible output directly or estimate the local Thevenin equivalent and find the outputs at the intersections of the Thevenin laws and the local droop law. We also established that the duration of the training phase for FD power talk in principle requires M more time slots to construct the detection space successfully than the TDMA variant.

As the sporadic load variations disturb and may alter the detection space, the detection space needs to be reconstructed by reactivation of the training phase. We review two potential techniques for activation of the training phase. A simple solution is to perform training periodically and update the detection spaces, disregarding whether the load has changed or not. A more involved approach is to employ a model change detector that tracks the bus voltage and if a change is detected, reinitiates the training phase. We investigate and compare these two approaches. For this purpose we model load changes via a Poisson process whose intensity λ is the expected number of changes per time slot, where $\lambda \ll 1$. This model is general, taking into account only the instants when the loads change and not the nature of the changes. Also, we assume that whenever a load change occurs, the detection space becomes outdated and the detection becomes completely unreliable, which is the worst-case scenario.

For periodic training we assume that the training phase occurs periodically after each VSC transmits B bits of information. In TDMA power talk, when scheduled to transmit, VSC m transmits reliably one bit of information in a single slot with probability $1 - p$, where

$$p = 1 - e^{-\lambda} \tag{11.18}$$

is the probability that the load has changed (at least once) during a slot. In the absence of noise it can be shown by a simple analysis that the net transmission rate for TDMA power talk is

$$\eta_{\text{TDMA}} = \frac{(1 - p)^{L_{\text{TDMA}}+1}[1 - (1 - p)^{MB}]}{p(L_{\text{TDMA}} + MB)} \eta_{\text{TDMA}}^{\text{S}}, \tag{11.19}$$

where $\eta_{\text{TDMA}}^{\text{S}}$ is the corresponding stable rate. In the FD variant, when the Chang-Weldon coding solution for multiple access is used, each unit has to send a block of bits of length $\frac{1}{\eta_{\text{FD}}^{\text{S}}}$ to deliver one bit of information. In the absence of noise, it can be shown that the corresponding net transmission rate is

$$\eta_{\text{FD}} = \frac{(1 - p)^{L_{\text{FD}}+\frac{1}{\eta_{\text{FD}}^{\text{S}}}}\left[1 - (1 - p)^{\frac{B}{\eta_{\text{FD}}^{\text{S}}}}\right]}{\left[1 - (1 - p)^{\frac{1}{\eta_{\text{FD}}^{\text{S}}}}\right]\left(L_{\text{FD}} + \frac{B}{\eta_{\text{FD}}^{\text{S}}}\right)}, \tag{11.20}$$

where $\eta_{\text{FD}}^{\text{S}}$ is the net transmission rate when the load is stable. It can be shown that when $p \to 0$, then $\eta \to \frac{B}{L\eta^{\text{S}}+B}\eta^{\text{S}}$ for both the TDMA variant and the FD variant.

In an alternative approach, each VSC tracks the output voltage and current and decides whether the system state has changed. The state changes should be detected by all units simultaneously with high reliability; an option is to use a standard model change detector that tracks the voltage level in each slot. Assume that each VSC implements such a detector, operated in the following way: (1) if a change is detected, then the current transmission is stopped, (2) L_{BS} "blank slots" (e.g., nominal operation symbols) are inserted by all units to allow the system to reach the steady state after the load change, and (3) the training phase of length L is reinitiated. Under this approach, it can be shown that the effective net transmission rate is

$$\eta = \frac{1}{p + (1 - p)^{-(L+L_{BS})}} \eta^{S}, \tag{11.21}$$

which holds both for TDMA and FD power talk. When $p \to 0$, then $\eta \to \eta^{S}$.

For the purpose of illustration of the concepts introduced, we evaluate the power talk variants in a single-bus system in terms of the net reception rate μ, defined as the average number of bits observed by a single unit per slot and calculated as

$$\mu = (K - 1)\eta, \tag{11.22}$$

when the net transmission rates are equal; this is a suitable metric for applications that require aggregate information about the status of other units (e.g., optimal dispatch). Fig. 11.9 depicts μ for both variants.

For TDMA binary power talk, the asymptotic upper bound for the net reception rate is one bit per unit per slot, as could be expected. For the FD variant, μ is significantly larger and increases with K, although not monotonically because of the rates of the Chang-Weldon coding method (see [23] for a discussion on Chang-Weldon coding). Thus FD-based power talk is a better solution for microgrid applications in which information about the status of the rest of the system is required. Also, the protocol variant with a model change detector performs significantly better than the periodic training strategy, for both the TDMA variant and the FD variant; the price to pay is increased implementation complexity and sensitivity to potential misdetections and false alarms, which is not analyzed here. Further, for both protocols, the performance loss with respect to the stable operation, when $p \to 0$, is larger for the FD-based solution because of a longer training phase. This deviation becomes more apparent as the

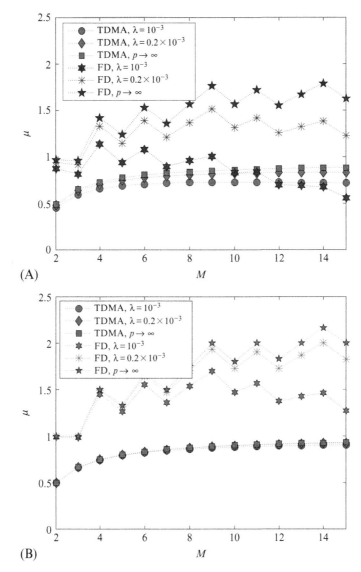

Fig. 11.9 Time division multiple access (TDMA) versus full-duplex (FD) power talk for single-bus systems in terms of the net reception rate per unit μ as a function of the number of transmitting units M and the load change intensity λ: (A) periodic training; (B) on-demand training. The effect of the observation noise has been suppressed by our choosing long power talk symbol slots $T_S \geq 50$ ms.

number of units M increases, as well as for larger values of the load change intensity λ. It can also be observed that the gain of the FD variant over the TDMA variant reduces with increasing M and increasing λ. In principle, from Eqs. (11.19) to (11.21) one can determine the values of M for which $\eta_{\text{TDMA}} > \eta_{\text{FD}}$, given λ and B. It can be shown that FD-based power talk is more efficient for single-bus systems with a smaller number of units.

5. CONSTRAINTS: SIGNALING SPACE

In this section we discuss the aspect of power talk input constraints by introducing the concept of the signaling space to capture the effects that the operating constraints of the microgrid have on the power talk schemes [14–16]. Every microgrid is subject to operational constraints that may not be violated and that are associated with system stability and power delivery quality, such as limits on the voltage, current, power dissipation, droop slope, etc. The *signaling space* is defined the set of all possible symbols $\mathbf{x}_m$, $m = 1, \ldots, M$, that satisfy all such constraints for any value of the loads R_{eq}. For simplicity, here we focus only on the constraints pertaining to the bus voltage and output current (as they are the most important): $V_{\min}^* \leq V_m^* \leq V_{\max}^*$, $I_{m,\min} \leq I_m \leq I_{m,\max}$, $m = 1, \ldots, M$, where $V_{\min}^* / V_{\max}^*$ are the minimum/maximum allowable bus voltages and where $I_{m,\min} / I_{m,\max}$ are the minimum/maximum output currents of VSC m; usually, $I_{m,\min} = 0$ and $I_{m,\max}$ is the current rating of the unit. Assuming that the loads in the system vary between some minimum and maximum power demand (i.e., $R_{eq} \in [R_{eq,\min}, R_{eq,\max}]$), it can be shown that the signaling space, for any power talk strategy, lies at the intersection of the M regions determined by the following inequalities:

$$R_{D,m} \left(\frac{V_{\min}^* - G_m(R_{eq,\min})}{H_m(R_{eq,\min})} \right) + V_{\min}^* \leq V_m$$

$$\leq R_{D,m} \left(\frac{V_{\max}^* - G_m(R_{eq,\max})}{H_m(R_{eq,\max})} \right) + V_{\max}^*, \tag{11.23}$$

$$G_m(R_{eq,\max}) \leq V_m \leq R_D I_{m,\max} + H_m(R_{eq,\min}) I_{m,\max} + G_m(R_{eq,\min}) \tag{11.24}$$

for $m = 1, \ldots, M$, where G_m and H_m are the voltage and the resistance of the Thevenin equivalent seen from VSC m. The signaling space region (11.23) stems from the bus voltage constraints $V_{\min}^* \leq V_m^* \leq V_{\max}^*$, while the region (11.24) comes from the output current constraints $I_{m,\min} \leq I_m \leq I_{m,\max}$. The calculation of G_m and H_m depends on the specific power talk communication strategy; however, from Eqs. (11.23), (11.24) it follows that the units do not need precise knowledge of the system to determine the signaling space when transmitting; they only need to estimate the equivalent response of the system at the minimum and maximum loads.

When the VSCs use $\mathbf{x}_m \neq \mathbf{x}_m^n$ to signal in power talk, the output power of each unit deviates from the nominal; that is, $P_m(\mathbf{x}_m) \neq P_m(\mathbf{x}_m^n)$, $m = 1, \ldots, M$, where $P_m(\mathbf{x}_m^n)$ is the output power of VSC m when *all* units operate nominally. To account for this effect, we introduce the relative power deviation of unit m with respect to the nominal output power P_m^n:

$$\delta_m(\mathbf{x}_1, \ldots, \mathbf{x}_M) = \frac{\sqrt{\mathbb{E}_{R_{eq}}\left\{(P_m - P_m^n)^2\right\}}}{\mathbb{E}_{R_{eq}}\left\{P_m^n\right\}} \tag{11.25}$$

for $m = 1, \ldots, M$, where the averaging is performed over the values of the equivalent load. Then the average relative power deviation per VSC is simply

$$\delta = \frac{1}{M} \sum_{m=1}^{M} \mathbb{E}_{\mathbf{X}_1, \ldots, \mathbf{X}_M}\{\delta_m(\mathbf{x}_1, \ldots, \mathbf{x}_M)\}, \tag{11.26}$$

where the averaging in Eq. (11.26) is performed over the combinations of all input symbols. Again, note in the TDMA case only a single input in $\mathbf{x}_1, \ldots, \mathbf{x}_M$ is an actual power talk symbol of the active unit, while the rest of them are nominal inputs. In the FD case, all inputs represent power talk symbols. Finally, we impose an average power deviation limit γ, requiring that

$$\delta \leq \gamma; \tag{11.27}$$

that is, the average power deviation per unit with respect to the nominal mode of operation is bounded by γ. In the TDMA case, Eq. (11.27) translates to a constraint that applies only to a currently transmitting VSC, whereas in the FD case it applies to all units jointly. We note that Eq. (11.27) is used only as a constellation design metric that constrains the average deviated power to be a fraction γ of the average supplied power $\mathbb{E}_{R_{eq}}\left\{P_k^n\right\}$ in nominal mode. The actual value of the deviated power depends on the state of the channel (i.e., the loads), while the total dissipated energy (i.e., power consumption) depends on the frequency and duration of power talk intervals.

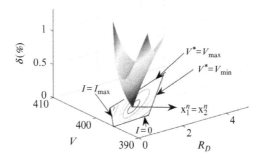

Fig. 11.10 Signaling space and average power deviation of voltage source converter (VSC) 1 and VSC 2 in a single-bus system with two VSC units and resistive load for the following constraints: $390\,\text{V} \leq V^* \leq 400\,\text{V}; 0\,\text{A} \leq I_m \leq 5\,\text{A}, m = 1, 2; 50\,\Omega \leq R_{eq} \leq 250\,\Omega; \mathbf{x}_m^n = (400\,\text{V}, 2\,\Omega), m = 1, 2$.

Fig. 11.10 depicts the effective signaling space as a set of allowable droop control combinations when the units engage in TDMA communication and the induced average power deviation as a result of the use of droop parameters different from the nominal ones; the calculation of the signaling space for FD power talk is straightforward. The signaling space is bounded by the bus voltage and output current constraints; evidently, the signaling space provides the support for the average power deviation function δ, which increases as we place the signaling symbol $\mathbf{x}$ far from $\mathbf{x}_m^n$. The increase in δ with the distance between $\mathbf{x}$ and $\mathbf{x}_m^n$ in the signaling space is nonisotropic as a result of the steady-state model, which is nonlinear in the droop control parameters as well as the loads. In practical systems with strict constraints that strive to minimize the power deviation introduced with power talk, one should choose symbols near pilot, making the region of small δ of practical importance.

6. CONCLUSION

The core idea of power talk is to modulate information using primary control loops of the VSCs that regulate the bus voltage. In this chapter we illustrated a set of techniques based on detection spaces and the Thevenin equivalent that enable reliable power talk communication amid the challenges imposed by the DC microgrid buses as communication channels. The main communication impairment of power talk is random load variations, leading to uncontrollable changes of the bus voltage. We investigated techniques to counter the effect of load changes, showing that it is possible to optimize the power talk operation given the statistics of the load changes.

The communication solution presented assumes discrete-time, slot-level synchronization among the communicating microgrid entities, as well as that issues pertaining to scheduling are resolved. To achieve these requirements in practice, one would have to use and adapt the related existing communication methods and algorithms and design a full-fledged communication protocol. These considerations are beyond the scope of the material presented here.

The achievable rates of power talk depend on the bandwidth of the primary control loops. In practice, it could be expected that power talk can achieve rates on the order of 100 baud to 1 kliobaud. Nevertheless, considering that the intermicrogrid communications are of machine type in nature, these rates may prove to be satisfactory. Moreover, when assessing the potential of the proposed technique, one should also take into account its inherent advantages, which are the use of existing microgrid power equipment, software implementation, and reliability and availability that derive from the reliability and availability of the microgrid.

REFERENCES

[1] R. Lasseter, Microgrids, in: Power Engineering Society Winter Meeting, vol. 1, IEEE, 2002, pp. 305–308.
[2] T. Dragicevic, X. Lu, J. Vasquez, J. Guerrero, DC microgrids—Part II: a review of power architectures, applications and standardization issues, IEEE Trans. Power Electron. 31 (5) (2016) 3528–3549.
[3] F. Blaabjerg, Z. Chen, S. Kjaer, Power electronics as efficient interface in dispersed power generation systems, IEEE Trans. Power Electron. 19 (5) (2004) 1184–1194.
[4] S.F. Bush, Smart Grid: Communication-Enabled Intelligence for the Electric Power Grid, John Wiley & Sons Ltd., 2014.
[5] J. Guerrero, J. Vasquez, J. Matas, L. de Vicuna, M. Castilla, Hierarchical control of droop-controlled AC and DC microgrids; a general approach toward standardization, IEEE Trans. Ind. Electron. 58 (1) (2011) 158–172.
[6] J. Guerrero, M. Chandorkar, T. Lee, P. Loh, Advanced control architectures for intelligent microgrids; part I: decentralized and hierarchical control, IEEE Trans. Ind. Electron. 60 (4) (2013) 1254–1262.
[7] I.U. Nutkani, P.C. Loh, P. Wang, F. Blaabjerg, Cost-prioritized droop schemes for autonomous AC microgrids, IEEE Trans. Power Electron. 30 (2) (2015) 1109–1119.
[8] H. Liang, B.J. Choi, A. Abdrabou, W. Zhuang, X. Shen, Decentralized economic dispatch in microgrids via heterogeneous wireless networks, IEEE J. Sel. Areas Commun. 30 (6) (2012) 1061–1074.

[9] S. Galli, A. Scaglione, Z. Wang, For the grid and through the grid: the role of power line communications in the smart grid, Proc. IEEE 99 (6) (2011) 998–1027.

[10] J. Schonberger, R. Duke, S. Round, DC-bus signaling: a distributed control strategy for a hybrid renewable nanogrid, IEEE Trans. Ind. Electron. 53 (5) (2006) 1453–1460.

[11] D. Chen, L. Xu, L. Yao, DC voltage variation based autonomous control of DC microgrids, IEEE Trans. Power Deliv. 28 (2) (2013) 637–648.

[12] K. Sun, L. Zhang, Y. Xing, J. Guerrero, A distributed control strategy based on DC bus signaling for modular photovoltaic generation systems with battery energy storage, IEEE Trans. Power Electron. 26 (10) (2011) 3032–3045.

[13] T. Dragicevic, J.M. Guerrero, J.C. Vasquez, A distributed control strategy for coordination of an autonomous LVDC microgrid based on power-line signaling, IEEE Trans. Ind. Electron. 61 (7) (2014) 3313–3326.

[14] M. Angjelichinoski, C. Stefanovic, P. Popovski, H. Liu, P. Loh, F. Blaabjerg, Power talk: how to modulate data over a DC micro grid bus using power electronics, in: IEEE GLOBECOM 2015, December, San Diego, CA, 2015.

[15] M. Angjelichinoski, C. Stefanovic, P. Popovski, F. Blaabjerg, Power talk in DC micro grids: constellation design and error probability performance, in: IEEE SmartGridComm 2015, November, Miami, FL, 2015.

[16] M. Angjelichinoski, Č. Stefanović, P. Popovski, H. Liu, P.C. Loh, F. Blaabjerg, Multiuser Communication Through Power Talk in DC MicroGrids, IEEE J. Sel. Areas Commun. 34 (7) (2016) 2006–2021. Special Issue on Powerline Communications and Its Integration with the Networking Ecosystem.

[17] T. Dragicevic, J. Guerrero, J. Vasquez, D. Skrlec, Supervisory control of an adaptivedroop regulated DC microgrid with battery management capability, IEEE Trans. Power Electron. 29 (2) (2014) 695–706.

[18] K. Iwanicki, M. van Steen, S. Voulgaris, Gossip-based clock synchronization for large decentralized systems, in: A. Keller, J.-P. Martin-Flatin (Eds.), Self-Managed Networks, Systems, and Services, Lecture Notes in Computer Science, vol. 3996, Springer, Berlin, 2006, pp. 28–42.

[19] P. Midya, P. Krein, Noise properties of pulse-width modulated power converters: openloop effects, IEEE Trans. Power Electron. 15 (6) (2000) 1134–1143.

[20] S. Mazumder, A. Nayfeh, D. Boroyevich, Theoretical and experimental investigation of the fast- and slow-scale instabilities of a DC-DC converter, IEEE Trans. Power Electron. 16 (2) (2001) 201–216.

[21] A. Sangswang, C. Nwankpa, Random noise in switching DC–DC converter: verification and analysis, in: Proceedings of IEEE ISCAS'03, May, Bangkok, Thailand, 2003.

[22] A. Sangswang, C. Nwankpa, Effects of switching-time uncertainties on pulsewidthmodulated power converters: modeling and analysis, IEEE Trans. Circuits Syst. I: Fundam. Theory Appl. 50 (8) (2003) 1006–1012.

[23] S.-C. Chang, E. Weldon, Coding for T-user multiple-access channels, IEEE Trans. Inf. Theory 25 (6) (1979) 684–691.

[24] S.-C. Chang, Further results on coding for T-user multiple-access channels (Corresp.), IEEE Trans. Inf. Theory 30 (2) (1984) 411–415.

[25] G. Khachatrian, S. Martirossian, Code construction for the T-user noiseless adder channel, IEEE Trans. Inf. Theory 44 (5) (1998) 1953–1957.

CHAPTER 12

Pilot-Scale Implementation of Coordinated Control for Autonomous Microgrids

M.S. Mahmoud, N.M. Alyazidi

King Fahd University of Petroleum and Minerals (KFUPM), Dhahran, Saudi Arabia

1. ELECTRONICALLY COUPLED DISTRIBUTED GENERATION UNITS

Autonomous operation of a microgrid requires sophisticated control strategies and protection systems. Depending on the electrical proximity of the distributed generation (DG) units and their dedicated loads, several topologies for microgrids can be defined: parallel connection, ring connection, and radial connection of DG units. Each DG unit within a microgrid is connected to the point of common coupling (PCC), where the dedicated loads are also connected. When all the PCCs are along a transmission line with nonzero impedance between the PCCs, the radial configuration is obtained [1].

Several control schemes have been proposed for the islanded operation of microgrids in the technical literature. The (dq) current control strategy for multiple DG units in an islanded microgrid, based on frequency-power and voltage-reactive power droop characteristics of each DG unit, is well known and extensively reported [2–5]. In this approach, each DG unit is equipped with two droop characteristics:
• frequency as a linear function of real power; and
• voltage magnitude as a linear function of reactive power.
On the basis of these droop characteristics, frequency is dominantly controlled by real power flow, and voltage magnitude is regulated by reactive power flow of the DG unit. This approach does not directly incorporate load dynamics in the control loop. Thus large and/or fast load changes can result in either poor dynamic response or even voltage/frequency instability. A control strategy for autonomous operation of a DG unit and its dedicated load is introduced in [6]. This method is intended for a fairly fixed load and cannot accommodate large perturbations in the load parameters.

Microgrid
http://dx.doi.org/10.1016/B978-0-08-101753-1.00012-7

Fig. 12.1 shows a microgrid system that consists of the parallel connection of several DG units. All DG units are connected to the PCC, where the main grid and the local load are also connected. Each DG unit is coupled to the main grid via a voltage source converter [7]. A single-line diagram of the three-phase microgrid system is depicted in Fig. 12.2.

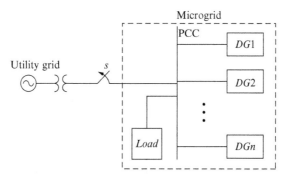

Fig. 12.1 Microgrid system with multiple parallel-connected DGs.

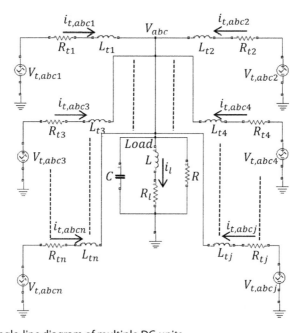

Fig. 12.2 Single-line diagram of multiple DG units.

1.1. System Layout

The study system is an islanded microgrid consisting of two electronically coupled DG units and a passive RLC load that are all connected to the PCC (see Fig. 12.2). Each DG unit is modeled by a DC voltage source, a three-phase voltage source converter, and a series RL filter (see Fig. 12.3). The microgrid parameters are given in Table 12.1.

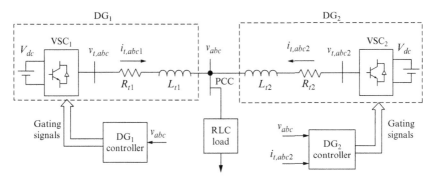

Fig. 12.3 Microgrid system consisting of two DG units.

Table 12.1 Parameters of the two distributed generation units

Quantity	Value	Comment
R_t	1.5 mΩ (0.01 pu)	Resistance of VSC filter
L_t	300 μH (0.785 pu)	Inductance of VSC filter
VSC rated power	2.5 MW (1 pu)	
VSC terminal voltage (line-line)	600 V (1 pu)	$V_{base} = 600$ V
f_{sw}	60 Hz	PWM carrier frequency
V_{dc}	2000 V	DC bus voltage
R	75 Ω (1 pu)	Load terminal resistance
L	111.9 mH (0.554 pu)	Load terminal inductance
C	62.86 μF (1.805 pu)	Load terminal capacitance
$Q = R\sqrt{\frac{C}{L}}$	1.8	Load quality factor
$f_{res} = \frac{1}{2a;\pi\sqrt{LC}}$	60 Hz	Load resonance frequency
$q_\ell = \frac{L\Omega_0}{R_\ell}$	120	Inductor quality factor
f_0	60 Hz	System nominal frequency
V_{dc}	230 V	DC bus voltage
R_s	1.0 Ω	Transmission line resistance
L_s	100 μH	Transmission line inductance

The system in Fig. 12.3 is to operate in islanded mode. In this case a feedback control system should be designed to regulate the voltage magnitude of the load at each PCC. In this chapter, the decentralized control strategy [7] is used to control the voltage magnitudes at the PCCs. Moreover, to control the power exchanges between the two DG units and to provide a constant frequency for the microgrid, an internal oscillator provides the phase-angle for both DG units. The internal oscillator is placed in the control system of one of the DG units and the phase angle is transmitted to the control system of the other DG unit by a communication method.

1.2. Mathematical Model

A state-space model for the islanded microgrid in Fig. 12.2 operating under balanced conditions and using a rotating reference frame (dq-frame) with frequency of ω_0 is provided in Eq. (12.1). By application of Kirchhoff's voltage and current laws, the equations of the microgrid system can be written as

$$v_{abc} = L\frac{di_{L,abc}}{dt} + R_l i_{L,abc},$$

$$v_{t,abc_1} = L_{t_1}\frac{di_{t,abc_1}}{dt} + R_{t_1} i_{t,abc_1} + v_{abc},$$

$$v_{t,abc_2} = L_{t_2}\frac{di_{t,abc_2}}{dt} + R_{t_2} i_{t,abc_2} + v_{abc}$$

$$\vdots$$

$$\vdots$$

$$v_{t,abc_n} = L_{t_n}\frac{di_{t,abc_n}}{dt} + R_{t_n} i_{t,abc_n} + v_{abc},$$

$$i_{t,abc_1} + i_{t,abc_2} + \cdots + i_{t,abc_n}$$
$$= \frac{v_{abc}}{R} + i_{L,abc} + C\frac{dv_{abc}}{dt}, \quad (12.1)$$

where v_{abc}, $v_{t,abc1}$, $v_{t,abc2}$, $i_{L,abc}$, $i_{t,abc1}$, and $i_{t,abc2}$ are 3×1 vectors containing the three-phase variables. Under the balanced conditions, the zero sequence of a three-phase variable x_{abc} in Eq. (12.1) is zero.

The model in Eq. (12.1) can be transformed to the $\alpha\beta$-frame with use of $x_{\alpha\beta} = x_a e^{j0} + x_b e^{j2\pi/3} + x_c e^{j4\pi/3}$, where $x_{\alpha\beta} = x_\alpha + jx_\beta$. Therefore the state-space equations of the system in the $\alpha\beta$-frame are

$$\frac{dv_{\alpha\beta}}{dt} = -\frac{1}{RC}v_{\alpha\beta} + \frac{1}{C}i_{t,\alpha\beta1} - \frac{1}{C}i_{L,\alpha\beta}$$

$$+ \frac{1}{C}i_{t,\alpha\beta2} + \cdots + \frac{1}{C}i_{t,\alpha\beta_n},$$

$$\frac{di_{t,\alpha\beta1}}{dt} = -\frac{1}{L_{t1}}v_{\alpha\beta} - \frac{R_{t1}}{L_{t1}}i_{t,\alpha\beta1} + \frac{1}{L_{t1}}v_{t,\alpha\beta1},$$

$$\frac{di_{L,\alpha\beta}}{dt} = \frac{1}{L}v_{\alpha\beta} - \frac{R_l}{L}i_{L,\alpha\beta},$$

$$\frac{di_{t,\alpha\beta2}}{dt} = -\frac{1}{L_{t2}}v_{\alpha\beta} - \frac{R_{t2}}{L_{t2}}i_{t,\alpha\beta2} + \frac{1}{L_{t2}}v_{t,\alpha\beta2}$$

$$\vdots$$

$$\vdots$$

$$\frac{di_{t,\alpha\beta_n}}{dt} = -\frac{1}{L_{t_n}}v_{\alpha\beta} - \frac{R_{t_n}}{L_{t_n}}i_{t,\alpha\beta_n} + \frac{1}{L_{t_n}}v_{t,\alpha\beta_n}. \tag{12.2}$$

To simplify the control design procedure, the dynamical equations (12.2) should be transformed to the dq-frame as

$$x_{\alpha\beta} = X_{dq}e^{j\theta} = (X_d + jX_q)e^{j\theta}, \tag{12.3}$$

where $\theta(t) = \int_0^t w(\xi)d\xi + \theta_0$ is the phase angle of any three-phase signal in the system. The dq-frame state-space equation can be written as

$$\frac{dV_{dq}}{dt} + jw_0 V_{dq} = -\frac{1}{RC}V_{dq} + \frac{1}{C}I_{t,dq1} - \frac{1}{C}I_{L,dq}$$

$$+ \frac{1}{C}I_{t,dq2} + \cdots + \frac{1}{C}I_{t,dqn},$$

$$\frac{dI_{t,dq_1}}{dt} + jw_0 I_{t,dq1} = -\frac{1}{L_{t_1}}V_{dq} - \frac{R_{t_1}}{L_{t_1}}I_{t,dq1} + \frac{1}{L_{t_1}}V_{t,dq_1},$$

$$\frac{dI_{L,dq}}{dt} + jw_0 I_{L,dq} = \frac{1}{L}V_{dq} - \frac{R_l}{L}L, dq,$$

$$\frac{dI_{t,dq2}}{dt} + jw_0 I_{t,dq2} = -\frac{1}{L_{t2}}V_{dq} - \frac{R_{t2}}{L_{t2}}I_{t,dq2} + \frac{1}{L_{t2}}V_{t,dq2}$$

$$\vdots$$

$$\frac{dI_{t,dq_n}}{dt} + jw_0 I_{t,dq2} = -\frac{1}{L_{t_n}}V_{dq} - \frac{R_{t_n}}{L_{t_n}}I_{t,dq_n} + \frac{1}{L_{t_n}}V_{t,dq_n}.$$

For state-space presentation,

$$\dot{x} = Ax + Bu,$$
$$y = Cx, \tag{12.4}$$

where x is the state vector defined as

$$x = [x_1, \; x_2]^T,$$
$$x_1 = [V_d, \; V_q, \; I_{td1}, \; I_{tq1}, \; I_{Ld}, \; I_{Lq}, \; I_{td2}, \; I_{tq2}],$$
$$x_2 = [\ldots, \; \ldots, \; I_{td_n}, \; I_{tq_n}]^T.$$

The control vector u and the output vector y are

$$u = \begin{bmatrix} V_{td_1} & V_{tq_1} & V_{td_2} & V_{tq_2} & \cdots & V_{td_n} & V_{tq_n} \end{bmatrix}^T,$$
$$y = \begin{bmatrix} V_d & V_q & I_{td_2} & I_{tq_2} & \cdots & I_{td_n} & I_{tq_n} \end{bmatrix}^T.$$

In islanded operation of the microgrid, the master DG unit handles the voltage and frequency control of the load, and the slave DG units regulate their power components by use of the conventional dq-current control strategy. It has been shown [7] that the parallel connection of n DG units in islanded mode forms an interconnected control system that is conveniently controlled in a two-level coordinating scheme.

2. LABORATORY-SCALE EXPERIMENT I

In what follows, we apply the two-level coordinating control strategy to a pilot-scale microgrid of the type depicted in Fig. 12.3 with $n = 2$. The dynamic performance of the two-DG microgrid is now verified in a simulation environment within MATLAB/SimPowerSystems and evaluated in terms of voltage tracking and perturbations in the load parameters.

2.1. Voltage Tracking Properties of Distributed Generation Unit 1

The voltage tracking properties of the DG unit 1 controller are evaluated in this simulation study. The entry data are as follows:
- The d and q components of the load voltage at PCC1 are initially set at 0.2 and 0.6 pu.
- The d and q components of the load voltage at PCC2 are initially set at 0.5 and 0.7 pu.

- The reference signals of the load voltage at PCC1, $V_{1d,ref}$ and $V_{1q,ref}$, are subjected to step changes.
- The d component of the load voltage steps up to 0.4 pu at $t = 0.7$ s, and the q component of the load voltage steps down to 0.5 pu at $t = 1$ s.

The ensuing simulation results are shown in Figs. 12.4 and 12.5, from which fast transient responses of both DG units to the step changes and decoupling of the control channels are seen.

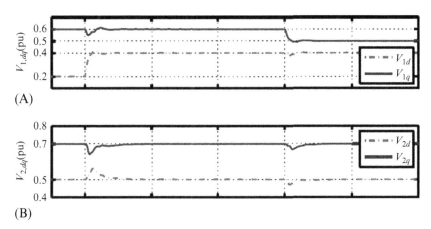

Fig. 12.4 Set point tracking of DG1: (d, q) components of the load voltage at (A) PCC1 and (B) PCC2.

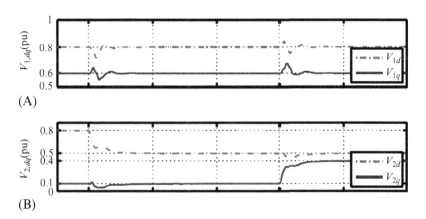

Fig. 12.5 Set point tracking of DG1: control signals of (A) DG1 and (B) DG2.

2.2. Voltage Tracking Properties of Distributed Generation Unit 2

In this case the performance of the proposed controller with respect to the step changes in the reference signals of the DG unit 2 is verified. The entry data are as follows:

- The dq components of the load voltage at PCC1 are regulated at 0.6 and 0.8 pu by DG unit 1.
- The d component of the reference signal of DG unit 2 steps down from 0.8 to 0.5 pu at $t = 0.7$ s.
- The q component of the reference signal of DG unit 2 steps up from 0.1 to 0.4 pu at $t = 1$ s.

The resulting dynamic responses of the microgrid to these step changes in the reference signals are plotted in Figs. 12.6 and 12.7.

2.3. Effect of Load Perturbations

In this case study the robust stability and the performance of the proposed decentralized controller with respect to the uncertainties in the load parameters are evaluated. While the microgrid supplies the rated loads, three load changes are imposed. In all cases, the d and q components of the load voltages at PCC1 and PCC2 are regulated at 0.8 and 0.3 pu, and at 0.5 and 0.9 pu, respectively.

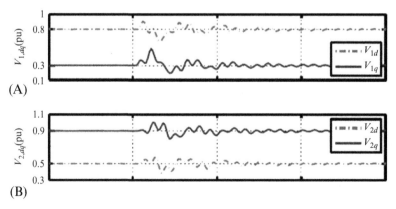

Fig. 12.6 Set point tracking of DG2: (d, q) components of the load voltage at (A) PCC1 and (B) PCC2.

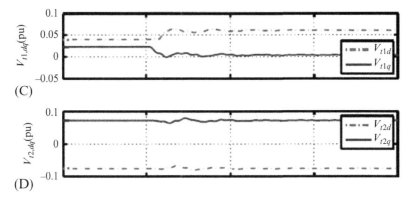

Fig. 12.7 Set point tracking of DG2: control signals of (C) DGl and (D) DG2.

2.3.1. The First Case

Here the load resistances in the three phases are changed equally such that the resultant load remains balanced. In this case the load resistance at PCC1 is suddenly stepped up from the nominal value $(76\,\Omega)$ to 25 times the nominal value $(1900\,\Omega)$ at $t = 1$ s.

- In Fig. 12.8 the d and q components of the load voltage at PCC1 and PCC2 are shown, and they demonstrate that both DG controllers successfully regulate the dedicated load voltages within about three cycles.

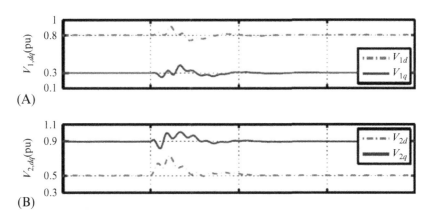

Fig. 12.8 Dynamic performance of the microgrid to a change in the load resistance at PCC1: (d, q) components of the load voltage at (A) PCC1 and (B) PCC2.

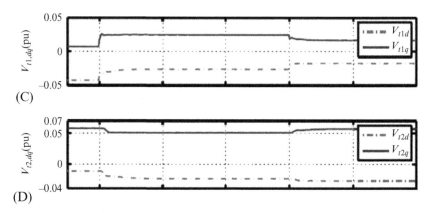

Fig. 12.9 Dynamic performance of the microgrid to a change in the load resistance at PCC1: control signals of (C) DG1 and (D) DG2.

- Fig. 12.9 indicates that the proposed controllers are capable of maintaining the load voltage despite uncertainties in the load resistance.

2.3.2. The Second Case

A similar load change is imposed in the load resistance at PCC2 from the nominal value (76 Ω) to 20 times the nominal value (1520 Ω) at $t = 1$ s.

- In Fig. 12.10 the d and q components of the load voltage at PCC1 and PCC2 are shown, and they demonstrate that both DG controllers successfully regulate the load voltages.

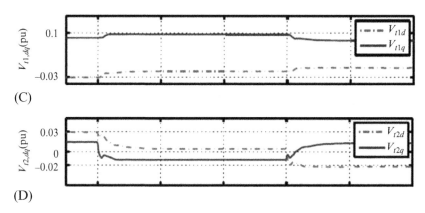

Fig. 12.10 Dynamic performance of the microgrid to a change in the load resistance at PCC2: (d, q) components of the load voltage at (C) PCC1 and (D) PCC2.

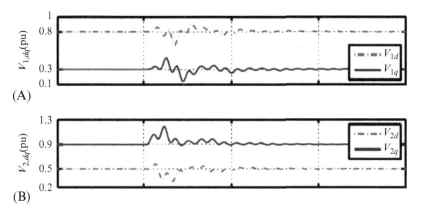

(A)

(B)

Fig. 12.11 Dynamic performance of the microgrid to a change in the load resistance at PCC2: control signals of (A) DG1 and (B) DG2.

• Fig. 12.11 indicates that the proposed controllers are capable of maintaining the load voltage despite uncertainties in the load resistance.

This case study also shows that the two-level coordinating control strategy is robust against load perturbations.

2.3.3. The Third Case

In this case the load resistance and capacitance at PCC1 are fixed at their rated values, and the load inductance is suddenly stepped down from its rated value (111.9 mH) to 0.65 times the rated value (72.735 mH) at $t = 1$ s.

• In Fig. 12.12 the d and q components of the load voltage at PCC1 and PCC2 are shown, and they confirm that the two-level coordinating

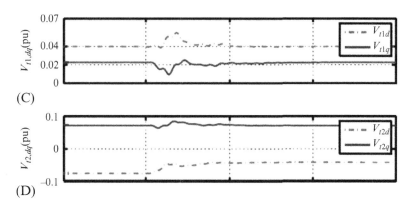

(C)

(D)

Fig. 12.12 Dynamic performance of the microgrid to a change in the load inductance at PCC1: (d, q) components of the load voltage atPCC1 (C) and PCC2 (D).

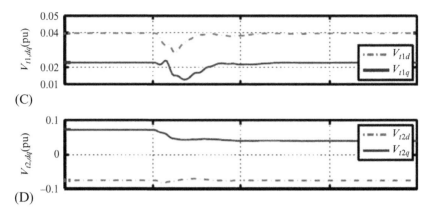

Fig. 12.13 Dynamic performance of the microgrid to a change in the load inductance at PCC1: control signals of DG1 (C) and DG2 (D).

control strategy successfully regulates the output signals to their desired values within about three cycles.

- Fig. 12.13 verifies that the controller is capable of maintaining the load voltages despite uncertainties in the load inductance.

This case study also shows that the two-level coordinating control strategy is robust with respect to the uncertainty in the load inductance.

3. LABORATORY-SCALE EXPERIMENT II

Control of the voltage and frequency during islanded operation of DG units is also a major challenge. A method for intentionally islanding a single DG unit to feed a local load was proposed in [8, 9]. With more than one DG unit on the island, it is necessary to regulate the voltage during microgrid operation, which can be achieved by use of a voltage versus reactive power droop controller [10]. To complete the resynchronization process once the grid is restored, a supervisory control mechanism will monitor the overall process and provide information to the local controller to respond accordingly.

The purpose of this section is to test the coordinated control scheme with use of a laboratory-based microgrid setup for validating the voltage and frequency control. The following are the assigned tasks:

1. setting up a microgrid having two synchronous machine based microsources;
2. development of local control for the machines; and

3. controlling the machine operating modes on the basis of the availability of grid power supply.

3.1. Microgrid Setup

An appropriate pilot experiment is developed in the system of systems (SoS) laboratory to demonstrate the controller function for synchronous machine-based DG units where two synchronous machines coupled to DC machines operating as microsources and are managed by MATLAB/Simulink. The DC machine represents a gas microturbine connected to the synchronous generator. A reactive load is composed of passive resistive loads along with an induction machine.

A one-line diagram for the laboratory setup is provided in Fig. 12.14. The ratings for the machines and loads are given in Table 12.2.

The control parameters for the synchronous generators are the torque that is provided to the shaft of the machines. This is equivalent to the output torque of the DC machine to which the synchronous machine is coupled. In grid-connected mode the torque controls the speed of the machine and hence the frequency of the supply; in islanding mode it controls the active power generated by the machine. The field voltage of the synchronous machine is also a control parameter. In islanding mode the field voltage controls the voltage, and in-grid connected mode it controls the reactive power generated by the machine.

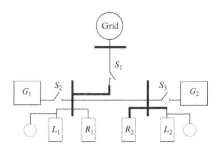

Fig. 12.14 Single-line diagram of laboratory experiment II.

Table 12.2 Synchronous machine and load ratings

Synchronous machines	Resistive load	Inductive load	Speed
230 V, 2.5 kVA	500 W	5000 W, 500 VAR	1800 rpm

3.2. Implementation

During this experiment it is assumed that the system starts with the grid serving the entire load. The generators start up and synchronize with the grid and start serving a share of the load. At a certain time when the grid support is lost, the machines are switched into droop mode and the load in the system is shared according to the ratings of the machines. The scenario of running the laboratory test is described by the following procedure, where every step has the approximate time instant presented to match with the voltage and frequency waveforms as well as active and reactive power waveforms.

- At $t = 10$ s, synchronous machine G_1 is enforced to run at a constant speed of 1800 rpm and generate grid voltage.
- At $t = 60$ s, when the synchronization lamps across (S_2) are all *off* at the same time, the synchronization switch (S_2) is closed.
- At $t = 70$ s, machine G_2 is commanded to run at a constant speed of 1800 rpm and generate grid voltage.
- At $t = 130$ s, when the synchronization lamps across (S_3) are *off*, the synchronization switch (S_2) is closed.
- At $t = 175$ s, machines G_1 and G_2 are commanded to generate a portion of the power drawn by the microgrid. During the experiment we command the machines to each generate 1000 W and 500 VAR each. Thus the grid is now supplying about 500 W and 250 VAR.
- At $t = 200$ s, at some selected moment, we assume that the grid support is lost. At this instant, switch S_1 is turned off and the controller is shifted to droop control. Now the machines share the load equally.
- At $t = 250$ s, we turn off the inductive load L_1 to demonstrate the sharing of the load. With R_1, R_2, and L_2 in the system, each machine can be seen to generate about 750 W and 300 VAR.
- At $t = 280$ s, at another selected instant, we assume that the fault preventing grid support to the microgrid was cleared and that the grid support is back.
- At $t = 375$ s, the grid frequency and voltage do not match the system voltage and frequency. Therefore the machines are commanded to increase the voltage and frequency to match the grid.
- At $t = 425$ s, when the synchronization lamps across S_1 are all off, synchronization switch S_1 is closed again and the droop controllers are turned off.

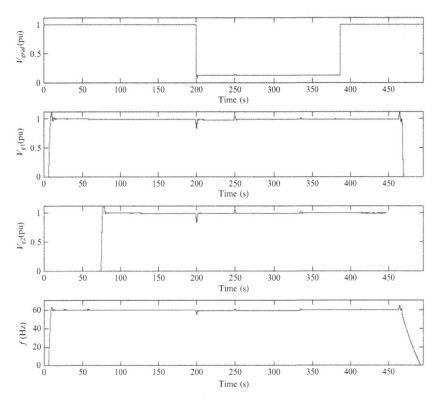

Fig. 12.15 Voltage and frequency waveforms from the test system.

- Now the loads in the microgrid are again served with the nominal voltage and frequency by the grid and partly by the machines in the microgrid, ensuring that machines G_1 and G_2 are stable.

  Fig. 12.15 shows the voltage and frequency waveforms during the entire experimentation period. It can be seen that:
- After 130 s, both generators are in the system and supplying power, and the load is served with the nominal voltage and frequency irrespective of the support from the grid. Figs. 12.16 and 12.17 depict the active and reactive power waveforms. Once machines G_1 and G_2 have both synchronized to the microgrid at 130 s, the load power is shared by the grid and the machines.
- At 200 s, the microgrid islands and machines G_1 and G_2 increase their generation to compensate for the loss of grid support and share the power according to their droops to continue to serve the load.

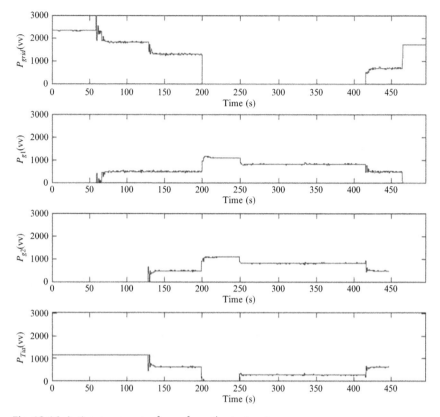

Fig. 12.16 Active power waveforms from the test system.

- When there is a step change in the load at 250 s, both generators decrease their output and continue serving the remaining load.
- At 375 s, the system voltage and frequency are increased by the shifting of the droop curve to facilitate the resynchronizing process.

4. LABORATORY-SCALE EXPERIMENT III

In an autonomous microgrid, DG units can be either dispatchable or nondispatchable. Generally, nondispatchable DG units are dependent on weather conditions. Alternatively, selection of the size of dispatchable DG units should be in accordance with the peak load demand since they can share the power according to their rating [11]. Normally, the microgrid operates in frequency droop control. To deal with an ever increasing load demand, it is considered an effective strategy to have some battery energy

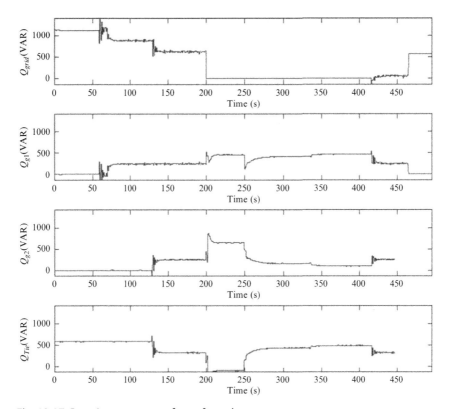

Fig. 12.17 Reactive power waveforms from the test system.

storage systems (BESSs) to aid the autonomous microgrid operation [12]. A BESS is usually expensive and its lifetime can be shortened by several charge–discharge cycles. Therefore such a unit must come online only when required and must supply only the amount of power that cannot be supplied by the DG units in an autonomous microgrid [13].

In an islanded microgrid, the performance of a BESS to provide energy balance during quick fluctuation in load demand is explained in [14]. Optimized economic operation by load transfer is discussed in [15] for a grid-connected microgrid with a BESS to reduce the microgrid operating cost. The centralized control is discussed in [16] for optimization of operation of a microgrid that consists of various DG units, storage devices, and controllable loads. Coordinated control is introduced in [17] for microresources with solar photovoltaic and battery storage to support the voltage and frequency in an islanded microgrid.

The purpose of this section is to assess the effectiveness of a hybrid droop control strategy to protect an autonomous microgrid against overloading by use of a BESS. To prevent any overloading, if the BESS is also controlled in the same frequency droop control as DG units, it will share the load power with the existing DG units according to its rating, thereby failing to supply only the required excess power [18]. Accordingly, DG units will not be utilized to their maximum capacity and the BESS will discharge faster. Therefore it might be completely discharged if the overload persists for longer. Hence, any BESS must supply only the excess amount of power.

4.1. Microgrid-Battery Energy Storage System Architecture

The architecture of an autonomous microgrid with a BESS is depicted in Fig. 12.18 and contains DG units, loads, and BESSs. Inertial-type DG

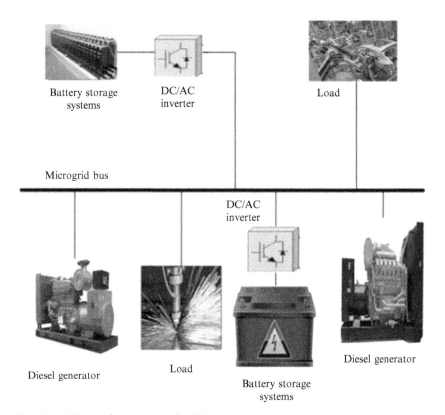

Fig. 12.18 Microgrid structure with BESS.

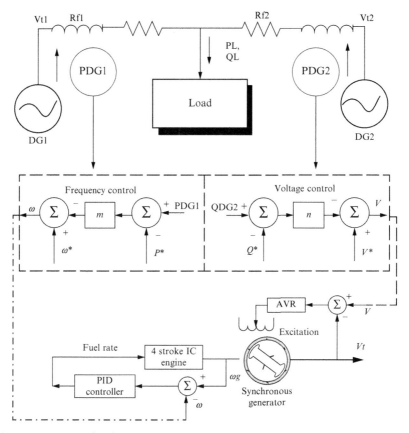

Fig. 12.19 Microgrid with two DGs.

units (diesel generators) with different ratings are considered as dispatchable sources. To examine the situation of providing power support during overloading through BESSs, only two diesel generators are considered (see Fig. 12.19). The diesel generators serve two cumulative loads: one constant PQ type and one R-L type. Two BESSs in angle droop control with different ratings are used to fill the gap between power generation and load, when the generation of the two diesel generators is at its maximum. The BESS converter switches are blocked during nominal operation and they can be brought online quickly during system stress.

The parameters of the selected DG units and BESSs are given in Tables 12.3 and 12.4.

In the sequel, we assess two distinct schemes: the first excludes BESSs and the second includes BESSs.

Table 12.3 Parameters of the diesel generator set

System data	Value
Rated voltage	11 kV
Rated power of diesel generator 1	250 kW
Rated power of diesel generator 2	500 kW
Rated frequency	50 Hz
Rated speed	1500 rpm
Reactance	**Value per unit**
X_d : Unsaturated d-axis synchronous reactance	0.116
X'_d : Unsaturated d-axis transient synchronous reactance	0.0074
X''_d : Unsaturated d-axis subtransient synchronous reactance	0.00294
X_q : Unsaturated q-axis synchronous reactance	0.00637
X''_q : Unsaturated q-axis subtransient synchronous reactance	0.00524
X_2 : Negative sequence reactance	0.044
X_0 : Zero sequence reactance	0.00245
Time constants	**Value (ms)**
t''_{d0} : d-axis subtransient open circuit time constant	25
t''_{q0} : q-axis subtransient open circuit time constant	4
t'_{d0} : d-axis transient open circuit time constant	368
t_a : Armature time constant	

Table 12.4 Parameters of the distributed generation units

System quantities	Values
Feeder impedance	$R_{f1} = 3.025\,\Omega,\ L_{f1} = 57.8\,\mathrm{mH}$
	$R_{f2} = 3.025\,\Omega,\ L_{f2} = 57.8\,\mathrm{mH}$
	$R_{f3} = 3.025\,\Omega,\ L_{f3} = 57.8\,\mathrm{mH}$
	$R_{f4} = 3.025\,\Omega,\ L_{f1} = 57.8\,\mathrm{mH}$
	$R_{f5} = 3.025\,\Omega,\ L_{f1} = 57.8\,\mathrm{mH}$
BESS rated power	**Battery 1 100 kW, battery 2 50 kW**
Load 2	$R_{La} = 1000\,\Omega,\ L_{La} = 100\,\mathrm{mH}$
	$R_{Lb} = 1000\,\Omega,\ L_{Lb} = 100\,\mathrm{mH}$
	$R_{Lc} = 1000\,\Omega,\ L_{Lc} = 100\,\mathrm{mH}$
Droop coefficient	**Frequency-voltage**
m_1	0.015 rad/MWs
m_2	0.0075 rad/MWs
n_1	0.04 kV/MVAR
n_2	0.02 kV/MVAR
Droop coefficient	**Angle droop**
m_{b1}	0.4 rad/MWs
m_{b2}	0.8 rad/MWs

4.2. Microgrid-Battery Energy Storage System Operation: First Scheme

Consider the microgrid shown in Fig. 12.19, in which only the DG units are supplying the load. The DG units operate in frequency droop control to share power according to their rating. The feeders and load are denoted by subscripts f and L, respectively. In a frequency droop control method, each DG unit uses its real power output to set the frequency at its point of connection. Also the voltage magnitude is related to the reactive power, and therefore a voltage droop controller is used to set the voltage magnitude as

$$V = V^* + n(Q^* - Q),$$

where V and V^* are the instantaneous and rated DG voltage, respectively. The rated and measured reactive power are denoted by Q and Q^*, respectively, and n is the droop coefficient.

Note in Fig. 12.19 that:

- Each DG unit is equipped with an automatic voltage regulator and an internal combustion engine plus a governor, which is a proportional-integral-derivative controller.
- The input to the governor is the frequency error, and its output controls the fuel rate of the internal combustion engine.
- The droop voltage V is given to the automatic voltage regulator such that the bus voltage is regulated.

The parameters of the governor (proportional-integral-derivative controller) and the automatic voltage regulator are given in Tables 12.5 and 12.6.

Table 12.5 Parameters of the governor

System data	Value
Proportional gain	5000
Integral gain	1000
Derivative gain	200

Table 12.6 Parameters of the proportional-integral-derivative controller

System data	Value
Proportional gain	−0.002
Integral gain	−50
Derivative gain	−0.01

Now consider an autonomous 11-kV microgrid with two DG units where the ratings of diesel generator 1 and diesel generator 2 are 250 and 500 kW, respectively, and the value of f_d is ±0.3 Hz with the droop gains set on the basis of this condition and given in Table 12.4.

At the beginning, the load demand is 350 kW, which is below half of the total rating of the DG units. Subsequently at 3 s, the load demand increases to 500 kW. The load power shared by the diesel generators according to their rating is shown in Fig. 12.20, while the frequencies of the DG units are shown in Fig. 12.21, and are above or below 50 Hz depending on the total load requirement.

For further demonstration, consider the case in which the two DG units are sharing 700-kW power, when the load power demand increases to 850 kW, which is beyond the cumulative rating of the DG units. The system response is shown in Fig. 12.22. It is readily seen that the DG units supply their maximum rated power for about 12 s, before the whole system collapses. Due to the inertia associated with the DG units, a serious failure does not occur instantaneously.

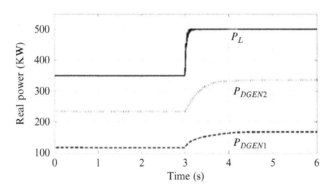

Fig. 12.20 Load power sharing between the DGs

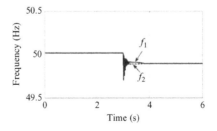

Fig. 12.21 DG frequencies before and after load change.

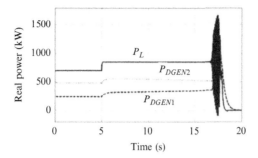

Fig. 12.22 Serious failure during overload.

4.3. Microgrid-Battery Energy Storage System Operation: Second Scheme

Consider the autonomous microgrid shown in Fig. 12.23, in which two BESSs are supplying a load. The angle droop control is given by

$$\delta_b = \delta_b^* + m_b^*(P_b^* - P_b),$$

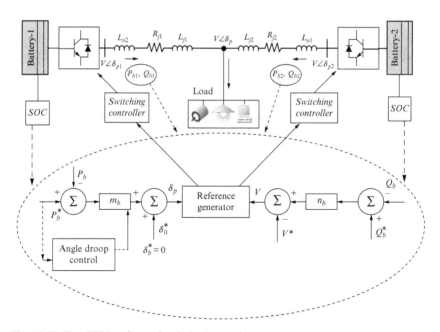

Fig. 12.23 Two BESS with an islanded microgrid.

where δ_b and δ_b^* are the actual voltage and its reference angle, respectively; the rated and actual real power of the BESSs are represented by P_b^* and P_b, respectively; and m_b indicates the coefficient of the droop line.

The requirement for an angle droop is that all the DG units must operate with respect to a single reference angle. If an autonomous microgrid contains only converter-interfaced DG units, the reference angle can be set arbitrarily.

Now consider the implementation of the microgrid in Fig. 12.23. The ratings of BESS 1 and BESS 2 are assumed to be 100 and 50 kW, respectively, while the load demand is assumed to be 90 kW. At the beginning, the reference is chosen as 0 degrees and is changed arbitrarily to 30 degrees at 1 s. The angle of the BESS 1 output voltage (δ_{b1}) is shown in Fig. 12.24. It can be seen that this angle jumps at 1 s. However, the load power and the power sharing ratio do not change, as is evident from Fig. 12.25.

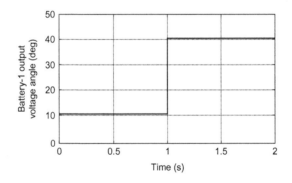

Fig. 12.24 Change in angle with a change in reference angle.

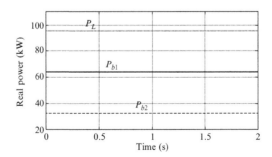

Fig. 12.25 Real power does not change with the change in reference angle.

5. CONCLUSIONS

The multilevel control strategy of islanded microgrid operation for n microgrids connected in parallel is shown to provide an efficient control framework where the parallel connection of the microgrids forms a comprehensive system including n microgrids. This coordinated control strategy verified that the islanded microgrid units cannot be controlled by their individual local controllers only, but also require some correction from the supervisory system controller.

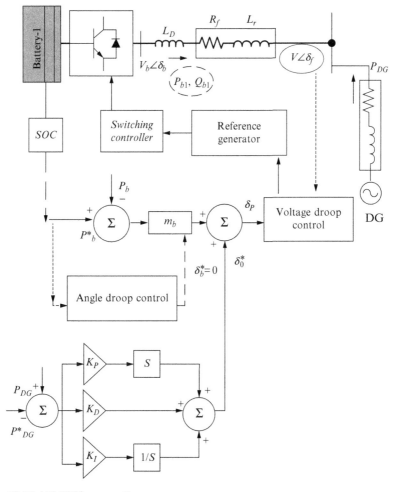

Fig. 12.26 MG-BESS connection.

In this chapter we have described laboratory-scale pilot experiments on the coordinated control strategy for different microgrid applications. It is interesting to observe that the supervisory system controller is not effected on the transient response of the load voltage and can accommodate any variations.

A major issue pertains to the microgrid structure and implementation is the transition between islanded and grid-connected modes of operation. This transition induces a change in the system dynamics, since on going from islanded to grid connection, the model set up will increase in complexity, leading to further requirements on the DG units. On the other hand, disconnection from the main grid to autonomous mode results in stability issues. Subsequently, examination of the transition behavior demands adequate stability analysis and simulation to provide careful guidelines for safe operation. It is represented by [7].

Further studies on the implementation will examine the dynamic system in Fig. 12.26.

ACKNOWLEDGMENTS

This work is supported by the Deanship of Scientific Research (DSR) at King Fahd University of Petroleum and Minerals (KFUPM) through book writing research project no. BW 151004.

REFERENCES

[1] R. Moradi, H. Karimi, M. Karimi-Ghartemani, Robust decentralized control for islanded operation of two radially connected DG systems, in: IEEE International Symposium on Industrial Electronics (ISIE), 2010, pp. 2272–2277.
[2] F. Katiraei, M.R. Iravani, P.W. Lehn, Microgrid autonomous operation during and subsequent to islanding process, IEEE Trans. Power Deliv. 20 (1) (2005) 248–257.
[3] H. Song, K. Nam, Dual current control scheme for PWM converter under unbalanced input voltage conditions, IEEE Trans. Ind. Electron. 46 (1999) 953–959.
[4] P. Piagi, R.H. Lasseter, Autonomous control of microgrids, in: IEEE Power Engineering Society General Meeting, June 18–22, 2006.
[5] J.A.P. Lopes, C.L. Moreira, A.G. Madureira, Defining control strategies for microgrids islanded operation, IEEE Trans. Power Syst. 21 (2) (2006) 916–924.
[6] H. Karimi, H. Nikkhajoei, M.R. Iravani, Control of an electronically-coupled distributed resource unit subsequent to an islanding event, IEEE Trans. Power Deliv. 23 (1) (2008), doi:10.1109/TPWRD.2007.911189.
[7] M.S. Mahmoud, O. Al-Buraiki, Two-level control for improving the performance of microgrid in islanded mode, in: Proceedings of the IEEE International Symposium on Industrial Electronics 2014, June 1–4, Istanbul, Turkey, 2014, pp. 254–259.
[8] H.W. Dommel, Digital computer solution of electromagnetic transients in single and multiphase, IEEE Trans. Power App. Syst. 88 (4) (1969) 388–399.

[9] Y. Deng, S. Foo, H. Li, Real time simulation of power flow control strategies for fuel cell vehicle with energy storage by using real time digital simulator (RTDS), in: Proceedings of the 6th IEEE International Conference on Power Electronics and Motion Control (IPEMC), 2009, pp. 2323–2327.

[10] B. Lasseter, Microgrids [distributed power generation], in: IEEE Power Engineering Society Winter Meeting, vol. 1, 2001, pp. 146–149.

[11] M. Goyal, A. Ghosh, F. Zare, Power sharing control with frequency droop in a hybrid microgrid, in: IEEE Power and Energy Society General Meeting, Vancouver, July, 2013.

[12] R. Lasseter, M. Erickson, Integration of Battery-Based Energy Storage Element in the CERTs Microgrid, Technical Report, University of Wisconsin, Madison, 2009.

[13] Altair Nanotechnologies Inc., Application for Advanced Batteries in Microgrid Environments, Altair Nanotechnologies Inc., 2012.

[14] Z. Haihua, T. Bhattacharya, T. Duong, T.S.T. Siew, A.M. Khambadkone, Composite energy storage system involving battery and ultracapacitor with dynamic energy management in microgrid applications, IEEE Trans. Power Electron. 26 (3) (2011) 923–930.

[15] Z. Jianping, G. Zhenyu, R. Yuliang, D. Xinhui, Y. Xiaohai, A economic operation optimization for microgrid with battery storage and load transfer, in: Systems and Informatics (ICSAI), November, 2014, pp. 186–191.

[16] A.G. Tsikalakis, N.D. Hatziargyriou, Centralized control for optimizing microgrids operation, IEEE Trans. Energy Convers. 23 (1) (2008) 241–248.

[17] S. Adhikari, L. Fangxing, Coordinated V-f and P-Q control of solar photovoltaic generators with MPPT and battery storage in microgrids, IEEE Trans. Smart Grid 5 (3) (2014) 1270–1281.

[18] M. Goyal, A. Ghosh, F. Shahnia, Distributed battery storage for overload prevention in an islanded microgrid, in: Australian Universities Power Engineering Conference (AUPEC), September, Perth, Australia, 2014, pp. 1–6.

INDEX

Note: Page numbers followed by b indicate boxes, f indicate figures and t indicate tables.

Printed in the United States
By Bookmasters